Fundamentals of Nanoscale Film Analysis

Fundamentals of Nanoscale Film Analysis

Terry L. Alford
Arizona State University
Tempe, AZ, USA

Leonard C. Feldman
Vanderbilt University
Nashville, TN, USA

James W. Mayer
Arizona State University
Tempe, AZ, USA

 Springer

Terry L. Alford
Arizona State University
Tempe, AZ, USA

Leonard C. Feldman
Vanderbilt University
Nashville, TN, USA

James W. Mayer
Arizona State University
Tempe, AZ, USA

Fundamentals of Nanoscale Film Analysis

ISBN 13: 978-1-4419-3980-7

ISBN 13: 978-0-387-29261-8(e-book)

ISBN 10: 0-387-29261-6 (e-book)

Printed on acid-free paper.

Printed in the United States of America.

springer.com

To our wives and children,
Katherine and Dylan,
Betty, Greg, and Dana,
and
Betty, Jim, John, Frank, Helen, and Bill

Contents

Preface

A major feature in the evolution of modern technologies is the important role of surfaces and near surfaces on the properties of materials. This is especially true at the nanometer scale. In this book, we focus on the fundamental physics underlying the techniques used to analyze surfaces and near surfaces. New analytical techniques are emerging to meet the technological requirements, and all are based on a few processes that govern the interactions of particles and radiation with matter. Ion implantation and pulsed electron beams and lasers are used to modify composition and structure. Thin films are deposited from a variety of sources. Epitaxial layers are grown from molecular beams and physical and chemical vapor techniques. Oxidation and catalytic reactions are studied under controlled conditions. The key to these methods has been the widespread availability of analytical techniques that are sensitive to the composition and structure of solids on the nanometer scale.

This book focuses on the physics underlying the techniques used to analyze the surface region of materials. This book also addresses the fundamentals of these processes. From an understanding of processes that determine the energies and intensities of the emitted energetic particles and/or photons, the application to materials analysis follows directly.

Modern materials analysis techniques are based on the interaction of solids with interrogating beams of energetic particles or electromagnetic radiation. These interactions and their resulting radiation/particles are based upon on fundamental physics. Detection of emergent radiation and energetic particles provides information about the solid's composition and structure. Identification of elements is based on the energy of the emergent radiation/particle; atomic concentration is based on the intensity of the emergent radiation. We discuss in detail the relevant analytical techniques used to uncover this information. Coulomb scattering from atoms (Rutherford backscattering spectrometry), the formation of inner shell vacancies in the electronic structure (X-ray photoelectron spectroscopy), transitions between levels (electron microprobe and Auger electron spectroscopies), and coherent scattering (X-ray and electron diffractometry) are fundamental to materials analysis. Composition depth profiles are obtained with heavy-ion sputtering in combination with surface-sensitive techniques (electron spectroscopies and secondary ion mass spectrometry). Depth profiles are also found from energy loss of light ions (Rutherford backscattering and prompt nuclear analyses). Structures of surface layers are characterized using diffraction (X-ray, electron, and low-energy electron diffraction), elastic scattering (ion channeling), and scanning probes (tunneling and atomic force microscopies).

Because this book focuses on the fundamentals of modern surface analysis at the nanometer scale, we have provided derivations of the basic parameters—energy and cross section or transition probability. The book is organized so that we start with the classical concepts of atomic collisions as applied to Rutherford scattering (Chapter 2), energy loss (Chapter 3), sputtering (Chapter 4), channeling (Chapter 5), and electron interactions (Chapter 6). An overview is given of diffraction techniques in both real space (X-ray diffraction, Chapter 7) and reciprocal space (electron diffraction, Chapter 8) for structural analysis. Wave mechanics is required for an understanding of photoelectric cross sections and fluorescence yields; we review the wave equation and perturbation theory in Chapter 9. We use these relations to discuss photoelectron spectroscopy (Chapter 10), radiative transitions (Chapter 11), and nonradiative transitions (Chapter 12). Chapter 13 discusses the application of nuclear techniques to thin film analysis. Finally, Chapter 14 presents a discussion of scanning probe microscopy.

The current volume is a significant expansion of the previous work, *Fundamentals of Thin Films Analysis*, by Feldman and Mayer. New chapters have been added reflecting the progress that has been made in analysis of ultra thin films and nanoscale structures.

All the authors have been engaged heavily in research programs centered on materials analysis; we realize the need for a comprehensive treatment of the analytical techniques used in nanoscale surface and thin film analysis. We find that a basic understanding of the processes is important in a field that is rapidly changing. Instruments may change, but the fundamental processes will remain the same.

This book is written for materials scientists and engineers interested in the use of spectroscopies and/or spectometries for sample characterization; for materials analysts who need information on techniques that are available outside their laboratory; and particularly for seniors and graduate students who will use this new generation of analytical techniques in their research.

We have used the material in this book in senior/graduate-level courses at Cornell University, Vanderbilt University, and Arizona State University, as well as in short courses for scientists and engineers in industry around the world. We wish to thank Dr. N. David Theodore for his review of Chapters 7 and 8. We also thank Timothy Pennycook for proofreading the manuscript. We thank Jane Jorgensen and Ali Avcisoy for their drawings and artwork.

1
An Overview: Concepts, Units, and the Bohr Atom

1.1 Introduction

Our understanding of the structure of atoms and atomic nuclei is based on scattering experiments. Such experiments determine the interaction of a beam of elementary particles—photons, electrons, neutrons, ions, etc.—with the atom or nucleus of a known element. (In this context, we consider all incident radiation as *particles*, including photons.) The classical example is Rutherford scattering, in which the scattering of incident alpha particles from a thin solid foil confirmed the picture of an atom as composed of a small positively charged nucleus surrounded by electrons in circular orbits. As these fundamental interactions became understood, the scientific community recognized the importance of the inverse process—namely, measuring the interaction of radiation with targets of unknown elements to determine atomic composition. Such determinations are called *materials analysis*. For example, alpha particles scatter from different nuclei in a distinct and well-understood manner. Measurements of the intensity and energy of the scattered particles provides a direct measure of elemental composition. The emphasis in this book is twofold: (1) to describe in a quantitative fashion those fundamental interactions that are used in modern materials analysis and (2) to illustrate the use of this understanding in practical materials analysis problems.

The emphasis in modern materials analysis is generally directed toward the structure and composition of the surface and outer few tens to hundred nanometers of the materials. The emphasis comes from the realization that the surface and near-surface regions control many of the mechanical and chemical properties of solids: corrosion, friction, wear, adhesion, and fracture. In addition, one can tailor the composition and structure of the outer layers by directed energy processes utilizing lasers or electron and ion beams, as well as by more conventional techniques such as oxidation and diffusion.

In modern materials analysis, one is concerned with the source beam (also referred to as the *incident beam* or the *probe beam* or *primary beam*) of radiation; the beam of particles—photons, electrons, neutrons, or ions; the interaction cross section; the emergent radiation; and the detection system. The primary interest of this book is the interaction of the beam with the material to be analyzed, with emphasis on the energies and intensities of emitted radiation. As we will show, the energy of the emitted particles provides the signature or identification of the atom, and the intensity tells the amount of atoms (i.e., sample composition). The radiation source and the detection system are important topics in their own right; however, the main emphasis in this book is on the ability to conduct quantitative materials analysis that depends upon interactions within the target.

1.2   Nomenclature

Materials characterization involves the quantification of the structure, composition, amount, and depth distribution of matter with the use of energetic particles (e.g., ions, neutrons, alpha particles, protons, and electrons) and energetic photons (e.g., infrared radiation, visible light, UV light, X-rays, and gamma rays). Any materials-characterization techniques can be described in the following manner. The incident *probe beam* of energetic photons or particles *interrogates* the solid. The incident particle or photon reacts with the solid in various manners; these reactions (R_x) induce the emission of a variety of detected beams in the form of energetic particles or photons, i.e., the detected beam (Fig. 1.1). Hence, the primary interest of this book is in using the reaction (between the beam and the solid) and the intensity and energy of the detected beam to analyze solids. Since the energy of the detected particle/photon is measured, the actual names of the various techniques have the prefix SPECTRO, meaning *energy measurement*. The suffix gives information about the relationship between the specific incidence *photon/particle* and the detected *photon/particle*. For example, if the incident species is the same as the emitted species, the technique is a SPECTROMETRY: Rutherford backscattering spectrometry and X-ray diffractometry. If the incident species is different from the emitted species, then the term SPECTROSCOPY is used: Auger electron spectroscopy and X-ray photoelectron spectroscopy.

There is an impressive array of experimental techniques available for the analysis of solids. Figure 1.2 gives the flavor of the possible combinations. In some cases, the same incident and emergent radiation is employed (we will use the general terms *radiation* and *particles* for photons, electrons, ions, etc.). Listed below are examples, with commonly used acronyms in parentheses.

Primary electron in, Auger electron out: Auger electron spectroscopy (AES)
Alpha particle in, alpha particle out: Rutherford backscattering spectrometry (RBS)
Primary X-ray in, characteristic X-ray out: X-ray fluorescence spectroscopy (XRF)

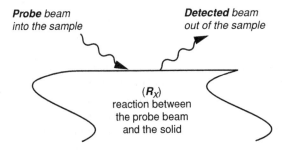

Probe *beam*
into the sample

Detected *beam*
out of the sample

(R_x)
reaction between
the probe beam
and the solid

FIGURE 1.1. Schematic of the fundamentals of materials characterization. The probe beam of energetic photons or particles interrogates the solid. The incident particle or photon reacts (R_x) and induces the emission of a variety of detected beams in the form of energetic particles or photons, i.e., the detected beam.

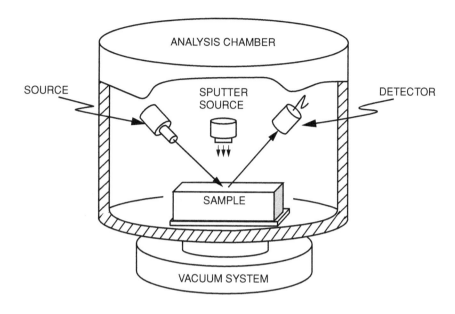

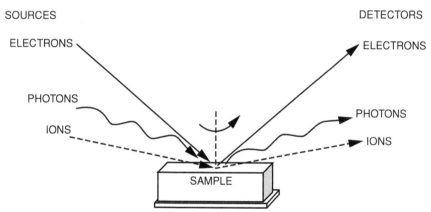

FIGURE 1.2. Schematic of radiation sources and detectors in thin film analysis techniques. Analytical probes are represented by almost any combination of source and detected radiation, i.e., photons in and photons out or ions in and photon out. Many chambers will also contain sample erosion facilities such as an ion sputtering as well as an evaporation apparatus for deposition of materials onto a clean substrate under vacuum.

In other cases, the incident and emergent radiation differ as indicated below:

X-ray in, electron out: X-ray photoelectron spectroscopy (XPS)
Electron in, X-ray out: electron microprobe analysis (EMA)
Ion in, target ion out: secondary ion mass spectroscopy (SIMS)

A beam of particles incident on a target either scatters elastically or causes an electronic transition in an atom. The scattered particle or the energy of the emergent radiation contains the signature of the atom. The energy levels in the transition are

TABLE 1.1. Nomenclature of many techniques available for the analysis of materials. The name of a given technique often provides a complete or partial description of the technique.

Method	In	Out	R_x
Energy Dispersive X-rays (EDX) Spectroscopy	e^-	$h\nu_{out}$	ΔE_{e^-}, $h\nu_{out} = \Delta E$
X-ray Fluorescence (XRF) Spectroscopy	$h\nu_{in}$	$h\nu_{out}$	ΔE, $h\nu_{in}$, $h\nu_{out} = \Delta E$
Particle Induced X-ray Emission (PIXE) Spectroscopy	H^+	$h\nu_{out}$	ΔE, α, $h\nu_{out} = \Delta E$
X-ray Photoelectron Spectroscopy (XPS)	$h\nu_{in}$	e^-	$h\nu_{in}$, $e^- E_B$
X-ray Diffractometry (XRD)	ν_{in}	ν_{out}	Coherent Scattering $\nu_{in} = \nu_{out}$ (particle characteristics)
Electron Diffractometry (ED)	e^-	e^-	Coherent Scattering $\nu_{in} = \nu_{out}$ (wave characteristic)
Rutherford Backscattering Spectrometry (RBS)	α	α	Elastic Scattering
Secondary Ion Mass Spectroscopy (SIMS)	Heavy ion	Matrix ion	Sputtered Ion (erosion due to momentum transfer)

characteristic of a given atom; hence, measurement of the energy spectrum of the emergent radiation allows identification of the atom. Table 1.1 gives a summary of various techniques based on the nomenclature of the incident *probe beam*, the induced emission, and the detected beam.

The number of atoms per cm^2 in a target is found from the relation between the number, I, of incident particles and the number of interactions. The term *cross section* is used as a quantitative measure of an interaction between an incident

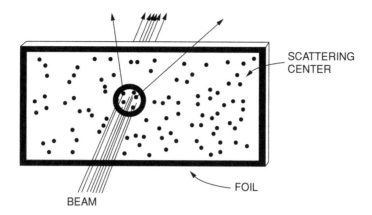

FIGURE 1.3. Illustration of the concept of cross section and scattering. The central circle defines a unit area of a foil containing a random array of scattering centers. In this example, there are five scattering centers per unit area. Each scattering center has an area (the cross section for scattering) of 1/20 unit area; therefore, the probability of scattering is 5/20, or 0.25. Then a fraction (0.25 in this example) of the incident beam will be scattered, i.e., 2 out of 8 trajectories in the drawing. A measure of the fraction of the scattered beam is a measure of the probability ($P = Nt\sigma$, Eq. 1.1). If the foil thickness and density are known, Nt can be calculated, yielding a direct measure of the cross section.

particle and an atom. The cross section σ for a given process is defined through the probability, P:

$$P = \frac{\text{Number of interactions}}{\text{Number of incident particles}}. \tag{1.1}$$

For a target containing Nt atoms per unit area perpendicular to an incident beam of I particles, the number of interactions is $I\sigma Nt$. From knowledge of detection efficiency for measuring the emergent radiation containing the signature of the transition, the number of atoms and ultimately the target composition can be found (Fig. 1.3).

The information required from analytical techniques is the species identification, concentration, depth distribution, and structure. The available analytical techniques have different capabilities to meet these requirements. The choice of analysis method depends upon the nature of the problem. For example, chemical bonding information can be obtained from techniques that rely upon transitions in the electronic structure around the atoms—the electron spectroscopies. Structural determination is found from diffraction or particle channeling techniques.

In the following chapters, we are mostly concerned with materials analysis in the outer microns of the sample's surface and near-surface region. We emphasize the energy of the emergent radiation as an identification of the element and the intensity of the radiation as a measure of the amount of material. These are the basic principles that provide the foundation for the different analytical techniques.

1.3   Energies, Units, and Particles

With few exceptions, the measurement of energy is the hallmark of materials analysis. Although the SI (or MKS) system of units gives the Joule (J) as the derived unit of energy, the electron volt (eV) is the traditional unit in materials analysis. The Joule is so large that it is inconvenient as a unit in atomic interactions. The electron volt is defined as the kinetic energy gained by an electron accelerated from rest through a potential difference of 1 V. Since the charge on the electron is 1.602×10^{-19} Coulomb and a Joule is a Coulomb-volt,

$$1\,\text{eV} = 1.602 \times 10^{-19}\,\text{J}. \tag{1.2}$$

Commonly used multiples of the eV are the keV (10^3 eV) and MeV (10^6 eV).

In determination of crystal structure by X-ray diffraction, the diffraction conditions are determined by atomic spacing and hence the wavelength of the photon. The wavelength λ is the ratio of c/ν, where c is the speed of light and ν is the frequency, so the energy E is

$$E(\text{keV}) = h\nu = \frac{hc}{\lambda} = \frac{1.24\,\text{keV-nm}}{\lambda\,(\text{nm})}, \tag{1.3}$$

where Planck's constant $h = 4.136 \times 10^{-15}$ eV-sec, $c = 2.998 \times 10^8$ m/sec, λ is in units of nm, and 1 nanometer is 10^{-9} m.

The energies of the emergent radiation provide the signature of the transition; the cross section determines the strength of the interaction. Although the MKS unit for cross sectional area is m^2, the measured values are often given in cm^2. It is convenient to use cgs units rather than SI units in relations involving the charge on the electron. The usefulness of cgs units is clear when considering the Coulomb force between two charged particles with Z_1 and Z_2 units of electronic charge separated by a distance r:

$$F = \frac{Z_1 Z_2 e^2 k_C}{r^2}, \tag{1.4}$$

where the Coulomb law constant $k_C = (1/4\pi\varepsilon_0) = 8.988 \times 10^9$ m/farad (F) in the SI system (where 1 F = 1 amp-s/V) and $k_c = 1$ in the cgs system. In cgs units, the value of e equals 4.803×10^{-10} stat C, which leads to a quick conversion factor for Coulomb interactions of

$$e^2 = 1.44 \times 10^{-13}\,\text{MeV-cm} = 1.44\,\text{eV-nm}. \tag{1.5}$$

In this book we use $k_C = 1$ and rely on Eq. 1.5 for e^2. The masses of particles, given in kg in SI units, are generally expressed in unified mass units (u), a measure that replaces the older atomic mass units, or amu. The neutral carbon atom with 6 protons, 6 neutrons, and 6 electrons is the reference for the unified mass unit (u), which is defined as $1/12^{\text{th}}$ the mass of the neutral ^{12}C carbon atom (where the superscript indicates the mass number 12). Avogadro's number N_A is the number of atoms or molecules in a mole (mol) of a substance and is defined as the number of atoms of an element needed to equal its atomic mass in grams. Avogadro's number of ^{12}C atoms is equivalent to a mass of exactly 12 g, and the mass of one ^{12}C atom is 12 mass units. The value of

Avogadro's number, the number of atoms/mol, is

$$N_A = 6.0220 \times 10^{23}, \tag{1.6}$$

and the unified mass unit u, the reciprocal of N_A, is

$$u = \frac{1}{N_A} = \frac{1g}{6.023 \times 10^{23}} = 1.661 \times 10^{-24} g . \tag{1.7}$$

A large part of this book is devoted to the extraction of depth profiles—the atomic composition or impurity concentration as a function of depth below the surface. In terms of length measurement, the natural unit is the nanometer (nm), where

$$1 \, nm = 10^{-9} \, m .$$

For example, the separation between atoms in a solid is about 0.3 nm. The measurement techniques give depth scales in terms of areal density, the number Nt of atoms per cm^2, where t is the thickness and N is the atomic density. For elemental solids, the atomic density and the mass density ρ in g/cm^3 are related by

$$N = N_A \, \rho / A, \tag{1.8}$$

where A is the atomic mass number and N_A is Avogadro's number. Another unit of thickness is the mass absorption coefficient, usually expressed as g/cm^2, the product of the mass density and linear thickness.

Each nucleus is characterized by a definite atomic number Z and mass number A. The atomic number Z is the number of protons and hence the number of electrons in the neutral atom; it reflects the atomic properties of the atom. The mass number gives the number of nucleons, protons, and neutrons; isotopes are nuclei (often called nuclides) with the same Z and different A. The current practice is to represent each nucleus by the chemical name with the mass number as a superscript, i.e., ^{12}C. The chemical atomic weight (or atomic mass) of elements as listed in the periodic table gives the average atomic mass, i.e., the average of the stable isotopes weighted by their abundance. Carbon, for example, has an atomic weight of 12.011, which reflects the 1.1% abundance of ^{13}C. Appendix 10 lists the elements and their relative abundance, atomic weight, atomic density, and specific gravity.

The masses of particles may be expressed in terms of energy through the Einstein relation

$$E = mc^2 , \tag{1.9}$$

which associates 1 J of energy with $1/c^2$ kg of mass. The mass of an electron is 9.11×10^{-31} kg, which is equivalent to an energy

$$E = (9.11 \times 10^{-31} \, kg)(2.998 \times 10^8 \, m/s)^2$$
$$= 8.188 \times 10^{-14} J = 0.511 \, MeV . \tag{1.10}$$

In materials analysis, the incident radiation is usually photons, electrons, neutrons, or low-mass ions (neutral atoms stripped of one or more electrons). For example, the proton is an ionized hydrogen atom, and the alpha particle is a helium atom with one or two electrons removed. The notation ^{4}He$^+$ and ^{4}He^{++} is often used to denote a helium atom with one or two electrons removed, respectively. The deuteron, ^{2}H$^+$, is a neutron

TABLE 1.2. Mass energies of particles and light nuclei.

Particle	Symbol	Mass energy (MeV)
Electron	e or $e-$	0.511
Proton	p	938.3
Neutron	n	939.6
Deuteron	d or $^2H^+$	1875.6
Alpha	α or $^4He^{++}$	3727.4

and proton bound together. The mass energies of some of these particles are given in Table 1.2. In analytical applications, the velocities of these particles are generally well below 10^7 m/sec; hence, relativistic effects do not enter, and the masses are independent of velocity.

1.4   Particle–Wave Duality and Lattice Spacing

In materials analysis, one tends to view the incident beam and emergent radiation as discrete particles—photons, electrons, neutrons, and ions. On the other hand, the interactions of radiation with matter and, in particular, the cross section for a transition is often based on the wave aspect of the radiation.

This wave–particle duality was of major concern in the early development of modern physics. The photon and the electron provide examples of the wave and particle nature of matter. For example, in the photoelectric effect, light behaves as if it were particle-like, that each photon interacting with an atom to give up its energy, $E = h\nu$, to an electron that can escape from the solid. The diffraction of X-rays from planes of atoms, on the other hand, satisfies wave interference conditions.

Electrons and their diffraction from crystal surfaces constitute a sensitive probe of surface structure. The classical, particle behavior of electrons, on the other hand, is illustrated in their deflection in electric and magnetic fields. One can associate both a wavelength λ and a momentum p with the motion of an electron. The De Broglie relation gives their connection:

$$\lambda = h/p , \tag{1.11}$$

where h is Planck's constant. Distances between lattice planes are on the order of a tenth of a nanometer (0.1 nm). For diffraction, the wavelengths of electrons are of comparable magnitude. The electron velocity, $v = p/m$, corresponding to a wavelength of 0.1 nm is

$$v = \frac{h}{m\lambda} = \frac{6.6 \times 10^{-34}}{9.1 \times 10^{-31} \times 10^{-10}} = 7.25 \times 10^6 \text{ m-sec}^{-1} ,$$

where MKS units are used with $h = 6.6 \times 10^{-34}$ J s. The energy is

$$E = \frac{1}{2}mv^2 = \frac{9.1 \times 10^{-31}(7.25 \times 10^6)^2}{2} = 2.39 \times 10^{-17} \text{J} = 150 \text{ eV} .$$

Electron diffraction studies of surfaces use electrons with low energies, between 40 eV and 150 eV, giving rise to the acronym LEED—low-energy electron diffraction.

Energies of 1.0–2.0 MeV He$^+$ are commonly used in materials analysis; here the wavelengths are orders of magnitude smaller than the lattice spacing, and the interactions of helium ions with solids are described on the basis of particle rather than wave behavior. For helium atoms, an energy of 2 MeV corresponds to a wavelength of 10^{-5} nm; whereas, distances between nearest-neighbor atoms in a solid are on the order of 0.2–0.5 nm.

The distances between atoms and atomic planes can be calculated from a known lattice constant and crystal structure. Aluminum, for example, contains $\sim 6 \times 10^{22}$ atoms/cm^3, has a lattice constant of 0.404 nm, and has a face-centered cubic (fcc) crystal structure. One monolayer of atoms on the (100) surface then contains an areal atom density of 2 atoms/(0.404 nm)^2 or 1.2×10^{15} atoms/cm^2. Almost all solids have monolayer density values of 5×10^{14}/cm^2 to 2×10^{15}/cm^2 on major crystallographic surfaces. In a loose way, a monolayer is usually thought of as 10^{15} atoms/cm^2. The spectroscopic sensitivity of various surfaces is often measured in units of monolayers or atoms/cm^2; bulk impurity determinations are usually given in atoms/cm^3.

1.5   The Bohr Model

The identification of atomic species from the energies of emitted radiation was developed from the concepts of the Bohr model of the hydrogen atom. Particle scattering experiments established that the atom could be treated as a positively charged nucleus surrounded by a cloud of electrons. Bohr assumed that the electrons could move in stable circular orbits called *stationary states* and would emit radiation only in the transition from one stable orbit to another. The energies of the orbits were derived from the postulate that the angular momentum of the electron around the nucleus is an integral multiple of h/2π (h/2π is written as $\hbar$). In this section, we give a brief review of the Bohr atom, which provides useful relations for simple estimates of atomic parameters.

For a single electron of mass m$_e$ in a circular orbit of radius r about a fixed nucleus of charge Ze, the balance between the Coulomb and centripetal forces leads to

$$\frac{Ze^2}{r^2} = m_e \frac{v^2}{r} .$$
(1.12)

Bohr assumed that the angular momentum, m$_e vr$, has values given by an integer n times $\hbar$,

$$m_e vr = n \hbar .$$

From the above equations, we have

$$v^2 = \frac{n^2 \hbar^2 e^2}{m_e^2 r^2} = \frac{Ze^2}{m_e r} ,$$

which can be rewritten to give the radii r_n of allowed orbits:

$$r_n = \frac{\hbar^2 n^2}{m_e Z e^2}.$$
(1.13)

For hydrogen, $Z = 1$, the radius a_o of the smallest orbit, $n = 1$, is known as the *Bohr radius* and is given by

$$a_o = \frac{\hbar^2}{m_e e^2} = 0.53 \times 10^{-10}\,\text{m} = 0.053\,\text{nm}, \tag{1.14}$$

and the Bohr velocity v_0 of the electron in this orbit is

$$v_0 = \frac{\hbar}{m_e a_o} = \frac{e^2}{\hbar} = 2.19 \times 10^8\,\text{cm-s}^{-1}. \tag{1.15}$$

The ratio of v_0 to the speed of light is known as the *fine-structure constant* α, given by:

$$\alpha = \frac{v_0}{c} = \frac{1}{137}. \tag{1.16}$$

The energy of the electron is defined here as zero when it is at rest at infinity. The potential energy, *PE*, of an electron in the Coulomb force field has a negative value, $-Ze^2/r$, in this convention, and the kinetic energy (*KE*) is $Ze^2/2r$ (Eq. 1.12), so the total energy E is

$$E = KE + PE = \frac{Ze^2}{2r} - \frac{Ze^2}{r} = -\frac{Ze^2}{2r}$$

or, for the *n*th orbital,

$$E_n = -\frac{Ze^2}{2r_n} = \frac{-m_e e^4 Z^2}{2\hbar^2 n^2} = \frac{E_o Z^2}{n^2}. \tag{1.17}$$

The electron bound to a positively charged nucleus has a discrete set of allowed energies,

$$E_n = -\frac{13.6Z^2}{n^2}\,\text{eV}. \tag{1.18}$$

The binding energy E_B of such an electron is the positive value $13.58Z^2/n^2$. The numerical value of the $n = 1$ state represents the energy required to ionize the atom by complete removal of the electron; for hydrogen, the ionization energy is 13.58 eV.

The Bohr theory does lead to the correct values for energy levels observed in H spectral lines. The nomenclature introduced by Bohr persists in the vocabulary of atomic physics: orbital, Bohr radius, and Bohr velocity. The quantities v_0 and α_0 are used repeatedly in this book, as they are the natural units with which to evaluate atomic processes.

Problems

1.1. Calculate the density of atoms in C (graphite), Si, Fe, and Au. Express your answer in atoms/cm^3.

1.2. Calculate the number of atoms/ cm^2 in one monolayer of Si(100), in Si(111), and in W(100).

1.3. Calculate the wavelength (in nm) of a 1 MeV He ion, a 150 eV electron, and a 1 keV Ar ion.

1.4. Show that $e^2 = 1.44\,\text{eV-nm}$.

1.5. Find the ratio of velocity of a 1 MeV ion to the Bohr velocity.

1.6. Use the literature and notes to state the incoming radiation (particles) in the following spectroscopies, and in each case, state the nature of the atomic transition involved:

AES – Auger electron spectroscopy
RBS – Rutherford backscattering spectrometry
SIMS – secondary ion mass spectroscopy
XPS – X-ray photoelectron spectroscopy
XRF – X-ray fluorescence spectroscopy
SEM – scanning electron μ probe
NRA – nuclear reaction analysis

1.7. In this book we repeatedly make estimates using the Bohr model of the atom. Test the validity of this approximation by calculating the K-shell binding energy, $E_K(n = 1)$; the L-shell binding energy, $E_L(n = 2)$; the wavelength at the K-shell absorption edge, ($\hbar\omega = E_K$), and the K X-ray energy ($E_K - E_L$) for Si, Ni, and W. Compare with the accurate values given in the appendices.

1.8. The Auger process, discussed in Chapter 12, corresponds to an electron transition involving the emission of an Auger electron with the energy ($E_K - E_L - E_L$), where K is for $n = 1$ and L is for $n = 2$. Show that, in the Bohr model, $a_o k = 1/\sqrt{2}$, where a_o is the K-shell radius a_o/Z and $\hbar k$ is the momentum of the outgoing electron.

1.9. An incident photon of sufficient energy can eject an electron from an inner shell orbit. Such an excited atom may relax by rearranging the outer electrons to fill the vacancy. This is said to occur in a time equivalent to the orbital time. Calculate this characteristic atomic time for Ni. In later chapters, we will show that the inverse of this time may be thought of as the rate for the Auger process.

References

1. L.C. Feldman and J.W. Mayer, *Fundamentals of Surface and Thin Film Analysis* (North-Holland, New York 1986).
2. J. D. McGervey, *Introduction to Modern Physics* (Academic Press New York, 1971).
3. F. K. Richtmyer, E. H. Kennard, and J. N. Cooper, *Introduction to Modern Physics*, 6th ed. (McGraw-Hill, New York, 1969).
4. R. L. Sproull and W. A. Phillips, *Modern Physics*, 3rd ed. (John Wiley and Sons, New York 1980).
5. P. A. Tipler, *Modern Physics* (Worth Publishers, New York, 1978).
6. R. T. Weidner and R. L. Sells, *Elementary Modern Physics*, 3rd ed. (Allyn and Bacon, Boston, MA, 1980).
7. J. C. Willmott, *Atomic Physics* (John Wiley and Sons, New York, 1975).
8. John Taylor, Chris Zafiratos, and Michael A. Dubson, *Modern Physics for Scientists and Engineers*, 2nd ed. (Prentice-Hall, New York, 2003).
9. D.C. Giancoli, *Physics for Scientists and Engineers with Modern Physics*, 3rd ed. (Prentice-Hall, New York, 2001).

2
Atomic Collisions and Backscattering Spectrometry

2.1 Introduction

The model of the atom is that of a cloud of electrons surrounding a positively charged central core—the nucleus—that contains Z protons and $A - Z$ neutrons, where Z is the atomic number and A the mass number. Single-collision, large-angle scattering of alpha particles by the positively charged nucleus not only established this model but also forms the basis for one modern analytical technique, Rutherford backscattering spectrometry. In this chapter, we will develop the physical concepts underlying Coulomb scattering of a fast light ion by a more massive stationary atom.

Of all the analytical techniques, Rutherford backscattering spectrometry is perhaps the easiest to understand and to apply because it is based on classical scattering in a central-force field. Aside from the accelerator, which provides a collimated beam of MeV particles (usually $^4\text{He}^+$ ions), the instrumentation is simple (Fig. 2.1a). Semiconductor nuclear particle detectors are used that have an output voltage pulse proportional to the energy of the particles scattered from the sample into the detector. The technique is also the most quantitative, as MeV He ions undergo close-impact scattering collisions that are governed by the well-known Coulomb repulsion between the positively charged nuclei of the projectile and target atom. The kinematics of the collision and the scattering cross section are independent of chemical bonding, and hence backscattering measurements are insensitive to electronic configuration or chemical bonding with the target. To obtain information on the electronic configuration, one must employ analytical techniques such as photoelectron spectroscopy that rely on transitions in the electron shells.

In this chapter, we treat scattering between two positively charged bodies of atomic numbers Z_1 and Z_2. The convention is to use the subscript 1 to denote the incident particle and the subscript 2 to denote the target atom. We first consider energy transfers during collisions, as they provide the identity of the target atom. Then we calculate the scattering cross section, which is the basis of the quantitative aspect of Rutherford backscattering. Here we are concerned with scattering from atoms on the sample surface or from thin layers. In Chapter 3, we discuss depth profiles.

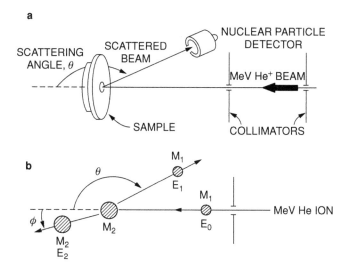

FIGURE 2.1. Nuclear particle detector with respect to scattering angle courtesy of MeV He$^+$ electron beam.

2.2 Kinematics of Elastic Collisions

In Rutherford backscattering spectrometry, monoenergetic particles in the incident beam collide with target atoms and are scattered backwards into the detector-analysis system, which measures the energies of the particles. In the collision, energy is transferred from the moving particle to the stationary target atom; the reduction in energy of the scattered particle depends on the masses of incident and target atoms and provides the signature of the target atoms.

The energy transfers or kinematics in elastic collisions between two isolated particles can be solved fully by applying the principles of conservation of energy and momentum. For an incident energetic particle of mass M_1, the values of the velocity and energy are v and $E_0 (= {}^1\!/_2 M_1 v^2)$, while the target atom of mass M_2 is at rest. After the collision, the values of the velocities v_1 and v_2 and energies E_1 and E_2 of the projectile and target atoms are determined by the scattering angle θ and recoil angle ϕ. The notation and geometry for the laboratory system of coordinates are given in Fig. 2.1b.

Conservation of energy and conservation of momentum parallel and perpendicular to the direction of incidence are expressed by the equations

$$\frac{1}{2}M_1 v^2 = \frac{1}{2}M_1 v_1^2 + \frac{1}{2}M_2 v_2^2, \tag{2.1}$$

$$M_1 v = M_1 v_1 \cos\theta + M_2 v_2 \cos\phi, \tag{2.2}$$

$$0 = M_1 v_1 \sin\theta - M_2 v_2 \sin\phi. \tag{2.3}$$

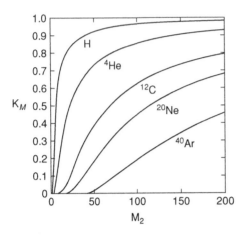

FIGURE 2.2. Representation of the kinematic factor K_{M_2} (Eq. 2.5) for scattering angle $\theta = 170°$ as a function of the target mass M_2 for ^{1}H, ^{4}He$^+$, ^{12}C, ^{20}Ne, and ^{40}Ar.

Eliminating ϕ first and then v_2, one finds the ratio of particle velocities:

$$\frac{v_1}{v} = \left[\frac{\pm(M_2^2 - M_1^2 \sin^2 \theta)^{1/2} + M_1 \cos \theta}{M_2 + M_1} \right]. \tag{2.4}$$

The ratio of the projectile energies for $M_1 < M_2$, where the plus sign holds, is

$$\frac{E_1}{E_0} = \left[\frac{(M_2^2 - M_1^2 \sin^2 \theta)^{1/2} + M_1 \cos \theta}{M_2 + M_1} \right]^2. \tag{2.5}$$

The energy ratio, called the *kinematic factor* $K = E_1/E_0$, shows that the energy after scattering is determined only by the masses of the particle and target atom and the scattering angle. A subscript is usually added to K, i.e., K_{M_2}, to indicate the target atom mass. Tabulations of K values for different M_2 and θ values are given in Appendix 1 and are shown in Fig. 2.2 for $\theta = 170°$. Such tables and figures are used routinely in the design of backscattering experiments. A summary of scattering relations is given in Table 3.1.

For direct backscattering through 180°, the energy ratio has its lowest value given by

$$\frac{E_1}{E_0} = \left(\frac{M_2 - M_1}{M_2 + M_1} \right)^2 \tag{2.6a}$$

and at 90° given by

$$\frac{E_1}{E_0} = \frac{M_2 - M_1}{M_2 + M_1}. \tag{2.6b}$$

In collisions where $M_1 = M_2$, the incident particle is at rest after the collision, with all the energy transferred to the target atom, a feature well known in billiards. For $\theta = 180°$, the energy E_2 transferred to the target atom has its maximum value given by

$$\frac{E_2}{E_0} = \frac{4 M_1 M_2}{(M_1 + M_2)^2}, \tag{2.7}$$

with the general relation given by

$$\frac{E_2}{E_0} = \frac{4 M_1 M_2}{(M_1 + M_2)^2} \cos^2 \phi. \tag{2.7'}$$

In practice, when a target contains two types of atoms that differ in their masses by a small amount ΔM_2, the experimental geometry is adjusted to produce as large a change ΔE_1 as possible in the measured energy E_1 of the projectile after the collision. A change of ΔM_2 (for fixed $M_1 < M_2$) gives the largest change of K when $\theta = 180°$. Thus $\theta = 180°$ is the preferred location for the detector ($\theta \cong 170°$ in practice because of detector size), an experimental arrangement that has given the method its name of *backscattering spectrometry*.

The ability to distinguish between two types of target atoms that differ in their masses by a small amount ΔM_2 is determined by the ability of the experimental energy measurement system to resolve small differences ΔE_1 in the energies of backscattered particles. Most MeV ^{4}He$^+$ backscattering apparatuses use a surface-barrier solid-state nuclear-particle detector for measurement of the energy spectrum of the backscattered particles. As shown in Fig. 2.3, the nuclear particle detector operates by the collection of the hole–electron pairs created by the incident particle in the depletion region of the reverse-biased Schottky barrier diode. The statistical fluctuations in the number of electron–hole pairs produce a spread in the output signal resulting in a finite resolution.

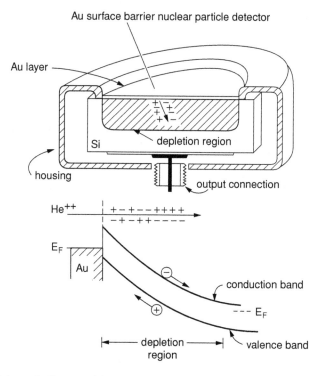

FIGURE 2.3. Schematic diagram of the operation of a gold surface barrier nuclear particle detector. The upper portion of the figure shows a cutaway sketch of silicon with gold film mounted in the detector housing. The lower portion shows an alpha particle, the He$^+$ ion, forming holes and electrons over the penetration path. The energy-band diagram of a reverse biased detector (positive polarity on *n*-type silicon) shows the electrons and holes swept apart by the high electric field within the depletion region.

Energy resolution values of 10–20 keV, full width at half maximum (FWHM), for MeV ^{4}He$^+$ ions can be obtained with conventional electronic systems. For example, backscattering analysis with 2.0 MeV ^{4}He$^+$ particles can resolve isotopes up to about mass 40 (the chlorine isotopes, for example). Around target masses close to 200, the mass resolution is about 20, which means that one cannot distinguish among atoms between ^{181}Ta and ^{201}Hg.

In backscattering measurements, the signals from the semiconductor detector electronic system are in the form of voltage pulses. The heights of the pulses are proportional to the incident energy of the particles. The pulse height analyzer stores pulses of a given height in a given voltage bin or channel (hence the alternate description, *multichannel analyzer*). The channel numbers are calibrated in terms of the pulse height, and hence there is a direct relationship between channel number and energy.

2.3   Rutherford Backscattering Spectrometry

In backscattering spectrometry, the mass differences of different elements and isotopes can be distinguished. Figure 2.4 shows a backscattering spectrum from a sample with approximately one monolayer of ^{63,65}Cu, ^{107,109}Ag, and ^{197}Au. The various elements are well separated in the spectrum and easily identified. Absolute coverages can be determined from knowledge of the absolute cross section discussed in the following section. The spectrum is an illustration of the fact that heavy elements on a light substrate can be investigated at coverages well below a monolayer.

The limits of the mass resolution are indicated by the peak separation of the various isotopes. In Fig. 2.4, the different isotopic masses of ^{63}Cu and ^{65}Cu, which have a natural abundance of 69% and 31%, respectively, have values of the energy ratio, or kinematic factor K, of 0.777 and 0.783 for $\theta = 170°$ and incident ^{4}He$^+$ ions ($M_1 = 4$).

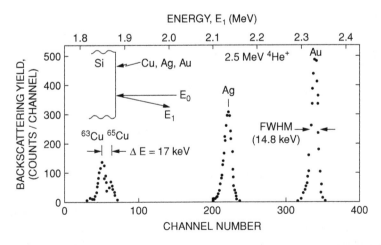

FIGURE 2.4. Backscattering spectrum for $\theta = 170°$ and 2.5 MeV ^{4}He$^+$ ions incident on a target with approximately one monolayer coverage of Cu, Ag, and Au. The spectrum is displayed as raw data from a multichannel analyzer, i.e., in counts/channel and channel number.

For incident energies of 2.5 MeV, the energy difference of particles from the two masses is 17 keV, an energy value close to the energy resolution (FWHM = 14.8 keV) of the semiconductor particle-detector system. Consequently, the signals from the two isotopes overlap to produce the peak and shoulder shown in the figure. Particles scattered from the two Ag isotopes, ^{107}Ag and ^{109}Ag, have too small an energy difference, 6 keV, and hence the signal from Ag appears as a single peak.

2.4   Scattering Cross Section and Impact Parameter

The identity of target atoms is established by the energy of the scattered particle after an elastic collision. The number N_s of target atoms per unit area is determined by the probability of a collision between the incident particles and target atoms as measured by the total number Q_D of detected particles for a given number Q of particles incident on the target in the geometry shown in Fig. 2.5. The connection between the number of target atoms N_s and detected particles is given by the scattering cross section. For a thin target of thickness t with N atoms/cm^3, $N_s = Nt$.

The differential scattering cross section, $d\sigma/d\Omega$, of a target atom for scattering an incident particle through an angle θ into a differential solid angle $d\Omega$ centered about θ is given by

$$\frac{d\sigma(\theta)}{d\Omega}d\Omega \cdot N_s = \frac{\text{Number of particles scattered into } d\Omega}{\text{Total number of incident particles}}.$$

In backscattering spectrometry, the detector solid angle Ω is small (10^{-2} steradian or less), so that one defines an average differential scattering cross section $\sigma(\theta)$,

$$\sigma(\theta) = \frac{1}{\Omega} \int_\Omega \frac{d\sigma}{d\Omega} \cdot d\Omega, \tag{2.8}$$

where $\sigma(\theta)$ is usually called the *scattering cross section*. For a small detector of area A, at distance l from the target, the solid angle is given by A/l^2 in steradians. For the geometry of Fig. 2.5, the number N_s of target atoms/cm^2 is related to the yield Y or the number Q_D of detected particles (in an ideal, 100%-efficient detector that subtends a solid angle Ω) by

$$Y = Q_D = \sigma(\theta)\Omega Q N_s, \tag{2.9}$$

where Q is the total number of incident particles in the beam. The value of Q is

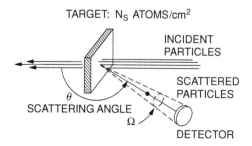

FIGURE 2.5. Simplified layout of a scattering experiment to demonstrate the concept of the differential scattering cross section. Only primary particles that are scattered within the solid angle $d\Omega$ spanned by the detector are counted.

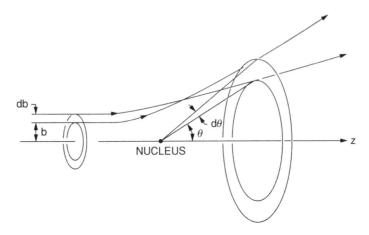

FIGURE 2.6. Schematic illustrating the number of particles between b and $b + db$ being deflected into an angular region $2\pi \sin\theta \, d\theta$. The cross section is, by definition, the proportionality constant; $2\pi b \, db = -\sigma(\theta)2\pi \sin\theta \, d\theta$.

determined by the time integration of the current of charged particles incident on the target. From Eq. 2.9, one can also note that the name *cross section* is appropriate in that $\sigma(\theta)$ has the dimensions of an area.

The scattering cross section can be calculated from the force that acts during the collision between the projectile and target atom. For most cases in backscattering spectrometry, the distance of closest approach during the collision is well within the electron orbit, so the force can be described as an unscreened Coulomb repulsion of two positively charged nuclei, with charge given by the atomic numbers Z_1 and Z_2 of the projectile and target atoms. We derive this unscreened scattering cross section in Section 2.5 and treat the small correction due to electron screening in Section 2.7.

The deflection of the particles in a one-body formulation is treated as the scattering of particles by a center of force in which the kinetic energy of the particle is conserved. As shown in Fig. 2.6, we can define the impact parameter b as the perpendicular distance between the incident particle path and the parallel line through the target nucleus. Particles incident with impact parameters between b and $b + db$ will be scattered through angles between θ and $\theta + d\theta$. With central forces, there must be complete symmetry around the axis of the beam so that

$$2\pi b \, db = -\sigma(\theta) \, 2\pi \, \sin\theta \, d\theta. \tag{2.10}$$

In this case, the scattering cross section $\sigma(\theta)$ relates the initial uniform distribution of impact parameters to the outgoing angular distribution. The minus sign indicates that an increase in the impact parameter results in less force on the particle so that there is a decrease in the scattering angle.

2.5   Central Force Scattering

The scattering cross section for central force scattering can be calculated for small deflections from the impulse imparted to the particle as it passes the target atom. As

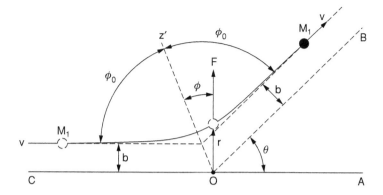

FIGURE 2.7. Rutherford scattering geometry. The nucleus is assumed to be a point charge at the origin O. At any distance r, the particle experiences a repulsive force. The particle travels along a hyperbolic path that is initially parallel to line OA a distance b from it and finally parallel to line OB, which makes an angle θ with OA. The scattering angle θ can be related to impact parameter b by classical mechanics.

the particle with charge Z_1e approaches the target atom, charge Z_2e, it will experience a repulsive force that will cause its trajectory to deviate from the incident straight line path (Fig. 2.7). The value of the Coulomb force F at a distance r is given by

$$F = \frac{Z_1 Z_2 e^2}{r^2}. \tag{2.11}$$

Let $\mathbf{p}_1$ and $\mathbf{p}_2$ be the initial and final momentum vectors of the particle. From Fig. 2.8, it is evident that the total change in momentum $\Delta\mathbf{p} = \mathbf{p}_2 - \mathbf{p}_1$ is along the z' axis. In this calculation, the magnitude of the momentum does not change. From the isosceles triangle formed by $\mathbf{p}_1$, $\mathbf{p}_2$, and $\Delta\mathbf{p}$ shown in Fig. 2.8, we have

$$\frac{\frac{1}{2}\Delta p}{M_1 v} = \sin\frac{\theta}{2}$$

or

$$\Delta p = 2M_1 v \sin\frac{\theta}{2}. \tag{2.12}$$

FIGURE 2.8. Momentum diagram for Rutherford scattering. Note that $|\mathbf{p}_1| = |\mathbf{p}_2|$, i.e., for elastic scattering the energy and the speed of the projectile are the same before and after the collision.

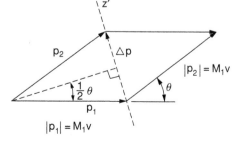

We now write Newton's law for the particle, $\mathbf{F} = d\mathbf{p}/dt$, or

$$d\mathbf{p} = \mathbf{F}\,dt.$$

The force F is given by Coulomb's law and is in the radial direction. Taking components along the z' direction, and integrating to obtain Δp, we have

$$\Delta p = \int (dp)_{z'} = \int F \cos\phi\,dt = \int F \cos\phi\frac{dt}{d\phi}d\phi, \qquad (2.13)$$

where we have changed the variable of integration from t to the angle ϕ. We can relate $dt/d\phi$ to the angular momentum of the particle about the origin. Since the force is central (i.e., acts along the line joining the particle and the nucleus at the origin), there is no torque about the origin, and the angular momentum of the particle is conserved. Initially, the angular momentum has the magnitude $M_1 vb$. At a later time, it is $M_1 r^2\,d\phi/dt$. Conservation of angular momentum thus gives

$$M_1 r^2\frac{d\phi}{dt} = M_1 vb$$

or

$$\frac{dt}{d\phi} = \frac{r^2}{vb}.$$

Substituting this result and Eq. 2.11 for the force in Eq. 2.13, we obtain

$$\Delta p = \frac{Z_1 Z_2 e^2}{r^2}\int \cos\phi\frac{r^2}{vb}d\phi = \frac{Z_1 Z_2 e^2}{vb}\int \cos\phi\,d\phi$$

or

$$\Delta p = \frac{Z_1 Z_2 e^2}{vb}(\sin\phi_2 - \sin\phi_1). \qquad (2.14)$$

From Fig. 2.7, $\phi_1 = -\phi_0$ and $\phi_2 = +\phi_0$, where $2\phi_0 + \theta = 180°$. Then $\sin\phi_2 - \sin\phi_1 = 2\sin(90° - {}^1\!/_2\theta)$. Combining Eqs. 2.12 and 2.14 for Δp, we have

$$\Delta p = 2M_1 v \sin\frac{\theta}{2} = \frac{Z_1 Z_2 e^2}{vb}2\cos\frac{\theta}{2}. \qquad (2.15a)$$

This gives the relationship between the impact parameter b and the scattering angle:

$$b = \frac{Z_1 Z_2 e^2}{M_1 v^2}\cot\frac{\theta}{2} = \frac{Z_1 Z_2 e^2}{2E}\cot\frac{\theta}{2}. \qquad (2.15b)$$

From Eq. 2.10, the scattering cross section can be expressed as

$$\sigma(\theta) = \frac{-b}{\sin\theta}\frac{db}{d\theta}, \qquad (2.16)$$

and from the geometrical relations $\sin\theta = 2\sin(\theta/2)\cos(\theta/2)$ and $d\cot(\theta/2) = -\frac{1}{2}d\theta/\sin^2(\theta/2)$,

$$\sigma(\theta) = \left(\frac{Z_1 Z_2 e^2}{4E}\right)^2\frac{1}{\sin^4\theta/2}. \qquad (2.17)$$

This is the scattering cross section originally derived by Rutherford. The experiments by Geiger and Marsden in 1911–1913 verified the predictions that the amount of scattering was proportional to $(\sin^4 \theta/2)^{-1}$ and E^{-2}. In addition, they found that the number of elementary charges in the center of the atom is equal to roughly half the atomic weight. This observation introduced the concept of the atomic number of an element, which describes the positive charge carried by the nucleus of the atom. The very experiments that gave rise to the picture of an atom as a positively charged nucleus surrounded by orbiting electrons has now evolved into an important materials analysis technique.

For Coulomb scattering, the distance of closest approach, d, of the projectile to the scattering atom is given by equating the incident kinetic energy, E, to the potential energy at d:

$$d = \frac{Z_1 Z_2 e^2}{E}. \tag{2.18}$$

The scattering cross section can be written as $\sigma(\theta) = (d/4)^2/\sin^4 \theta/2$, which for $180°$ scattering gives $\sigma(180°) = (d/4)^2$. For 2 MeV He$^+$ ions ($Z_1 = 2$) incident on Ag ($Z_2 = 47$),

$$d = \frac{(2)(47) \cdot (1.44\,\text{eV nm})}{2 \times 10^6\,\text{eV}} = 6.8 \times 10^{-5}\,\text{nm},$$

a value much smaller than the Bohr radius $a_0 = \hbar^2/m_e e^2 = 0.053\,\text{nm}$ and the K-shell radius of Ag, $a_0/47 \cong 10^{-3}\,\text{nm}$. Thus the use of an unscreened cross section is justified. The cross section for scattering to $180°$ is

$$\sigma(\theta) = (6.8 \times 10^{-5}\,\text{nm})^2/16 = 2.89 \times 10^{-10}\,\text{nm}^2,$$

a value of $2.89 \times 10^{-24}\,\text{cm}^2$ or 2.89 barns, where the barn $= 10^{-24}\,\text{cm}^2$.

2.6   Scattering Cross Section: Two-Body

In the previous section, we used central forces in which the energy of the incident particle was unchanged through its trajectory. From the kinematics (Section 2.2), we know that the target atom recoils from its initial position, and hence the incident particle loses energy in the collision. The scattering is elastic in that the total kinetic energy of the particles is conserved. Therefore, the change in energy of the scattered particle can be appreciable; for $\theta = 180°$ and ^{4}He$^+$($M_1 = 4$) scattering from Si ($M_2 = 28$), the kinematic factor $K = (24/32)^2 = 0.56$ indicates that nearly one-half the energy is lost by the incident particle. In this section, we evaluate the scattering cross section while including this recoil effect. The derivation of the center of mass to laboratory transformation is given in Section 2.10.

The scattering cross section (Eq. 2.17) was based on the one-body problem of the scattering of a particle by a fixed center of force. However, the second particle is not fixed but recoils from its initial position as a result of the scattering. In general, the two-body central force problem can be reduced to a one-body problem by replacing

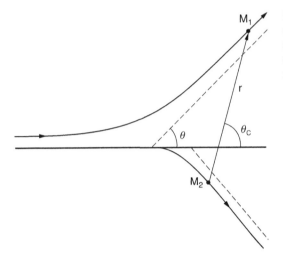

FIGURE 2.9. Scattering of two particles as viewed in the laboratory system, showing the laboratory scattering angle θ and the center of mass scattering angle θ_c.

M_1 by the reduced mass $\mu = M_1 M_2/(M_1 + M_2)$. The matter is not quite that simple as indicated in Fig. 2.9. The laboratory scattering angle θ differs from the angle θ_c calculated from the equivalent, reduced-mass, one-body problem. The two angles would only be the same if the second remains stationary during the scattering (i.e., $M_2 \gg M_1$).

The relation between the scattering angles is

$$\tan \theta = \frac{\sin \theta_C}{\cos \theta_C + M_1/M_2},$$

derived in Eq. 2.24. The transformation gives

$$\sigma(\theta) = \left(\frac{Z_1 Z_2 e^2}{4E}\right)^2 \frac{4}{\sin^4 \theta} \frac{\left(\{1 - [(M_1/M_2)\sin \theta]^2\}^{1/2} + \cos \theta\right)^2}{\{1 - [(M_1/M_2)\sin \theta]^2\}^{1/2}}, \qquad (2.19)$$

which can be expanded for $M_1 \ll M_2$ in a power series to give

$$\sigma(\theta) = \left(\frac{Z_1 Z_2 e^2}{4E}\right)^2 \left[\sin^{-4}\frac{\theta}{2} - 2\left(\frac{M_1}{M_2}\right)^2 + \cdots\right], \qquad (2.20)$$

where the first term omitted is of the order of $(M_1/M_2)^4$. It is clear that the leading term gives the cross section of Eq. 2.17, and that the corrections are generally small. For $He^+ (M_1 = 4)$ incident on Si ($M_2 = 28$), $2(M_1/M_2)^2 \cong 4\%$, even though appreciable energy is lost in the collision. For accurate quantitative analysis, this correction should be included, as the correction can be appreciable for scattering from light atoms such as carbon or oxygen. Cross section values given in Appendix 2 are based on Eq. 2.19. A summary of scattering relations and cross section formulae are given in Table 3.1.

2.7 Deviations from Rutherford Scattering at Low and High Energy

The derivation of the Rutherford scattering cross section is based on a Coulomb inter-action potential $V(r)$ between the particle Z_1 and target atom Z_2. This assumes that the particle velocity is sufficiently large so that the particle penetrates well inside the or-bitals of the atomic electrons. Then scattering is due to the repulsion of two positively charged nuclei of atomic number Z_1 and Z_2. At larger impact parameters found in small-angle scattering of MeV He ions or low-energy, heavy ion collisions (discussed in Chapter 4), the incident particle does not completely penetrate through the electron shells, and hence the innermost electrons screen the charge of the target atom.

We can estimate the energy where these electron screening effects become important. For the Coulomb potential to be valid for backscattering, we require that the distance of closest approach d be smaller than the K-shell electron radius, which can be estimated as a_0/Z_2, where $a_0 = 0.053$ nm, the Bohr radius. Using Eq. 2.18 for the distance of closest approach d, the requirement for d less than the radius sets a lower limit on the energy of the analysis beam and requires that

$$E > Z_1 Z_2^2 \frac{e^2}{a_0}.$$

This energy value corresponds to ~ 10 keV for He$^+$ scattering from silicon and ~ 340 keV for He$^+$ scattering from Au ($Z_2 = 79$). However, deviations from the Rutherford scattering cross section occur at energies greater than the screening limit estimate given above, as part of the trajectory is always outside of the electron cloud.

In Rutherford-backscattering analysis of solids, the influence of screening can be treated to the first order (Chu et al., 1978) by using a screened Coulomb cross section σ_{sc} obtained by multiplying the scattering cross section $\sigma(\theta)$ given in Eqs. 2.19 and 2.20 by a correction factor F,

$$\sigma_{sc} = \sigma(\theta)F, \tag{2.21}$$

where $F = (1 - 0.049 \, Z_1 Z_2^{4/3}/E)$ and E is given in keV. Values of the correction factor are given in Fig. 2.10. With 1 MeV ^{4}He$^+$ ions incident on Au atoms, the correction factor corresponds to only 3%. Consequently, for analysis with 2 MeV ^{4}He$^+$ ions, the screening correction can be neglected for most target elements. At lower analysis energies or with heavier incident ions, screening effects may be important.

At higher energies and small impact parameter values, there can be large departures from the Rutherford scattering cross section due to the interaction of the incident particle with the nucleus of the target atom. Deviations from Rutherford scattering due to nuclear interactions will become important when the distance of closest approach of the projectile-nucleus system becomes comparable to R, the nuclear radius. Although the size of the nucleus is not a uniquely defined quantity, early experiments with alpha-particle scattering indicated that the nuclear radius could be expressed as

$$R = R_0 A^{1/3}, \tag{2.22}$$

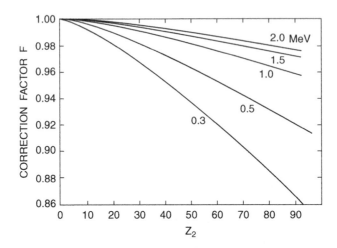

FIGURE 2.10. Correction factor F, which describes the deviation from pure Rutherford scattering due to electron screening for He^+ scattering from the atoms Z_2, at a variety of incident kinetic energies. [Courtesy of John Davies]

where A is the mass number and $R_0 \cong 1.4 \times 10^{-13}$ cm. The radius has values from a few times 10^{-13} cm in light nuclei to about 10^{-12} cm in heavy nuclei. When the distance of closest approach d becomes comparable with the nuclear radius, one should expect deviations from the Rutherford scattering. From Eqs. 2.18 and 2.22, the energy where $R = d$ is

$$E = \frac{Z_1 Z_2 e^2}{R_0 A^{1/3}}.$$

For $^4He^+$ ions incident on silicon, this energy is about 9.6 MeV. Consequently, nuclear reactions and strong deviations from Rutherford scattering should not play a role in backscattering analyses at energies of a few MeV.

One of the exceptions to the estimate given above is the strong increase (resonance) in the scattering cross section at 3.04 MeV for $^4He^+$ ions incident on ^{16}O, as shown in Fig. 2.11. This reaction can be used to increase the sensitivity for the detection of oxygen. Indeed, many nuclear reactions are useful for element detection, as described in Chapter 13.

2.8   Low-Energy Ion Scattering

Whereas MeV ions can penetrate on the order of microns into a solid, low-energy ions (~keV) scatter almost predominantly from the surface layer and are of considerable use for *first monolayer analysis*. In low-energy scattering, incident ions are scattered, via binary events, from the atomic constituents at the surface and are detected by an electrostatic analyzer (Fig. 2.12). Such an analyzer detects only charged

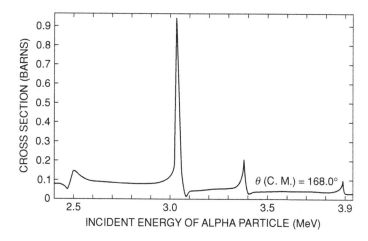

FIGURE 2.11. Cross section as a function of energy for elastic scattering of $^4He^+$ from oxygen. The curve shows the anomalous cross section dependence near 3.0 MeV. For reference, the Rutherford cross section 3.0 MeV is ∼0.037 barns.

particles, and in this energy range ($\cong 1$ keV), particles that penetrate beyond a mono-layer emerge nearly always as neutral atoms. Thus this experimental sensitivity to only charged particles further enhances the surface sensitivity of low-energy ion scattering. The main reasons for the high surface sensitivity of low-energy ion scattering is the charge selectivity of the electrostatic analyzer as well as the very large cross section for scattering.

The kinematic relations between energy and mass given in Eqs. 2.5 and 2.7 re-main unchanged for the 1 keV regime. Mass resolution is determined as before by the energy resolution of the electrostatic detector. The shape of the energy spectrum is, however, considerably different than that with MeV scattering. The spectrum consists of a series of peaks corresponding to the atomic masses of the atoms in the surface layer.

Quantitative analysis in this regime is not straightforward for two primary reasons: (1) uncertainty in the absolute scattering cross section and (2) lack of knowledge of the probability of neutralization of the surface scattered particle. The latter factor is minimized by use of projectiles with a low neutralization probability and use of detection techniques that are insensitive to the charge state of the scattered ion.

Estimates of the scattering cross section are made using screened Coulomb potentials, as discussed in the previous section. The importance of the screening correction is shown in Fig. 2.13, which compares the pure Rutherford scattering cross section to two different forms of the screened Coulomb potential. As mentioned in the previous section, the screening correction for ∼1 MeV He ions is only a few percent (for He$^+$ on Au) but is 2–3 orders of magnitude at ∼1 keV. Quantitative analysis is possible if the scattering potential is known. The largest uncertainty in low-energy ion scattering is not associated with the potential but with the neutralization probability, of relevance when charge sensitive detectors are used.

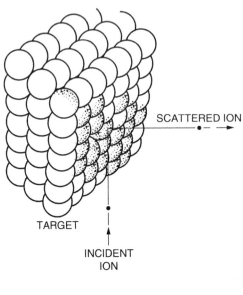

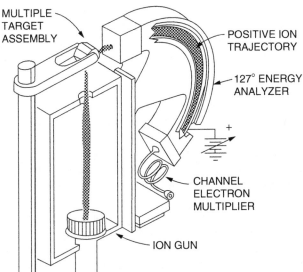

FIGURE 2.12. Schematic of self-contained electrostatic analyzer system used in low-energy ion scattering. The ion source provides a beam of low-energy ions that are scattered (to 90°) from samples held on a multiple target assembly and analyzed in a 127° electrostatic energy analyzer.

Low-energy spectra for ^{3}He and ^{20}Ne ions scattered from an Fe–Re–Mo alloy are shown in Fig. 2.14. The improved mass resolution associated with heavier mass projectiles is used to clearly distinguish the Mo from the Re. This technique is used in studies of surface segregation, where relative changes in the surface composition can readily be obtained.

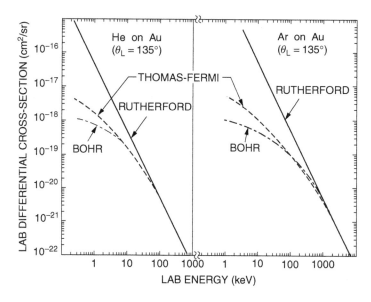

FIGURE 2.13. Energy dependence of the Rutherford, Thomas–Fermi, and Bohr cross sections for a laboratory scattering angle of 135°. The Thomas–Fermi and Bohr potentials are two common approximations to a screened Coulomb potential: (a) He$^+$ on Au and (b) Ar on Au. From J.M. Poate and T.M. Buck, *Experimental Methods in Catalytic Research*, Vol. 3. [Academic Press, New York, 1976, vol. 3.]

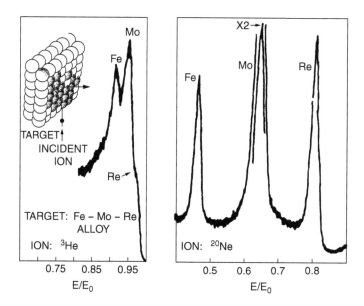

FIGURE 2.14. Energy spectra for ^{3}He$^+$ scattering and ^{20}Ne scattering from a Fe–Mo–Re alloy. Incident energy was 1.5 keV. [From J.T. McKinney and J.A. Leys, 8th *National Conference on Electron Probe Analysis*, New Orleans, LA, 1973.]

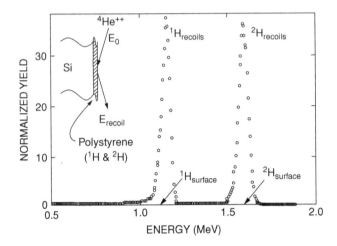

FIGURE 2.15. The forward recoil structures of ^{1}H and ^{2}H (deuterium) from 3.0 MeV ^{4}He$^+$ ions incident on a thin ($\approx$200 nm) polystyrene film on silicon. The detector is placed so that the recoil angle $\phi = 30°$. A mylar film 10 micron (μm) thick mylar film is mounted in front of the detector.

2.9   Forward Recoil Spectrometry

In elastic collisions, particles are not scattered in a backward direction when the mass of the incident particle is equal to or greater than that of the target atom. The incident energy is transferred primarily to the lighter target atom in a recoil collision (Eq. 2.7). The energy of the recoils can be measured by placing the target at a glancing angle (typically 15°) with respect to the beam direction and by moving the detector to a forward angle ($\theta = 30°$), as shown in the inset of Fig. 2.15. This scattering geometry allows detection of hydrogen and deuterium at concentration levels of 0.1 atomic percent and surface coverages of less than a monolayer.

The spectrum for ^{1}H and ^{2}H (deuteron) recoils from a thin polystyrene target are shown in Fig. 2.15. The recoil energy from 3.0 MeV ^{4}He$^+$ irradiation and recoil angle ϕ of 30° can be calculated from Eq. 2.7' to be 1.44 MeV and 2.00 MeV for ^{1}H and ^{2}H, respectively. Since ^{2}H nuclei recoiling from the surface receive a higher fraction ($\sim^2\!/_3$) of the incident energy E_0 than do ^{1}H nuclei ($\sim^1\!/_2$), the peaks in the spectrum are well separated in energy. The energies of the detected recoils are shifted to lower values than the calculated position due to the energy loss in the mylar film placed in front of the detector to block out He$^+$ ions scattered from the substrate.

The application of forward recoil spectrometry to determine hydrogen and deuterium depth profiles is discussed in Chapter 3. The forward recoil geometry can also be used to detect other light-mass species as long as heavy-mass analysis particles are used.

2.10   Center of Mass to Laboratory Transformation

The derivation of the Rutherford cross section assumes a fixed center of force. In practice, the scattering involves two bodies, neither of which is fixed. In general, any

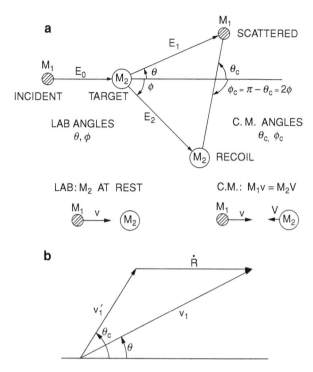

FIGURE 2.16. (a) The relationship between the scattering angles in the laboratory system, namely, θ and ϕ, and the scattering angles in the center of the mass system, θ_c, ϕ_c. (b) Vector diagram illustrating the relationship between the velocity of particle 1 in the laboratory system, $\mathbf{v}_1$, and the velocity in the center of mass system, $\mathbf{v}_1'$.

two-body central force problem can be reduced to a one-body problem. However, since actual measurements are done in the laboratory, one must be aware of the appropriate transformation. The transformation equations yield finite and important *corrections* that must be incorporated in careful analytical work. These corrections are most important when the mass of the projectile, M_1, becomes comparable to the mass of the target M_2. Under these conditions, the recoil effects (nonfixed scattering center) become largest.

The relationship between the scattering angles in the laboratory system, namely, θ and ϕ, and the angles in the center of mass (CM) system are illustrated in Fig. 2.16a. The first step is to determine an analytical relation between the scattering angles in the two systems.

We use the following notation: $\mathbf{r}_1$ and $\mathbf{v}_1$ are the position and velocity vectors of the incident particle in the laboratory system; $\mathbf{r}_1'$ and $\mathbf{v}_1'$ are the position and velocity vectors of the incident particle in the center-of-mass system; and $\mathbf{R}$ and $\dot{\mathbf{R}}$ are the position and velocity vectors of the center of mass in the laboratory system.

By definition,

$$\mathbf{r}_1 = \mathbf{R} + \mathbf{r}_1',$$

so

$$\mathbf{v}_1 = \dot{\mathbf{R}} + \mathbf{v}_1'.$$

The geometrical relationship between vectors and scattering angles shown in Fig. 2.16b indicates the relation

$$\tan\theta = \frac{v_1' \sin\theta_c}{v_1' \cos\theta_c + |\dot{\mathbf{R}}|}. \tag{2.23}$$

The definition of the center-of-mass vector, R, is

$$(M_1 + M_2)\mathbf{R} = M_1\mathbf{r}_1 + M_2\mathbf{r}_2,$$

so that

$$(M_1 + M_2)\dot{\mathbf{R}} = M_1\dot{\mathbf{r}}_1 + M_2\dot{\mathbf{r}}_2,$$

where M_2, $\mathbf{r}_2$ refers to the target atom. From the vector diagram,

$$\mathbf{v}_1' = \mathbf{v}_1 - \dot{\mathbf{R}},$$

or

$$\mathbf{v}_1' = \frac{M_2}{M_1 + M_2}(\dot{\mathbf{r}}_1 - \dot{\mathbf{r}}_2).$$

Since the system is conservative, the relative velocity, $\dot{\mathbf{r}}_1 - \dot{\mathbf{r}}_2$, is the same before and after the collision. Initially, $\dot{\mathbf{r}}_2 = 0$, so

$$v_1' = \frac{M_2}{M_1 + M_2}v,$$

where v is the initial velocity of the particle. The constant velocity of the CM can also be derived from the definition:

$$(M_1 + M_2)\dot{\mathbf{R}} = M_1\mathbf{v}.$$

Substituting the relations for $\dot{\mathbf{R}}$ and v_1' in Eq. 2.23, we have

$$\tan\theta = \frac{\sin\theta_c}{\cos\theta_c + \frac{M_1}{M_2}}. \tag{2.24}$$

When $M_1 \ll M_2$, the angles in the two systems are approximately equal; the massive scatterer M_2 suffers little recoil. A useful form of Eq. 2.24 is written as

$$\cot\theta = \cot\theta_c + x\csc\theta_c,$$

where $x = M_1/M_2$. This can be rearranged to yield

$$\cot\theta - \cot\theta_c = x\csc\theta_c,$$

or

$$\sin\theta_c \cos\theta - \cos\theta_c \sin\theta = x\sin\theta$$

so that

$$\sin(\theta_c - \theta) = x\sin\theta. \tag{2.25}$$

For simplicity, we let $\xi = \theta_c - \theta$. From Eq. 2.25, we have

$$\cos\xi \, d\xi = x\cos\theta \, d\theta$$

and

$$\frac{d\xi}{d\theta} = \frac{d\theta_c}{d\theta} - 1 = \frac{x \cos \theta}{\cos \xi},$$

or

$$\frac{d\theta_c}{d\theta} = \frac{\sin \theta_c}{\sin \theta \cos \xi}.$$

Then

$$\frac{d\sigma}{d\Omega} = \left(\frac{Z_1 Z_2 e^2}{2E}\right)^2 \left[\frac{(1+x)\sin \theta_c}{2 \sin \theta \sin^2 \theta_c/2}\right]^2 / \cos \xi,$$

where E, the energy in lab coordinates, is given by

$$E = E_c(1+x).$$

It is useful to derive an expression for the cross section simply in terms of θ and x. We make use of the fact that

$$1 + x = (\sin \xi + \sin \theta)/ \sin \theta,$$

$$\sin \left(\theta_c/2\right) \sin^2 \left(\theta_c/2\right) = \cot \left(\theta_c/2\right),$$

$$\frac{\sin \theta + \sin \xi}{\cos \theta + \cos \xi} = \tan \frac{\theta_c}{2},$$

so that

$$\frac{(1+x)\sin \xi_c}{2 \sin^2 \left(\theta_c/2\right)} = \frac{\cos \theta + \cos \xi}{\sin \theta},$$

and

$$\frac{d\sigma}{d\Omega} = \left(\frac{Z_1 Z_2 e^2}{2E}\right)^2 \frac{(\cos \theta + \cos \xi)^2}{\sin^4 \theta \cos \xi}.$$

Noting that $\cos \xi = (1 - \sin^2 \xi)^{1/2} = (1 - x^2 \sin^2 \theta)^{1/2}$, we obtain

$$\frac{d\sigma}{d\Omega} = \left(\frac{Z_1 Z_2 e^2}{2E}\right)^2 \frac{[\cos \theta + (1 - x^2 \sin^2 \theta)^{1/2}]^2}{\sin^4 \theta (1 - x^2 \sin^2 \theta)^{1/2}}, \tag{2.26}$$

which is the form given in Eq. 2.19.

Problems

2.1. ^{4}He^{++} particles are scattered from a thin foil of an elemental material with atomic number Z_1, mass density ρ_1, number A_1, and thickness t_1, and are observed at some fixed angle θ. The first foil is replaced with a second one (Z_2, ρ_2, A_2, t_2). What is the ratio of the number of particles observed at θ for the first and second foils?

2.2. A beam of 2 MeV helium ions is incident on a silver foil 10^{-6} cm thick and undergoes Coulomb scattering in accordance with the Rutherford formula.

(a) What is the distance of closest approach?
(b) Find the impact parameter for He^+ ions scattered through $90°$.
(c) What fraction of the incident 2 MeV He^+ ions will be backscattered (i.e., $\theta > 90°$)?

The density of silver is $10.50 \, g/cm^3$, and its atomic weight is $107.88 \, g$ mol. [Hint: The integrated cross section for scattering through angles $0°$ to $90°$ is $\int_0^{\pi/2} d\sigma$.]

2.3. An α particle, $^4He^+$, makes a head-on collision with (a) a gold nucleus, (b) a carbon nucleus, (c) an α particle, and (d) an electron, each initially at rest. What fraction of the α particle's initial kinetic energy is transferred to the struck particle in each instance?

2.4. (a) Using the formula for the Rutherford scattering cross section in center-of-mass coordinates (CM) and the relations for the recoil energy (Eq. 2.7), and noting that $\phi = \pi/2 - \theta_c/2$ (Fig. 2.16), calculate an expression for $d\sigma/dE_2$, the cross section for transferring an energy E_2 to a nucleus. Hint:

$$\frac{d\sigma}{d\theta_c} \frac{d\theta_c}{dE_2} = \frac{d\sigma}{dE_2}.$$

(b) Using the result of part (a), integrate $d\sigma/dE_2$ from E_{min} to E_{max} to find the *total* cross section for transferring an energy greater than E_{min}.

(c) Evaluate the result of part (b) in cm^2 for the case of 1.0 MeV He ions bombarding Si. Use $E_{min} = 14 \, eV$, the displacement energy of a Si atom bound in an Si lattice. Compare this cross section to $\sigma_{Ruth}(\theta = 180°)$.

(d) Use the result of (c) to calculate the fraction of atoms displaced (i.e., undergoing an energy transfer greater than 14 eV) for 1 μC of He^+ ions incident on a target where the He^+ beam diameter $= 1$ mm. This result is only a lower limit to the displacements, since we have ignored displacements due to recoiling Si atoms.

2.5. A carbon film is known to contain surface contaminants of Au, Ag, and Si. Sketch the backscattering spectrum, indicating the energies of the various peaks and their relative heights.

2.6. An accelerator produces an He^+ ion current of 50 nA at 1.0 MeV. Using a 1 cm^2 detector 5 cm from the target at a scattering angle of $170°$, determine the smallest amount of Au (atoms/cm^2) that can be detected. Detectability is arbitrarily defined as 100 counts in 1 hr. Under similar conditions, what is the detection limit for oxygen? Compare these limits with the number of atoms/cm^2 in a monolayer ($\cong 10^{15}$ atoms/cm^2).

2.7. Derive the expression for E_2 (Eq. 2.7'), the energy transferred to the target atom using conservation of energy and momentum relations. Give the expression for E_2/E_0 for $\theta = 90°$.

2.8. Use the small angle approximation ($\sin \theta = \theta$) to show that the scattering cross section can be expressed as $\sigma(\theta) = (Z_1 Z_2 e^2/E)^2(\theta)^{-4}$. Derive this expression using the impulse approximation in which the force of $Z_1 Z_2 e^2/b^2$ acts on the particle for an effective time $t = l/v$, where $l = 2b$. [Hint: An intermediate step in the derivation is to show that $b = Z_1 Z_2 e^2/E\theta$.]

References

1. W. K. Chu, J. W. Mayer, and M.-A. Nicolet, *Backscattering Spectrometry* (Academic Press, New York, 1978).
2. H. Goldstein, *Classical Mechanics* (Addison-Wesley, Reading, MA, 1959).
3. J. W. Mayer and E. Rimini, *Ion Beam Handbook for Material Analysis* (Academic Press, New York, 1977).
4. F. K. Richtmyer, E. H. Kennard, and T. N. Cooper, *Introduction to Modern Physics*, 6th ed. (McGraw-Hill, New York, 1969).
5. P. A. Tipler, *Modern Physics* (Worth Publishers, New York, 1978).

3
Energy Loss of Light Ions and Backscattering Depth Profiles

3.1 Introduction

In the previous chapter, it was tacitly assumed that the atoms to be identified were at the surface of the materials. In this chapter, we consider composition depth profiles that can be obtained from Rutherford backscattering spectrometry (RBS) and nuclear reaction analysis. In this case, the depth scale is established by the energy loss dE/dx of light (H^+, d^+, and He^+) ions at high energies (0.5–5 MeV) during their passage through the solid (Fig. 3.1). The energy lost in penetration is directly proportional to the thickness of material traversed, so a depth scale can be assigned directly and quantitatively to the energy spectra of detected particles. The yield of backscattered particles or reaction products is proportional to the scattering or reaction cross sections, so the composition depth profile can be found from knowledge of energy loss and cross sections.

3.2 General Picture of Energy Loss and Units of Energy Loss

For light ions such as $^4He^+$ penetrating a solid, the energetic particles lose energy primarily through excitation and ionization in inelastic collisions with atomic electrons—termed *electronic-energy loss*. Microscopically, energy loss due to excitation and ionization is a discrete process. Macroscopically, however, it is a good assumption that the moving ions lose energy continuously. All we are concerned with here is the average energy loss during the penetration of ions into a given material.

To measure energy loss, we must determine two quantities: the distance Δt that the ions traverse in the target, and the energy loss ΔE in this distance. The mass density ρ or the atomic density N are frequently combined with the distance, in the form $\rho \Delta t$ or $N \Delta t$, to express the amount of material per unit area or the number of atoms per unit area that the projectiles have traversed in losing energy ΔE to the target material. Energy loss can be expressed in several different ways. Some frequently used units are

$$dE/dx : \text{eV/nm},$$
$$(1/\rho) \, dE/dx : \text{eV/}(\mu\text{g/cm}^2),$$
$$\varepsilon = (1/N) \, dE/dx : \text{eV/}(\text{atoms/cm}^2), \text{ eV-cm}^2.$$

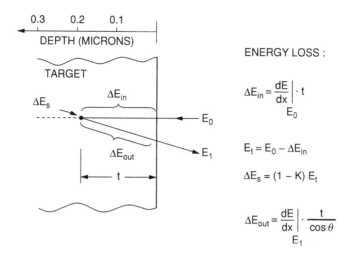

DEPTH (MICRONS)

TARGET

ENERGY LOSS :

$$\Delta E_{in} \simeq \left.\frac{dE}{dx}\right|_{E_0} \cdot t$$

$$E_t = E_0 - \Delta E_{in}$$

$$\Delta E_s = (1 - K)\, E_t$$

$$\Delta E_{out} \simeq \left.\frac{dE}{dx}\right|_{E_1} \cdot \frac{t}{\cos\theta}$$

FIGURE 3.1. Energy-loss components for a projectile that scatters from depth t. The sequence is: energy lost via electronic stopping on inward path ΔE_{in}; energy lost in the elastic scattering process, ΔE_{in}; and energy lost to electronic stopping in the outward path, ΔE_{out}. Then $E_1 = E_0 - \Delta E_{in} - \Delta E_s - \Delta E_{out}$.

Recently, most authors have adopted $(1/N)\, dE/dx$ (eV-cm^2) as the stopping cross section ε; we give the ^{4}He$^+$ stopping cross section in these units in Appendix 3.

3.3 Energy Loss of MeV Light Ions in Solids

3.3.1 Applicable Energy Ranges

When a He$^+$ or H ion moves through matter, it loses energy through interactions with electrons that are raised to excited states or ejected from atoms. The radii of atomic nuclei are so small compared with atomic dimensions that nuclear scattering is rare compared with interactions with electrons; therefore, in a first approximation, nuclear interactions may be neglected in the slowing down process.

Theoretical treatments of inelastic collisions of charged particles with target atoms or molecules are separated into fast collisions and slow collisions. The criterion is the velocity of the projectile relative to the mean orbital velocity of the atomic or molecular electrons in the shell or subshell of a given target atom. When the projectile velocity v is much greater than that of an orbital electron (fast-collision case), the influence of the incident particle on an atom may be regarded as a sudden, small external perturbation. This picture leads to Bohr's theory of stopping power. The collision produces a sudden transfer of energy from the projectile to the target electron. The energy loss of a fast particle to a stationary nucleus or electron can be calculated from scattering in a central-force field. The stopping cross section decreases with increasing velocity because the particle spends less time in the vicinity of the atom. In the low-energy (slow velocity) regime, this argument does not hold, and it is found that the stopping power is proportional to velocity. The maximum in the stopping cross section

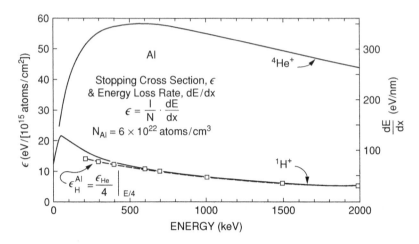

FIGURE 3.2. Stopping cross section ϵ and energy loss rate dE/dx ^{4}He$^+$ and ^{1}H$^+$ in Al. The open squares for the hydrogen data are scaled from the He$^+$ data by evaluating the stopping powers at the same velocity $(E/4)$ and scaling by 4 for the Z_1 dependence. The atomic density of Al is 6×10^{22} atoms/cm^3.

is found at the energy separating these two regions. In backscattering spectrometry, we are concerned with the region near and above the maximum.

One useful estimate of the lower energy limit of the fast-collision case is to compare the particle velocity or energy with the Bohr velocity v_0 of an electron in the innermost orbit of a hydrogen atom:

$$v_0 = \frac{e^2}{\hbar} = \frac{c}{137} = 2.2 \times 10^8 \text{cm/s}. \tag{3.1}$$

This velocity is equivalent to that of a 0.1 MeV He ion or 25 keV H ion. As shown in Fig. 3.2, the values of the energy loss reach a maximum around 0.5 MeV for He in Al. In this high-energy, fast-collision regime, values of dE/dx are proportional to Z_1^2 at the same velocity. The dashed curve and squares in Fig. 3.2 are the values for the energy loss of ^{4}He ions in Al reduced by a factor of four $(Z_{1H}^2/Z_{1He}^2 = 1/4)$ and plotted at an energy of one-quarter of that for the He ion $(M_H/M_{He}) = 1/4)$, i.e., at the same velocity. The incident He or H particle is considered as fully ionized (He$^+$ or α particle, H$^+$ or proton) in its passage through matter—the velocity of the particle is sufficiently great so that it is stripped of its electrons. At lower velocities, the average charge of the projectile becomes lower and the number of electrons available for excitation decreases; therefore, the energy loss decreases.

3.3.2   Derivation of dE/dx

In 1913 Bohr derived an expression for the rate of energy loss of a charged particle on the basis of classical considerations. He considered a heavy particle, such as a particle or a proton, of charge Z_1e, mass M, and velocity v passing an atom electron of mass m_e at a distance b (Fig. 3.3). As the heavy particle passes, the Coulomb force

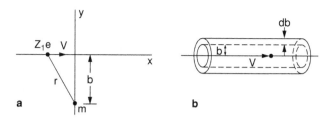

FIGURE 3.3. (a) A heavy particle of charge $Z_1 e$ passing an electron at distance b (b) A shell of radius b and thickness db with its axis the path of the heavy charged particle.

acting on the electron changes direction continuously. If the electron moves negligibly during the passage of the heavy particle, the impulse, $\int F dt$, parallel to the path is zero by symmetry, since for each position of the incident particle in the $-x$ direction there is a corresponding position in the $+x$ direction that makes an equal and opposite contribution to the x component of the momentum. Throughout the passage, however, there is a force in the y direction, and momentum Δp is transferred to the electron. This problem of energy transfer is similar to the Coulomb force scattering used to derive the Rutherford scattering law in Chapter 2, if we think of the electron moving towards the *stationary* projectile at the same velocity v. The momentum transferred to the electron during the full passage is therefore

$$\Delta p = \frac{2 Z_1 e^2}{b v},\tag{3.2}$$

where we have used Eq. 2.15a with $\theta \cong 0$, a small angle approximation. If the electron has not achieved a relativistic velocity, its kinetic energy is given by

$$\frac{\Delta p^2}{2m} = \frac{2 Z_1^2 e^4}{b^2 m v^2} = T,\tag{3.3}$$

where T is the energy transfer in the collision.

The differential cross section, $d\sigma(T)$, for an energy transfer between T and $T + dT$ is

$$d\sigma(T) = -2\pi b \, db,\tag{3.4}$$

and the energy loss per unit path length, dE/dx, is

$$-\frac{dE}{dx} = n \int_{T_{\min}}^{T_{\max}} T \, d\sigma,\tag{3.5}$$

where n is the number of electrons per unit volume. In terms of impact parameter b,

$$-\frac{dE}{dx} = n \int_{b_{\min}}^{b_{\max}} T \, 2\pi b \, db,\tag{3.6}$$

which reduces to

$$-\frac{dE}{dx} = \frac{4\pi Z_1^2 e^4 n}{mv^2} \ln \frac{b_{max}}{b_{min}}. \tag{3.7}$$

To choose a meaningful value for b_{min}, we observe that if the heavy particle collides head on with the electron, the maximum velocity transferred to a stationary electron is $2v$. The corresponding maximum kinetic energy (for a nonrelativistic v) is $T_{max} = \frac{1}{2}m(2v)^2 = 2mv^2$. If this value of T_{max} is inserted in Eq. 3.3, the corresponding b_{min} becomes

$$b_{min} = \frac{Z_1 e^2}{mv^2}. \tag{3.8}$$

If b_{max} is allowed to become infinite, $-dE/dx$ goes to infinity because of the contribution of an unlimited number of small energy transfers given to distant electrons. But the smallest energy an atomic electron can accept must be sufficient to raise it to an allowed excited state. If I represents the average excitation energy of an electron, we choose $T_{min} = I$, and find

$$b_{max} = \frac{2Z_1 e^2}{\sqrt{2mv^2 I}}. \tag{3.9}$$

When Eqs. 3.8 and 3.9 are substituted in Eq. 3.7, we obtain

$$-\frac{dE}{dx} = \frac{2\pi Z_1^2 e^4 n}{mv^2} \ln \frac{2mv^2}{I}.$$

This calculation is based on direct collisions with electrons in the solid. There is another term of comparable magnitude due to distant resonant energy transfer. The derivation is outside the scope of this book but leads, in its simplest form, to a total stopping power twice that shown above, i.e.,

$$-\frac{dE}{dx} = \frac{4\pi Z_1^2 e^4 n}{mv^2} \ln \frac{2mv^2}{I}, \tag{3.10}$$

or

$$-\frac{dE}{dx} = \frac{2\pi Z_1^2 e^4}{E} N Z_2 \left(\frac{M_1}{m}\right) \ln \frac{2mv^2}{I},$$

where $E = M_1 v/2$ and $n = N Z_2$, with N given by the atomic density in the stopping medium.

Thus we can regard the electronic interactions as composed of two contributions: (1) close collisions with large momentum transfers, where the particle approaches within the electronic orbits, and (2) distant collisions with small momentum transfers, where the particle is outside the orbits. The two contributions are nearly equal (equipartition rule) for the particle velocities used in Rutherford backscattering.

The average excitation energy I for most elements is roughly $10 Z_2$ in eV, where Z_2 is the atomic number of the stopping atoms. Experimental and calculated values of I are given in Fig. 3.4. The description of stopping power so far ignores the shell structure of the atoms and variations in electron binding. Experimentally, these effects show up as small deviations (except for the very light elements) from the $10 Z_2$ approximations shown in Fig. 3.4.

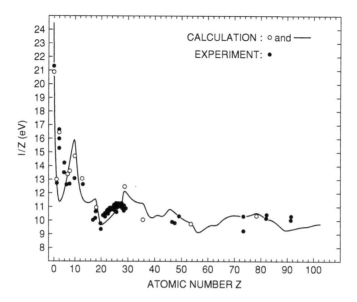

FIGURE 3.4. Calculation of mean excitation energy by Lindhard and Scharff's theory with a Hartree–Fock–Slater charge distribution. The calculation $1/Z$ vs. atomic number Z reveals structure, as has been observed in many experimental measurements. [From W.K. Chu and D. Powers, *Phys. Lett.* 40A, 23 (1972), with permission from Elsevier.]

The complete energy loss formula (often referred to as the *Bethe formula*) contains corrections that include relativistic terms at high velocities and corrections for the nonparticipation of the strongly bound inner-shell electrons. For helium ions in the energy regime of a few MeV, relativistic effects are negligible and nearly all the target electrons participate ($n = NZ_2$) in the stopping process. Consequently, Eq. 3.10 can be used to estimate values of dE/dx. In analysis, it is preferable to use tabulated or numerical values such as those listed in Appendix 3.

For example, the electronic energy loss of 2 MeV ^{4}He$^+$ ions in Al has a value (calculated from Eq. 3.10) of 315 eV/nm using values of $n = NZ_2 = 780/\mathrm{nm}^3$ and $I = 10Z_2 = 130\,\mathrm{eV}$. The value given in Appendix 3 is $\varepsilon = 44.25\,\mathrm{eV}/(10^{15}\,\mathrm{atoms/cm}^2)$ or a value of $dE/dx = 266\,\mathrm{eV/nm}$ ($dE/dx = \varepsilon N$). Thus the first-order treatment gives values to within 20% of the experimental values.

3.3.3 Comparison of Energy Loss to Electrons and to Nuclei

A penetrating He ion can also transfer energy to the nuclei of the solid through small-angle scattering events. This component to the total energy loss of an energetic ion is termed *nuclear energy loss*. Nuclear energy loss is much smaller than the electronic loss. If the derivation of Eq. 3.7 is repeated for collisions with target atoms, we find

$$-\left.\frac{dE}{dx}\right|_n = \frac{4\pi Z_2^2 Z_1^2 e^4 N}{M_2 v^2} \ln \frac{b_{\max}}{b_{\min}}, \tag{3.11}$$

where b_{min} corresponds to the maximum energy transfer to a target atom in a head-on collision,

$$T_{max} = \frac{4M_1 M_2}{(M_1 + M_2)^2} E, \tag{3.12}$$

so that b_{min} becomes

$$b_{min} = \frac{2Z_1 Z_2 e^2}{vp}, \tag{3.13}$$

and b_{max} can be approximated by Eq. 3.9, with I now representing a displacement energy.

In a comparison of Eqs. 3.7 and 3.11, the major differences are the mass (m or M_2) in the denominator and the charge Z_2 of the target atom. For protons, neglecting the ratio of log terms, the ratio of nuclear to electronic energy loss per atom is

$$\frac{dE/dx|_n}{dE/dx|_e} \cong N\frac{Z_2^2}{M_2} \times \frac{m}{n} = \frac{Z_2}{M_2}m \cong \frac{1}{3600}, \tag{3.14}$$

where the number of electrons per volume $n \cong Z_2 N$, $M_2 \cong 2\,Z_2 m_p$, and $m_p \cong 1836\,m_e$ is the mass of the proton.

3.4 Energy Loss in Compounds—Bragg's Rule

The process by which a particle loses energy when it moves swiftly through a medium consists of a random sequence of independent encounters between the moving projectile and an electron attached to an atom of the solid. For a target that contains more than one element, the energy loss is the sum of the losses of the constituent elements weighted by the abundance of the elements. This postulate is known as *Bragg's rule* and states that the stopping cross section $\varepsilon^{A_m B_n}$ of a solid of composition $A_m B_n$ is given by

$$\varepsilon^{A_m B_n} = m\varepsilon^A + n\varepsilon^B, \tag{3.15}$$

where ε^A and ε^B are the stopping cross sections of the atomic constituents A and B.

To take the specific example of SiO_2 on a molecular basis,

$$\varepsilon^{SiO_2} = \varepsilon^{Si} + 2\varepsilon^O, \tag{3.16}$$

where ε^{SiO_2} is now the stopping power/molecule, so $dE/dx = N\varepsilon^{SiO_2}$, where N is the number of molecules/volume. Figure 3.5 shows the stopping cross section for SiO_2 on a molecular basis.

The energy loss value, dE/dx, for 2.0 MeV He is 283 eV/nm, close to the value of elemental Si, namely, 246 eV/nm.

3.5 The Energy Width in Backscattering

As MeV He ions traverse the solid, they lose energy along their incident path at a rate dE/dx between 300 and 600 eV/nm. In thin film analysis, to a good approximation,

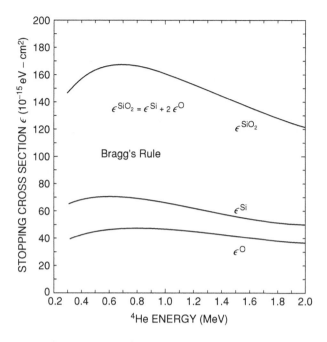

FIGURE 3.5. Stopping cross sections for ^{4}He$^+$ ions in Si, O, and SiO$_2$. The oxide stopping cross section was determined on the molecular basis ε^{SiO_2}, on the assumption that Bragg's rule of linear additivity holds, with 2.3×10^{22} SiO$_2$ molecules/cm^2.

the total energy loss ΔE into a depth t is proportional to t. That is,

$$\Delta E_{in} = \int_0^t \frac{dE}{dx} dx \cong \left. \frac{dE}{dx} \right|_{in} \cdot t, \tag{3.17}$$

where $dE/dx|_{in}$ is evaluated at some average energy between the incident energy E_o and $E_o - t(dE/dx)$. The energy of a particle at depth t is

$$E(t) = E_o - t \, dE/dx|_{in}. \tag{3.18}$$

After large-angle scattering, the particle energy is $KE(t)$, where K is the kinematic factor defined in Eq. 2.5. The particle loses energy along the outward path and emerges with an energy

$$E_1(t) = KE(t) - \frac{t}{|\cos \theta|} \left. \frac{dE}{dx} \right|_{out} = -t \left(K \left. \frac{dE}{dx} \right|_{in} + \frac{1}{|\cos \theta|} \left. \frac{dE}{dx} \right|_{out} \right) + KE_o, \tag{3.19}$$

where θ is the scattering angle. The energy width ΔE of the signal from a film of thickness Δt is

$$\Delta E = \Delta t \left(K \left. \frac{dE}{dx} \right|_{in} + \frac{1}{|\cos \theta|} \left. \frac{dE}{dx} \right|_{out} \right) = \Delta t [S]. \tag{3.20a}$$

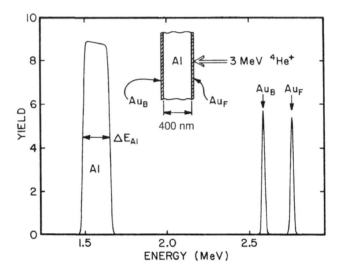

FIGURE 3.6. The backscattering spectrum ($\theta = 170°$) for 3.0 MeV He$^+$ ions incident on a 400 nm Al film with thin Au markers on the front and back surfaces.

The subscripts *in* and *out* refer to the energies at which dE/dx is evaluated, and $[S]$ is often referred to as the *backscattering energy loss factor*. The backscattering spectrum at $\theta = 170°$ for 3 MeV ^{4}He incident on a 400 nm Al film with thin Au markers ($\cong 3$ monolayers of Au) on the front and back surfaces is shown in Fig. 3.6. The energy loss rate dE/dx along the inward path in Al is $\cong 220$ eV/nm at energies of 3 MeV and is $\cong 290$ eV/nm on the outward path at energies of about 1.5 MeV ($K_{Al} \cong 0.55$). Inserting these values into Eq. 3.20, we obtain an energy width ΔE_{Al} of 165 keV. The energy separation between the two Au peaks is slightly larger, 175 keV, because one uses K_{Au} in Eq. 3.20 along with dE/dx (Al).

The assumption of constant values for dE/dx or ε along the inward and outward tracks leads to a linear relation between ΔE and the depth t at which scattering occurs. For thin films, $\Delta t \leq 100$ nm, the relative change in energy along the paths is small. In evaluating dE/dx, one can use the *surface-energy approximation*, in which $(dE/dX)_{in}$ is evaluated at E_0 and $(dE/dx)_{out}$ is evaluated at ΔE_0. In this approximation, the energy width ΔE_0 from a film of thickness Δt is

$$\Delta E_0 = \Delta t[S_0] = \Delta t \left[K \frac{dE}{dx} \bigg|_{E_0} + \frac{1}{|\cos \theta|} \frac{dE}{dx} \bigg|_{KE_0} \right], \qquad (3.20b)$$

where the subscripts denote the surface-energy approximation. When the film thickness or the path length becomes appreciable, a better approximation can be made by selecting a constant value of dE/dx at a mean energy $\bar{E}$ intermediate between that at the end points of each track. For the inward track, the incident particle enters at energy E_0 and has an energy $E(\Delta t)$ before scattering at Δt so that $\bar{E}_{in} = \frac{1}{2}[E(\Delta t) + E_0]$. After scattering, the particle has an energy $KE(\Delta t)$ so that $\bar{E}_{out} = \frac{1}{2}[E_1 + KE(\Delta t)]$. In this *mean-energy approximation*, the energy $E(\Delta t)$ before scattering can be calculated from

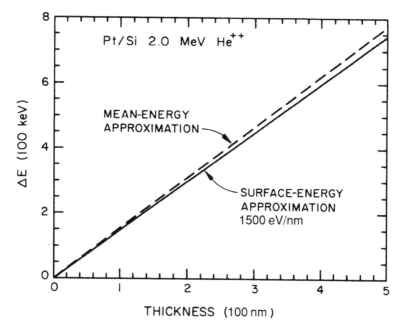

FIGURE 3.7. Comparison of the calculated relationship between the energy width, ΔE, and thickness for a Pt film. In the surface-energy approximation, the values of dE/dx are evaluated at the incident energy, E_0 on the inward path and KE_0 on the outward path. In the mean-energy approximation, dE/dx is evaluated at appropriate average energies as discussed in the text.

values of dE/dx or can be further approximated by assuming that the energy difference ΔE is measured or known and that this loss is subdivided equally between the incident and the outward path so that E is approximately $E_0 - \frac{1}{2}\Delta E$. Then $\bar{E}_{\text{in}} = E_0 - \frac{1}{4}\Delta E$ and $\bar{E}_{\text{out}} = E_1 + \frac{1}{4}\Delta E$.

A comparison between the surface-energy and the mean-energy approximations is shown in Fig. 3.7 for 2.0 MeV He scattering from a Pt film. In the surface-energy approximation, the conversion between energy width ΔE and thickness Δt is 1500 eV/nm. In the mean-energy approximation, the ΔE versus Δt relation deviates from a straight line relation, and the value of ΔE for a 500 nm film exceeds by about 3% the value from the surface-energy approximation. The comparison between the mean-energy and surface-energy approximations serves as a quick estimate of the probable error introduced by using the surface-energy approximation. The main point here is that a backscattering spectrum can be thought of as a linear depth profile of the elements within the sample.

3.6   The Shape of the Backscattering Spectrum

The energy spectrum from an infinitely thick target has a characteristic slope (Fig. 3.8) that can be understood from these relations between depth and energy loss and the energy dependence of the Rutherford cross section. In backscattering measurements,

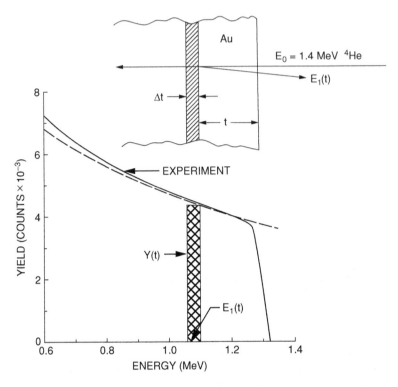

FIGURE 3.8. Backscattering spectrum for 1.4 MeV He$^+$ ions incident on a thick Au sample. The dashed line is calculated based on Eq. 3.25 and normalized to an experimental curve at 1.3 MeV.

the detector subtends a solid angle Ω so that the total number of detected particles Q_D or yield Y from a thin layer of atoms, Δt, is

$$Y = \sigma(\theta)\,\Omega QN\Delta t, \tag{3.21}$$

where Q is the measured number of incident particles and $(N\Delta t)$ is the number of target atoms/cm^2 in the layer.

For thicker layers or bulk targets, projectiles can scatter from any depth t, resulting in a continuous energy spectrum to low energy. The yield from a slice of width Δt at depth t is given (for $\theta = 180°$) by

$$Y(t) = \left(\frac{Z_1 Z_2 e^2}{4E(t)}\right)^2 NQ\Omega\,\Delta t, \tag{3.22}$$

where $E(t)$ is the energy of the particle at depth t Eq. 3.18 and N is the atomic density. In backscattering, one measures the spectrum of particles emerging with energy E_1. To convert Eq. 3.22 to a spectrum $Y(E_1)dE_1$ of the measured energy E_1, we note that $E(t)$ is an intermediate energy between E_0 and E_1. If we denote ΔE_{in} as the energy lost on the inward path, $\Delta E_{in} = E_0 - E(t)$, and ΔE_{out} as the energy lost on the outgoing path, $\Delta E_{out} = KE(t) - E_1$, the ratio

$$A = \frac{\Delta E_{out}}{\Delta E_{in}} = \frac{KE(t) - E_1}{E_0 - E(t)} \approx \frac{dE/dx|_{out}}{dE/dx|_{in}}. \tag{3.23}$$

This ratio is approximately constant for slowly varying energy-loss values, as is the case for 2.0 MeV He ions. Then the energy E at depth t is

$$E(t) = \frac{E_1 + AE_0}{K + A}. \tag{3.24}$$

The value of A can be determined explicitly, but for medium to heavy mass targets where $K \cong 1$ and $A \cong 1$, the value of $E(t) \cong (E_0 + E_1)/2$ and

$$Y(E_1) \propto \frac{1}{(E_0 + E_1)^2}. \tag{3.25}$$

This spectral shape for $E_0 = 1.4$ MeV is indicated in Fig. 3.7.

The shape of backscattering spectra and depth profiles can be obtained from computer programs (i.e., RUMP, Doolittle, 1985), which are used in both simulation and analysis of RBS data. Simulations of energy widths and signal heights, for example, guides in the design of sample configuration and scattering geometry.

3.7   Depth Profiles with Rutherford Scattering

The energy loss of light ions follows a well-behaved pattern in the MeV energy range. The values of dE/dx or ε can be used to obtain composition depth profiles from the energy spectra of backscattered particles or particles emitted in nuclear reactions. We illustrate the technique with backscattering spectra from an implanted Si substrate and a thin film on Si.

For dilute concentrations of an impurity, ≤ 1 atomic %, the stopping power is simply determined by the host. Figure 3.9 shows a spectrum of As implanted into Si. The conversion of the energy scale to a depth scale is given by Eq. 3.20 using $K = K_{As}$ and dE/dx for silicon. The shift, ΔE_{As}, indicates that the As is implanted below the surface of the Si.

The upper section in Fig. 3.10 shows a 100 nm Ni film on Si. Nearly all of the incident $^4He^+$ beam penetrates microns into the target before it is stopped. Particles scattered from the front surface of the Ni have an energy given by the kinematic equation, $E_1 = E_0 K$, where the kinematic factor K for $^4He^+$ backscattered at a laboratory angle of 170° is 0.76 for Ni and 0.57 for Si.

As particles traverse the solid, they lose energy along their incident path at a rate of about 640 eV/nm (assuming a bulk density for Ni of 8.9 g-cm^{-3}). In thin film analysis, to a good approximation, energy loss is linear with thickness. Thus, a 2 MeV particle will lose 64 keV penetrating to the Ni–Si interface. Immediately after scattering from the interface, particles scattered from Ni will have an energy of 1477 keV derived from $K_{Ni} \times (E_0 - 64)$. On their outward path, particles will have slightly different energy loss due to the energy dependence of the energy loss processes, in this case 690 eV nm^{-1}. On emerging from the surface, the $^4He^+$ ions scattered from Ni at the interface will have an energy of 1408 keV. The total energy difference ΔE between particles

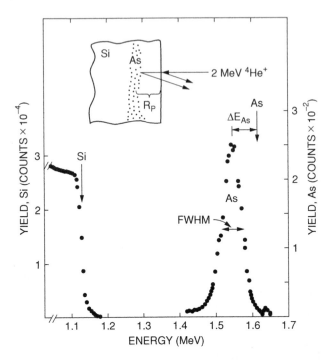

FIGURE 3.9. Energy spectrum of 2 MeV ^{4}He$^+$ ions backscattered from a silicon crystal implanted with a nominal dose of 1.2×10^{15} As ions/cm^2 at 250 keV. The vertical arrows indicate the energies of particles scattered from surface atoms of ^{28}Si and ^{75}As.

scattered at the surface and near the interface is 118 keV, a value that can be derived from Eq. 3.20.

In general, one is interested in reaction products or interdiffusion profiles, and the lower portion of Fig. 3.10 shows schematically a Ni film reacted to form Ni$_2$Si. After reaction, the Ni signal ΔE_{Ni} has spread slightly, owing to the presence of Si atoms contributing to the energy loss. The Si signal exhibits a step corresponding to Si in the Ni$_2$Si. It should be noted that the ratio of the heights H_{Ni}/H_{Si} of Ni to Si in the silicide layer gives the composition of the layer. To a first approximation, the expression of the concentration ratio is given by

$$\frac{N_{Ni}}{N_{Si}} = \frac{H_{Ni}}{H_{Si}} \frac{\sigma_{Si}}{\sigma_{Ni}} \simeq \frac{H_{Ni}}{H_{Si}} \left(\frac{Z_{Si}}{Z_{Ni}} \right)^2, \tag{3.26}$$

where we have ignored the difference in stopping cross sections along the outward path for particles scattered from Ni and Si atoms. The yield from the Ni or Si in the silicide is given closely by the product of signal height and energy with ΔE. Therefore, a better approximation to the concentration ratio of two elements A and B uniformly distributed within a film is

$$\frac{N_A}{N_B} = \frac{H_A \Delta E_A \sigma_B}{H_B \Delta E_B \sigma_A}. \tag{3.27}$$

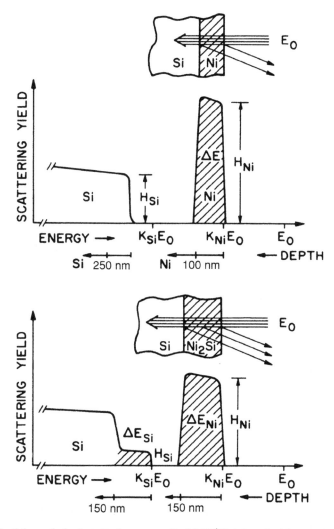

FIGURE 3.10. Schematic backscattering spectra for MeV $^4He^+$ ions incident on 100 nm Ni film on Si (top) and after reaction to form Ni_2Si (bottom). Depth scales are indicated below the energy axes.

In this case of Ni_2Si, the difference between applications of Eqs. 3.26 and 3.27 corresponds to a 5% difference in the determination of the stoichiometry of the silicide.

3.8 Depth Resolution and Energy-Loss Straggling

With backscattering spectrometry, one can determine composition changes with depth. In this section, we consider the limits to depth resolution δt in backscattering spectrometry. The relation between the energy resolution δE_1 and depth resolution is given by Eq. 3.20 as

$$\delta t = \delta E_1 / [S]. \tag{3.28}$$

3.8.1 Grazing Angle Techniques

Eqs. 3.20a and 3.28 give the following formula for the depth resolution:

$$\delta t = \frac{\delta E_1}{K (dE/dx)_{\text{in}} + \dfrac{(dE/dx)_{\text{out}}}{|\cos\theta|}}. \tag{3.29}$$

This equation corresponds to the case of a flat planar sample with the incident beam normal to the surface and a scattering angle θ. For a given detector energy resolution, depth sensitivity is optimum by maximizing the energy loss associated with scattering from a depth into the sample. In general, this is done by grazing-angle techniques so that the path length and hence the energy loss is a maximum. Using Eq. 3.29, the depth resolution is improved by observing the scattering at angles close to $90°$, so that the $\cos\theta$ term approaches zero. With this geometry, depth resolutions as small as 2 nm have been observed using standard solid-state detectors. There are three factors to consider that influence grazing-angle methods of improving depth resolution.

1. *Finite detector acceptance angle.* In any useful system, the detector possesses a finite detector angle that constitutes a broadening to the scattering angle set by the geometry. A common grazing exit-angle configuration consists of a slit aperture to define the detector angle, with a width of 1 mm in the scattering plane and a height of ~ 1 cm perpendicular to the plane, yielding a total area of 0.1 cm^2. If such a detector is 6 cm from the target, an additional uncertainty of approximately $1°$ is added to the scattering angle. This constitutes an appreciable broadening factor as the grazing angle becomes substantially less than $5°$. Smaller acceptance angles become impractical when considering the total charge (time) required for an experiment.
2. *Surface roughness.* One of the most difficult parameters for the analyst to control is the surface roughness of the unknown sample. It is clear, however, that surface roughness will set a limit to any grazing-angle technique. We note that polished semiconductor grade material is usually extraordinarily flat on the scale of a few degrees and lends itself to these types of analysis. In general, it is surface roughness that will set a final limit to depth resolution, even if acceptance angles are made extremely small.
3. *Straggling.* This contribution is discussed in the following section.

3.8.2 Straggling

The energy resolution is normally composed of two contributions: detector resolution δE_{d} and energy straggling δE_{s}. If the two contributions are independent and satisfy Poisson's statistics, the total resolution, δE_1, is given by

$$(\delta E_1)^2 = (\delta E_{\text{d}})^2 + (\delta E_{\text{s}})^2. \tag{3.30}$$

An energetic particle that moves through a medium loses energy via many individual encounters. Such a discrete process is subject to statistical fluctuations. As a result, identical energetic particles, which have the same initial velocity, do not have exactly the same energy after passing through a thickness Δt of a homogeneous medium. The energy loss ΔE is subject to fluctuations. This phenomenon is called *energy straggling*. Energy straggling places a finite limit on the precision with which energy losses, and hence depths, can be resolved.

Light particles such as $^1H^+$ or $^4He^+$ in the MeV energy range lose energy primarily by encounters with the electrons in the target, and the dominant contribution to energy straggling is the statistical fluctuations in these electronic interactions. The distribution of energy loss ΔE for many particles passing through a foil gives a distribution that is approximately Gaussian, when ΔE is small compared with the incident energy E_0. In the Gaussian region, the probability of finding an energy loss between ΔE and $d\Delta E$ is

$$P(\Delta E)\,d\Delta E = \frac{\exp\left[-\Delta E^2/2\Omega_B^2\right]}{\left(2\Omega_B^2\pi\right)^{1/2}}\,d\,\Delta E.$$

where Ω_B^2 is the mean square derivation. If we consider $d\sigma$ as the cross section for an energy transfer T, in a foil of thickness t containing n electrons/cm^3, then similar to Eq. 3.5,

$$\Delta E = nt \int T\,d\sigma$$

and

$$\Omega_B^2 = nt \int T^2 d\sigma. \tag{3.31}$$

From Eqs. 3.3 and 3.4, we have

$$d\sigma = \frac{2\pi Z_1^2 e^4}{mv^2 T^2}dT,$$

so

$$\Omega_B^2 = \frac{2\pi Z_1^2 e^4 nt}{mv^2}(T_{max} - T_{min}),$$

where $T_{max} = 2mv^2$ and $T_{min} = I$. For these swift particles, $T_{max} \gg I$ and $n = NZ_2$, giving

$$\Omega_B^2 = 4\pi Z_1^2 e^4 NZ_2 t, \tag{3.32}$$

an expression often referred to as the *Bohr value of energy straggling*. To determine the energy resolution in Eq. 3.30, we note that the full width at half maximum is $2(2\ln 2)^{1/2}$ times the standard deviation, $\delta E_S = 2.35\Omega_B$.

Bohr's theory predicts that energy straggling does not depend on the energy of the projectile and that the value of the energy variation increases with the square root of the electron density per unit area $NZ_2 t$ in the target. The quantity Ω_B^2/Nt for He ions is numerically equal to Z_2 within 4% when expressed in units of 10^{-12} (eV-cm^{-2}). This rule of thumb allows one to construct a simple estimate of the thickness of target material (atoms/cm^2) that produces an energy straggling of 15 keV for 2 MeV $^4He^+$.

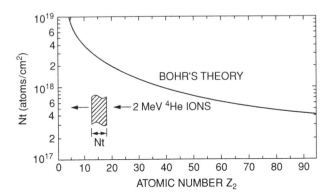

FIGURE 3.11. The amount of target material Nt (number of atoms/cm^2) required to produce 15 keV (FWHM) of energy straggling in a transmission experiment of 21 keV (FWHM) in a backscattering experiment. The projectiles were ^{4}He$^+$ at 2 MeV. [Adapted from Feldman and Mayer, 1986].

Figure 3.11 shows a calculation of straggling for elements throughout the periodic chart. In Si, for example ($Z_2 = 14$ and $N = 5 \times 10^{22}$ at. cm^{-3}), films approximately 500 nm thick can be analyzed before straggling becomes comparable to normal detector resolution of 15–20 keV.

Straggling sets a fundamental limit to depth resolution, possible with ion beam energy-loss techniques. Since Ω_B^2 is proportional to t, the straggling is a function of the depth of penetration. For He$^+$ ions incident on layers <100 nm, the straggling is small compared with the resolution of a solid-state detector and hence plays no role in the obtainable depth resolution. For depths greater than 200 nm, energy-loss straggling sets the limit in depth resolution. With a scattering angle of 95° (grazing exit angle of 5°), a particle scattering from a depth t in the sample corresponds to an energy-loss path length of 10t. Thus straggling, which contributes to the energy resolution at depths on the order of 200 nm in conventional backscattering, now becomes important at depths of approximately 20 nm. The point is that improvement in depth resolution by grazing-angle methods represents a genuine improvement only in the near-surface region.

3.9   Hydrogen and Deuterium Depth Profiles

Forward recoil spectrometry (Section 2.9) is a method for nondestructively obtaining depth profiles of light elements in solids. For the geometry shown in Fig. 3.12a, the technique can be used to determine hydrogen and deuterium concentration profiles in solid materials to depths of a few microns by using ^{4}He$^+$ ions at energies of a few MeV. The forward recoil technique is similar to backscattering analysis, but instead of measuring the energy of the scattered helium ion, the energies of the recoiling ^{1}H or ^{2}H nuclei are measured. Hydrogen is lighter than helium, and both particles are emitted in the forward direction. A mylar foil ($\cong$10 μm) is placed in front of the detector to block the penetration of the abundantly scattered helium ions while permitting the passage of the H ions. The stopping power of ^{1}H ions is sufficiently low compared with that

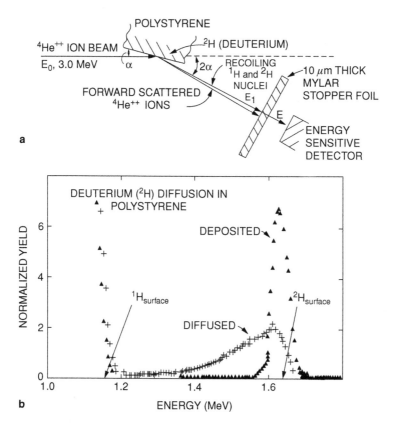

FIGURE 3.12. (a) Experimental geometry for forward recoil spectrometry experiments to determine depth profiles of ^{1}H and ^{2}H in solids. (b) Recoil spectrum of ^{2}H diffused in a sample of polystyrene for 1 hr at 170° C. The sample consisted of a bilayer film consisting of 12 nm of deuterated polystyrene on a large-molecular-weight ($M_w = 2 \times 10^7$) film of polystyrene.

of He ions (Fig. 3.2), so a 1.6 MeV ^{1}H ion only loses 300 keV in penetrating a film that completely stops 3 MeV ^{4}He$^+$ ions. The mylar absorber does introduce energy straggling that combined with the energy resolution of the detector, results in an energy resolution at the sample surface of about 40 keV.

Depth profiles are determined by the energy loss of the incident He$^+$ ion along the inward path and the energy loss of the recoil ^{1}H or ^{2}H ion along the outward path. The diffusion of deuterium (^{2}H) in polystyrene can be determined from spectra such as those shown in Fig. 3.12b. In that case, a ^{2}H ion detected at an energy of 1.4 MeV corresponds to a collision that originated a ^{2}H recoil from a depth about 400 nm below the surface. The use of forward recoil spectrometry allows the determination of hydrogen and deuterium diffusion coefficients in the range of 10^{-12}–10^{-14} cm^2/sec, a range that is difficult to determine by conventional techniques.

As an example of depth measurement, consider a layer of hydrocarbon on both sides of a self-supported 400 nm Al film (instead of the Au markers shown in Fig. 3.6). We will use a symmetrical scattering geometry with the sample inclined at an angle α to the beam and the detector at an angle 2α so that the path length to the back surface (sample thickness t) is the same, $t/\sin\alpha$, for both the inward He ion and the outward proton. Hydrogen recoils originating from the front surface will have an energy $E_2 = K'E_0$, where $K' = 0.480$ for $2\alpha = 30°$ (Eq. 2.7′, where K' denotes the recoil kinematic factor). Hydrogen recoils originating from the back surface at t will have an energy $E_2(t)$ given by

$$E_2(t) = K'E_0 - K'\Delta E_{He} - E_H, \tag{3.33}$$

where ΔE_{He} is the energy loss of the He^+ on the inward path and ΔE_H is the energy loss of the hydrogen on the outward path:

$$\Delta E_{He} = \left.\frac{dE}{dx}\right|_{He} \frac{t}{\sin\alpha}$$

and

$$\Delta E_H = \left.\frac{dE}{dx}\right|_H \frac{t}{\sin\alpha},$$

where $dE/dx\big|_{He}$ is evaluated at E_0 and $dE/dx\big|_H$ at E_2 or for simplicity at $K'E_0$.

The energy width ΔE between recoils from the front E_1 and back $E_1(t)$ surfaces is

$$\Delta E = \frac{t}{\sin\alpha}\left.\frac{dE}{dx}\right|_{He}\left\{K' + \left.\frac{dE}{dx}\right|_H\bigg/\left.\frac{dE}{dx}\right|_{He}\right\}, \tag{3.34}$$

where the stopping power ratio is about one-sixth. For a 400 nm Al film, the energy width ΔE from the H recoils at the front and back side is about 250 keV for 2 MeV He ions. This width is sufficiently large to make forward recoil spectrometry a useful technique for hydrogen analysis. Hydrogen depth profiles can also be determined from nuclear reaction analysis (Chapter 13) or by use of secondary ion mass spectroscopy, discussed in the following chapter (Chapter 4).

3.10   Ranges of H and He Ions

Hydrogen depth profile analysis by forward recoil spectroscopy (Section 3.9) requires the use of mylar foils to stop the scattered He ions while permitting the hydrogen ions to penetrate the foil into the detector (Fig. 3.12a). Mylar foils are also used in prompt radiation analysis using nuclear reactions (Chapter 13) in order to block the elastically scattered particles from reaching the detector (Fig. 13.11). In this section, we will discuss ranges of H and He ions in solids.

The penetration of alpha particles and protons in matter was a subject of great interest in the 1930s. The energies of these charged particles could be determined by measurements of their absorption in matter—particularly in air. These measurements were made by placing a collimated α-emitting source in air on a movable slide whose distance from a particle detector could be varied. The number of detected α particles stayed practically constant with increased separation between source and detector up to a distance R, and then the number of detected particles dropped to zero. The distance R is the range of the particles. Values of R can be correlated with the initial energy of the particle; for example, the mean range of 5.3 MeV α particles (^{210}Po) is about 3.8 cm in air, while that of 8.78 MeV particles (^{212}Po) is about 6.9 cm.

Here, we are interested in the stopping of 1–5 MeV charged particles (protons, deuterons, and α particles) in foils and can use many of the approximations used in evaluating α-particle ranges in air. The range R of a particle is given by

$$R = \int_0^{E_0} \left(\frac{dE}{dx}\right)^{-1} dE, \tag{3.35}$$

where E_0 is the initial kinetic energy and dE/dx is the energy lost per unit path length. For these swift nonrelativistic particles where electronic energy loss dominates, the rate of energy loss is given by Eq. 3.10.

For our purposes, the ranges of particles can be found in Northcliffe and Schilling (1970) or Ziegler (1977). We can use the scaling parameter, for which the range is given, using velocity as a parameter, as

$$R = \frac{M_1}{Z_1^2} F(v), \tag{3.36}$$

where $F(v)$ is a function of velocity. This expression is not exact for the neutralization phenomena at the end of the range, and other corrections are neglected; but it is sufficiently accurate for most cases, excluding very low energies. From Eq. 3.36, we conclude for the same incident velocities that $R(^4\text{He}^+) = R(^1\text{H}^+)$ (see Problem 3.1).

The range–energy relations for α particles ($^4\text{He}^+$) and protons ($^1\text{H}^+$) are given for mylar and silicon in Fig. 3.13a. The data show that for $^4\text{He}^+$ energies above 4 MeV, the ranges are equal to those of protons at the same velocity [$E(\text{He})/4$]. The scaling only

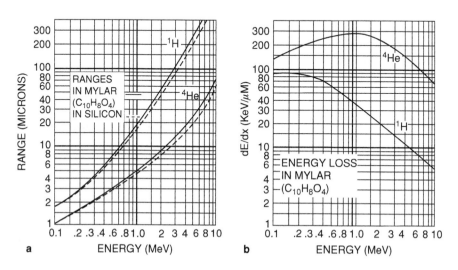

FIGURE 3.13. (a) The range in microns for protons and $^4\text{He}^+$ ions of energies of 0.1 to 10 MeV in mylar, $C^{10}H^8O^4$, and silicon. (b) The energy loss, in keV/μM, for protons and $^4\text{He}^+$ ions in mylar.

holds at energies above the maximum in the dE/dx curves shown in Figure 3.13b. In these curves, dE/dx values at the same velocity scale as Z_1^2, so the ^{1}H values are one-fourth of the ^{4}He$^+$ values.

A 10 μm mylar film is sufficient to stop a 2.5 MeV He ion. The energy loss of a 2.5 Me V proton in the film can be estimated from the range–energy relation. The range of a 2.5 MeV proton in mylar is 80 μm, and a 70 μm film has a thickness equivalent to the range of a 2.3 MeV proton; hence, the energy lost by a 2.5 MeV proton in traversing the film is about 0.2 MeV. This loss could also be estimated from the dE/dx curve in Fig. 3.13b by noting that a 2.5 MeV proton loses 18 keV/μm in mylar.

3.11   Sputtering and Limits to Sensitivity

The Z^2 dependence of the Rutherford scattering cross section clearly indicates a large sensitivity to heavy elements. It is of interest to ask for the ultimate sensitivity, i.e., the smallest amount of material detectable by this technique.

The limit for the ion scattering technique is set by sputtering. Sputtering is a process in which an energetic ion impinging on a solid creates a collision cascade due to small angle nuclear events. Some fraction of the secondary ions acquire the correct momentum to escape the solid, giving rise to an erosion process. This erosion technique is an important process in surface analysis and is described completely in Chapter 4. In the case of Rutherford scattering, this erosion process is an undesirable phenomenon and sets the limit on sensitivity. The basic question is under what conditions will the erosion of the material will occur before a measurement is complete. Sputtering is defined in terms of a sputtering yield Y, which is the number of atoms ejected from the solid per incident ion. In the following, we calculate the limit to sensitivity set by the sputtering process.

We consider a thin layer of material (possibly submonolayer) containing N_S atoms/cm^2. The yield of scattered ions, Q_D, is given by the usual equation in a Rutherford backscattering measurement:

$$Q_D = \sigma(\theta)\Omega Q N_S,$$

where $\sigma(\theta)$ is the differential cross section, Ω is the solid angle of the detector, and Q is the number of incident ions.

For the same number of incident ions, the loss of atoms from the layer, ΔN_S, due to sputtering by the incident ions is

$$\Delta N_S = YQ/a, \tag{3.37}$$

where a is the area of the probing beam spot. We require that the amount of erosion be less than the original film thickness, i.e.,

$$\Delta N_S < N_S, \tag{3.38}$$

which can be expressed as a limit on the value of Q given by

$$Q < \left(\frac{Q_D a}{Y \sigma(\theta) \Omega} \right)^{1/2},$$

(3.39)

and a corresponding minimum value of N_S given by

$$N_S > \left(\frac{Q_D Y}{\sigma(\theta) a \Omega} \right)^{1/2}.$$

(3.40)

In evaluating this quantity, we use the following standard numbers for the case of a layer of gold ($Z = 79$): $\sigma(\theta)$, the cross section for He scattering at 2 MeV to $170°$, is 10^{-23} cm^2/steradian; Ω, the solid angle of the detector corresponding to a 1 cm^2 detector 5 cm from the target, is 4×10^{-2} steradians; Y, the sputtering yield, is 10^{-3} (see Chapter 4); Q, the area of the probing beam, is 10^{-2} cm^2; and Q_D is arbitrarily taken as 10^2, the minimum number of counts required for a statistically significant measurement. With these values, we find a minimum layer thickness of 5×10^{12} Au atoms/cm^2 or close to 1/1000 of a monolayer. Using these expressions, we see that the scattering parameters and geometry could be optimized in a number of ways to further increase the sensitivity. Experience suggests that the absolute best sensitivity realizable for this favorable case of a heavy scatterer is 5×10^{11} atoms/cm^2. Note that the incident charge required is not prohibitive. Equation 3.39 corresponds to an incident particle dose of 5×10^{13} ions or about 10 μ Coulombs.

3.12   Summary of Scattering Relations

Throughout this book, and particularly in Chapters 2 and 3, we have developed and used a number of kinematic relations and cross sections and the backscattering factor. For the reader's convenience, we have collected these scattering relations in Table 3.1 on page 57.

Problems

3.1. Derive

$$R = \frac{M_1}{Z_1^2} F(v)$$

(Eq. 3.36), using $dE = Mv \, dv$ and $dx = mv^2 \, dE \, (4\pi Z_1^2 e^4 n \ln(2mv^2/I))^{-1}$, and estimate the range of 1.0 MeV deuterons in Si using Eq. 3.36 and Fig. 3.13.

3.2. Using Eq. 3.10:

    (a) Evaluate dE/dx in eV/nm for 1 MeV He in Si.

    (b) Convert to units of eV (10^{15} atoms/cm^2) and compare your answer with Appendix 3.

    (c) Determine dE/dx in eV/nm for 4 MeV C ions in silicon.

3.3. Show that the maximum in the stopping power is given by

$$E = \frac{M Z_2 I_0}{4m} e,$$

where e is 2.718 and $I_0 = 10\,\text{eV}$. Evaluate for He in Si.

3.4. Draw the backscattering spectrum ($\theta = 180°$) for 2.0 MeV He^+ expected for

(a) a sample of Si containing a uniform 1% Au impurity;
(b) a sample of Au containing a uniform 1% Si impurity;
(c) a sample of Si covered with a 100 nm Pt film; and
(d) a sample of Si with a 100 nm PtSi film on Si.

In all spectra, show specific energies for the leading edge (and trailing edge in thin film cases) of the spectral features, and indicate the relative heights.

3.5. For the scattering geometry in Fig. 3.12a, what is the energy width of the recoiling hydrogen atoms for a 200-nm-thick mylar target? Use the energy loss values in mylar given in Fig. 3.13 and ignore the influence of the stopper foil in front of the detector.

3.6. Determine the backscattering energy loss factor (S) (as in Eq. 3.20) for the case of nonnormal incidence: incident beam at angle θ_1 with respect to the sample normal and scattering angle θ_2.

3.7. Consider an experiment involving 2.0 MeV ions backscattered to 180° from a 200 nm SiO_2 target.

(a) Sketch the backscattering spectrum indicating the energies of the features in the spectrum.
(b) Assuming that the stopping power is energy independent, show that the ratio of the energy widths of scattering from oxygen and Si can be written as

$$(K_{Si} + 1)/(K_0 + 1),$$

where K denotes the kinematic factor.

(c) Recognizing that energy loss is not energy independent, derive a precise form for the ratio of the energy widths from the two elements. Evaluate the ratio and compare this to the expression given in part (b).
(d) Using the stopping powers, the stoichiometry, and the cross section, derive an equation for the heights of the Si and oxygen peaks and evaluate.
(e) Assume the oxide is of unknown stoichiometry, $Si_x O_y$. Derive an expression for the ratio of the heights in terms of the scattering cross section, the elemental stopping powers, and x/y. [Hint: Express the stopping power in terms of x and y using Bragg's rule.]

3.8. Assume in backscattering measurements that energy straggling is given by $\Omega_{tot}^2 = (K \Omega_B|_{in})^2 + (K \Omega_B|_{out})^2$. Calculate the amount of energy straggling in an RBS signal from a thin Cr layer underneath an Al film 400 nm thick with an analysis beam of 2 MeV ^4He ions ($\theta = 180°$). What is the total signal width (FWHM) if the detector resolution is 15 keV? To what thickness of Al does this correspond?

TABLE 3.1. Summary of Relations

Lab Energy of Scattered Particle; Kinematic Factor, K_{M_2}	$\sigma(\theta) = \left(\dfrac{Z_1 Z_2 e^2}{4E}\right)^2 \left[\sin^{-4}\dfrac{\theta}{2} - 2\left(\dfrac{M_1}{M_2}\right) + \cdots\right]$
Lab Energy of Recoil Nucleus	$E_2/E_0 = 1 - E_1/E_0 = \dfrac{4M_1 M_2}{(M_1+M_2)^2}\cos^2\phi; \phi < \pi/2$
Lab Angle of Recoil Nucleus	$\phi = \dfrac{1}{2}(\pi - \theta_c); \sin\phi = \left(\dfrac{M_1 E_1}{M_2 E_2}\right)^{1/2}\sin\theta; \quad \tan\theta = \dfrac{\sin 2\phi}{M_1/M_2 - \cos 2\phi}$
C, M, Angle of Scattered Particle	$\cos\theta_c = 1 - 2\cos^2\phi$
Lab Energy if Scattered Particle	$\dfrac{E_1}{E_0} = \dfrac{M_2 - M_1}{M_2 + M_1}, \theta = 90°; \quad \dfrac{E_1}{E_0} = \left(\dfrac{M_2 - M_1}{M_2 + M_1}\right)^2, \theta = 90°$
Rutherford Scattering Cross Section	$\dfrac{d\sigma}{d\Omega} = \left(\dfrac{Z_1 Z_2 e^2}{4E_c}\right)^2 \left[\dfrac{1}{\sin^4\theta_c/2}\right]$
	$\dfrac{d\sigma}{d\Omega} = \left(\dfrac{Z_1 Z_2 e^2}{4E_0}\right)^2 \left[\sin^{-4}\dfrac{\theta}{2} - 2\left(\dfrac{M_1}{M_2}\right) + \cdots\right]; M_1 < M_2$
Backscattering Factor, [S]	$\Delta E = \Delta t[S]$
Normal Incidence	$[S] = \Delta t\left[K\dfrac{dE}{dx}\Big|_E + \dfrac{1}{\lvert\cos\theta\rvert}\dfrac{dE}{dx}\Big|_{KE}\right]$

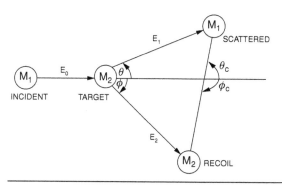

$M_1 Z_1$ = mass charge incident of particle
M, Z = mass, charge of target atom
θ, ϕ = scattered angle of scattered particle and recoil in laboratory system
θ_c, ϕ_c = angle of scattered particles and recoil in center-of-mass system
E_1, E_2 = the energy after the scattering of the projectile in the laboratory system
E = the energy of the incident projectile in the laboratory system
$E = (M_2/M_2 + M_1)E_0$ = incident energy in center-of-mass system
$\dfrac{dE}{dx}\Big|_{in}, \dfrac{dE}{dx}\Big|_{out}$ = the rate of energy loss on the inward, outward path

References

1. H. H. Anderson and J. F. Ziegler, *Hydrogen Stopping Power and Ranges in All Elements* (Pergamon Press, New York, 1977).
2. W. K. Chu, J. W. Mayer, and M. A. Nicolet, *Backscattering Spectrometry* (Academic Press, New York, 1978).
3. R. D. Evans, *The Atomic Nucleus* (McGraw-Hill Book Co., New York, 1955).
4. I. Kaplan, *Nuclear Physics* (Addison-Wesley, Reading, MA, 1964).
5. R. B. Leighton, *Principles of Modern Physics* (McGraw-Hill Book Co., New York, 1959).
6. L. C. Northcliffe and R. F. Schilling, "Range and Stopping Power Tables for Heavy Ions," *Nucl. Data Tables* 7(3–4), 733 (1970).
7. F. K. Richtmyer, E. H. Kennard, and J. N. Cooper, *Introduction to Modern Physics*, 6th ed. (McGraw-Hill Book Co., New York, 1978).
8. J. F. Ziegler, *Helium Stopping Powers and Ranges in All Elements* (Pergamon Press, New York, 1977).
9. L. R. Doolittle, *Nucl. Intr. and Meth.*, B9, 349 (1985).
10. L. C. Feldman and J. W. Mayer, *Fundaments of Surface and Thin Film Analysis* (Prentice Hall, NJ, 1986).

4
Sputter Depth Profiles and Secondary Ion Mass Spectroscopy

4.1 Introduction

This chapter deals with the erosion of the sample by energetic particle bombardment. In this process, called *sputtering*, surface atoms are removed by collisions between the incoming particles and the atoms in the near-surface layers of a solid. Sputtering provides the basis for composition depth profiling with surface analysis techniques, either by analysis of the remaining surface with electron spectroscopies or by analysis of the sputtered material. Here we describe the most widely used of these latter techniques, secondary ion mass spectroscopy (SIMS).

In previous chapters, we have been concerned with the energies and yields of particles scattered from the target material under analysis. With Rutherford backscattering spectrometery (RBS) using MeV He$^+$ ions, the energy loss along the inward and outward paths provides the depth information (Fig. 4.1). In other analytical techniques, the atoms to be identified must lie at the surface of the materials. For example, the observation depth in X-ray photoelectron spectroscopy and Auger-electron spectroscopy (XPS and AES) can be as small as 1.0–2.0 nm. In order to use XPS and AES to determine depth profiles, it is necessary to remove controlled thicknesses of the surface layer. This surface layer removal is carried out in materials analysis by bombarding the surface with low-energy (0.5–20 keV) heavy ions, such as O$^+$ or Ar$^+$, which eject or sputter target atoms from the surface. The yield of sputtered atoms, the number of sputtered atoms per incident ion, lies in the range of 0.5–20 depending upon ion species, ion energy, and target material. Surface-sensitive techniques can then be used after each layer is removed in order to determine the composition of the new surface and hence deduce the depth profile of the atomic composition. It is also possible to analyze the sputtered atoms, generally the ionized species, to determine the composition of the sputter-removed materials. This technique of secondary ion mass spectroscopy, or SIMS, has been used extensively in depth profiling. One can also measure the characteristic radiation emitted from excited sputtered ions or atoms to determine the composition of the sputter-removed material.

For sputtering, it is the energy lost in elastic collisions with the atomic cores—termed *nuclear energy loss*—that determines the energy transfer to and eventual ejection of surface atoms. For backscattering or nuclear analysis, the energetic particles lose energy primarily through electron excitation and ionization in inelastic collisions with atomic

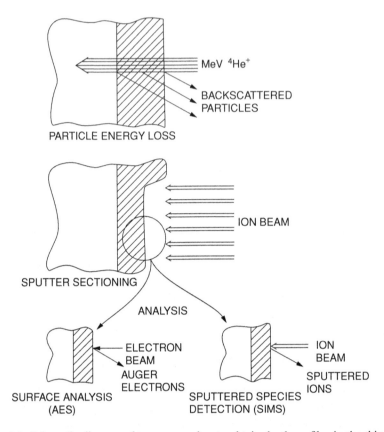

FIGURE 4.1. Schematic diagram of two approaches to obtain depth profiles in the thin films. With particle energy loss techniques, the thickness of the layer is determined from the energy loss of the energetic particles. With sputter sectioning techniques, the amount of material probed is determined by the sputtering yield. The surface composition can be directly analyzed either by electron spectroscopies or by the amount of material removed by sputter species detection.

electrons—termed *electronic energy loss*. A good assumption is that electronic energy loss and nuclear energy loss can be treated separately and independently. In Chapter 3, we described electronic energy loss and showed that the amount of nuclear energy loss was small. In the sputtering regime, the nuclear energy loss dominates.

4.2   Sputtering by Ion Bombardment—General Concepts

Surfaces of solids erode under ion bombardment. The erosion rates are characterized primarily by the sputtering yield Y, which is defined as

$$Y = \text{Sputtering yield} = \frac{\text{Mean number of emitted atoms}}{\text{Incident particle}}. \qquad (4.1)$$

The sputtering yield depends on the structure and composition of the target material, the parameters of the incident ion beam, and the experimental geometry. Measured values of Y cover a range of over seven decades; however, for the medium-mass ion species and keV energies of general interest in depth profiles, the values of Y lie between 0.5 and 20. The sputtering yields of MeV light ions for most materials are of the order of 10^{-3}. Consequently, Rutherford backscattering analysis will cause the sputtering of only a small fraction of a monolayer during a typical analysis (see Section 3.11).

Sputtering yields can be accurately predicted by theory for single-element materials. Figure 4.2 shows the energy and incident particle dependence of the sputtering yield

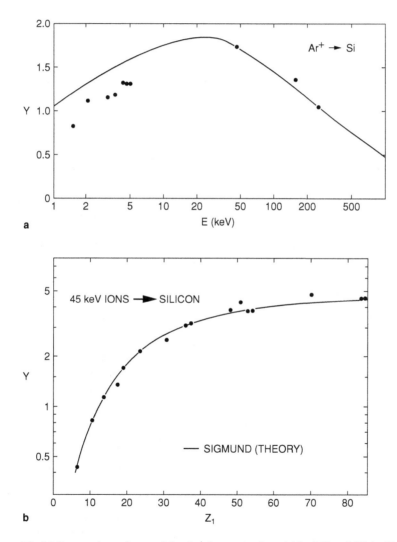

FIGURE 4.2. (a) Energy dependence of the Ar^+ ion sputtering yield of Si and (b) incident ion dependence of the Si sputtering yield. The solid line represents the calculations of Sigmund, and the data are from Andersen and Bay (1981). [With permission of Springer Science+Business Media.]

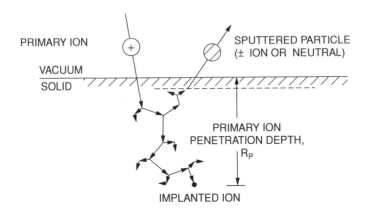

FIGURE 4.3. Schematic of the ion–solid interactions and the sputtering process.

of Si. The experimental values, in good agreement with calculations (solid line) by Sigmund (1981), are based on nuclear energy loss mechanisms and the sharing of this energy loss among the large number of atoms that define the collision cascade. For any given ion–target combination, it is desirable to refer to tabulated values or to determine the yield experimentally.

A number of review articles and books on the topic of sputtering are listed in the references at the end of this chapter. In the sputtering process, atoms are ejected from the outer surface layers. The bombarding ion transfers energy in collisions to target atoms that recoil with sufficient energy to generate other recoils (Fig. 4.3). Some of these backward recoils (about 1–2 atoms for a 20 keV Ar ion incident on Si) will approach the surface with enough energy to escape from the solid. It is these secondary recoils that make up most of the sputtering yield. For example, for the case of Ar ions on Si, target recoils in the backward direction toward the surface are kinematically forbidden, as is Ar backscattering (see Chapter 2). The sputtering process involves a complex series of collisions (the collision cascade) involving a series of angular deflections and energy transfers between many atoms in the solid. It is possible to simulate the sputtering process on a computer via a series of binary events, but such simulations do not readily yield the dependencies of the sputtering process on various experimental parameters. The problem has been approached based on transport theory, which considers the dynamics of the collision cascade and derives the total energy flux in the backward direction. Such a derivation is beyond the scope of this book. However, we do extract the important parameters based on nuclear energy loss concepts. Clearly, the most important parameter in the process is the energy deposited at the surface.

The sputtering yield should be proportional to the number of displaced or recoil atoms. In the linear cascade regime that is applicable for medium mass ions (such as Ar^+), the number of recoils is proportional to the energy deposited per unit depth in nuclear energy loss. We can then express the sputtering yield Y for particles incident normal to the surface as

$$Y = \Lambda \, F_D(E_0), \tag{4.2}$$

where Λ contains all the material properties such as surface binding energies, and $F_D(E_0)$ is the density of deposited energy at the surface and depends on the type, energy, and direction of the incident ion and the target parameters Z_2, M_2, and N.

The deposited energy at the surface can be expressed as

$$F_D(E_0) = \alpha N S_n(E_0), \tag{4.3}$$

where N is the atomic density of target atoms, $S_n(E)$ is the nuclear stopping cross section, and $NS_n(E) = dE\,dx|_n$ is the nuclear energy loss. In this equation, α is a correction factor that takes into account the angle of incidence of the beam to the surface and contributions due to large-angle scattering events that are not included in the development. The sputtering yield is calculated in Section 4.4.

The evaluation of $S_n(E)$ rests on the collision cross section for energy transfer to a substrate atom. In the keV sputtering regime where the particle velocity is much less than the Bohr velocity, screening of the nuclear charge by the electrons must be included in the description of the collisions. The procedure to obtain the sputtering yield is to first treat a screened potential from a description of the Thomas–Fermi approximations (Section 4.9), and then derive the collision cross section based on a screened potential to obtain the nuclear stopping cross section (Section 4.3).

4.3   Nuclear Energy Loss

A charged particle penetrating a solid loses energy through two processes: (1) energy transfer to electrons (electronic energy loss); and (2) energy transfer to the atoms of the solid (nuclear energy loss). In both cases, the interaction is basically of a Coulomb type; for the electronic case, it is pure Coulomb (see Chapter 3), while in the nuclear case, it is a form of screened Coulomb potential. The two mechanisms have different energy dependencies—in the electronic case, there is a peak in the cross section at projectile energies of the order of 0.1–1.0 MeV for light projectiles; in the nuclear case, the peak in the loss cross section is at much lower energy, of the order of 0.1–10 keV. In penetration theory, the electronic and nuclear energy losses are treated as uncorrelated and simply summed. In many cases, one or the other contribution is negligible and is simply ignored. Sputtering is governed by the energy deposited via nuclear energy loss at the surface of a solid. This mechanism transfers momentum and energy to the atoms of the solid, resulting in energetic secondaries and sputtering. In this section, we give a simple description of nuclear energy loss and compare it to the more sophisticated treatments. As in other sections of this text, our aim is to provide a simple mathematical description of the process in order to provide some insight into the quantitative understanding of the physics.

The derivation of nuclear energy loss uses two main assumptions: (1) a simple screened Coulomb potential and (2) the impulse approximation.

The interaction potential between two atoms Z_1 and Z_2 can be written in the form of a screened Coulomb potential (Section 4.9), using χ as the screening function:

$$V(r) = \frac{Z_1 Z_2 e^2}{r} \chi \left(\frac{r}{a} \right), \tag{4.4}$$

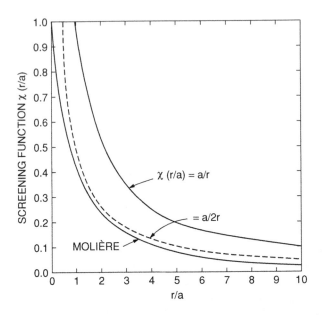

FIGURE 4.4. The screening function χ in the Molière approximation. At $r = a$, the value of the screening function has dropped to about 0.4 in support of the choice of a as an approximation for the atomic size. Shown also are the screening functions $\chi = a/r$ and $a/2r$.

where a is the Thomas–Fermi screening radius for the collision

$$a = \frac{0.885a_0}{(Z_1^{1/2} + Z_2^{1/2})^{2/3}}. \tag{4.5}$$

The values of a lie between 0.01 and 0.02 nm for most interactions. We take as a screening function

$$\chi(r/a) = a/2r, \tag{4.6}$$

leading to a potential of the form

$$V(r) = \frac{Z_1 Z_2 e^2 a}{2r^2}. \tag{4.7}$$

The screening functions a/r and $a/2r$ are illustrated in Fig. 4.4, and the latter reasonably follows the shape of the more accurate Molière potential for $r/a > 1$. The $1/r^2$ potential is not a good approximation for $r/a < 1$, since it does not go into the pure Coulomb form as $r \to 0$. For low-energy particles as used in sputtering, this problem is not severe, since the distance of closest approach is large, thus preventing interactions at small r.

The impulse approximation is appropriate for the small-angle, large-impact parameter collisions that dominate the sequence of scatterings which determine the charged particle trajectory. We have used this approximation in the derivation of electronic stopping given in Chapter 3. In the impulse approximation, the change in momentum is given by

$$\Delta p = \int_{-\infty}^{\infty} F_\perp \, dt, \tag{4.8}$$

or

$$\Delta p = \frac{1}{v} \int_{-\infty}^{\infty} F_\perp \, dx, \tag{4.9}$$

where $F_\perp$ is the component of the force acting on the ion perpendicular to its incident direction. By using the geometry of Fig. 3.3, the force may be written with $r = \sqrt{x^2 + b^2}$ as

$$F_\perp = \frac{\partial V(r)}{\partial y} = -\frac{\partial V\left(\sqrt{x^2 + b^2}\right)}{\partial b}. \tag{4.10}$$

Then

$$\Delta p = \frac{1}{v} \frac{\partial}{\partial b} \int_{-\infty}^{\infty} V\left(\sqrt{x^2 + b^2}\right) dx, \tag{4.11}$$

or, using Eq. 4.7,

$$\Delta p = \frac{1}{v} \frac{\partial}{\partial b} \int_{0}^{\infty} \frac{Z_1 Z_2 e^2 a}{(x^2 + b^2)} \, dx, \tag{4.12}$$

which reduces to

$$\Delta p = \frac{\pi Z_1 Z_2 e^2 a}{2 v b^2}. \tag{4.13}$$

The energy transferred, T, to the recoiling nucleus is

$$T = \frac{\Delta p^2}{2 M_2},$$

$$T = \frac{\pi^2 Z_1^2 Z_2^2 e^4 a^2}{8 M_2 v^2 b^4}. \tag{4.14}$$

The cross section $d\sigma(T)$ for transfer of energy between T and $T + dT$ is

$$d\sigma = -2\pi b \, db,$$

or

$$d\sigma = -\frac{\pi^2 Z_1 Z_2 e^2 a}{8\sqrt{(M_2/M_1)E}} T^{-3/2} \, dT, \tag{4.15}$$

where $E = M_1 v^2/2$. It is convenient to express this result in terms of the maximum energy transfer T_{max}, where

$$T_{max} = \frac{4M_1 M_2}{(M_1 + M_2)^2} E.$$

Then

$$d\sigma = -\frac{\pi^2 Z_1 Z_2 e^2 a}{4 T_{max}^{1/2}} \frac{M_1}{(M_1 + M_2)} T^{-3/2} \, dT. \tag{4.16}$$

The nuclear stopping cross section S_n is given by

$$S_n = - \int T \, d\sigma,$$

or

$$S_n = \frac{\pi^2 Z_1 Z_2 e^2 a}{2 T_{max}^{1/2}} \left(\frac{M_1}{M_1 + M_2} \right) T^{1/2} \Big|_0^{T_{max}}.$$

$$S_n = \frac{\pi^2 Z_1 Z_2 e^2 a M_1}{2(M_1 + M_2)}, \tag{4.17}$$

The nuclear energy loss is given by

$$\left. \frac{dE}{dx} \right|_n = N S_n, \tag{4.18}$$

where N is the number of atoms/volume in the solid. Note that, in this approximation, $dE/dx|_n$ is independent of energy, i.e.,

$$\left. \frac{dE}{dx} \right|_n = N \frac{\pi^2}{2} Z_1 Z_2 e^2 a \frac{M_1}{M_1 + M_2}. \tag{4.19}$$

Figure 4.5 compares this energy-independent value to the values of the nuclear energy loss using the Thomas–Fermi potential, and follows the description of Lindhard in which the nuclear energy loss is expressed in terms of a reduced energy ε'. This energy is given by the ratio of the Thomas screening distance to the distance of closest approach,

$$\varepsilon' = \frac{M_2}{M_1 + M_2} E \frac{a}{Z_1 Z_2 e^2}, \tag{4.20a}$$

and a reduced length ρ based on a cross section πa^2 and an energy ratio T_{max}/E,

$$\rho = x N M_2 4\pi a^2 M_1/(M_1 + M_2)^2. \tag{4.20b}$$

This form is generally useful in that the stopping power for any combination of projectile and target at any energy can be found. In this formalism, $d\varepsilon'/d\rho = S_n(\varepsilon')$, so if we use Eqs. 4.20a and 4.20b,

$$\left. \frac{dE}{dx} \right|_n = 4\pi a N Z_1 Z_2 e^2 \frac{M_1}{M_1 + M_2} S_n(\varepsilon'), \tag{4.21}$$

where $S_n(\varepsilon')$ depends on the form of $V(r)$. In terms of the value of $(dE/dx)_n$ derived in Eq. 4.19, the energy-independent value of the nuclear energy loss in reduced units of energy and length is 0.393. This value is slightly different than that given by Lindhard

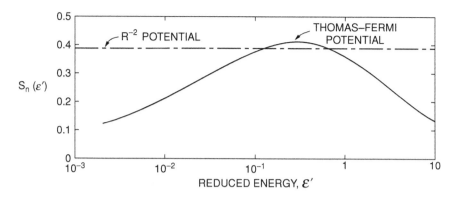

FIGURE 4.5. Reduced nuclear stopping cross section $S_n(\epsilon')$ (or $d\epsilon'/d\rho$) as a function of ϵ'. The Thomas–Fermi curve represents the most accurate value of S_n for the Thomas–Fermi potential; the horizontal line is the result for an R^{-2} potential [Eq. (4.17)].

et al. (1963), since we have used an impulse approximation, while these authors evaluate the scattering integral more completely.

Note that our approximation, Eq. 4.19, gives the correct order of magnitude for the stopping power but deviates considerably in the energy dependence. Most significantly, it does not display the $1/E$ dependence at high energy. This outcome is the result of using a $1/r^2$ potential rather than a $1/r$ potential. Clearly, the $1/r^2$ approximation is worst at high energies, where close collisions are important. Accurate values of the nuclear energy loss $(dE/dx)_n$ can be derived from Eq. 4.21 and Fig. 4.5, which give $S_n(\epsilon')$ for the more accurate Thomas–Fermi potential (Section 4.9).

For 1 keV Ar ions incident on a medium mass target, Cu, $a = 0.0103$ nm and $\epsilon' = 0.008$, and, for 10 keV O ions on Cu, $a = 0.115$ and $\epsilon' = 0.27$. (Ar and O are generally used in sputter profiling.) Thus, for ion energies from 1 to 10 keV, the values of ϵ' are in the range of 0.01–0.3; this is a range just below the plateau of dE/dx. As an approximation to estimate the magnitude of $dE/dx|_n$, an energy-independent value (a rough average) of $S_n(\epsilon') \cong 0.39$ can be used. For Ar^+ ions incident on Cu, $dE/dx|_n = 12.4$ eV/nm, and for oxygen ions incident on Cu, $dE/dx|_n \cong 3.2$ eV/nm.

4.4   Sputtering Yield

The yield Y of sputtered particles from single-element amorphous targets was expressed in Eq. 4.2 as the product of two terms: one, Λ, containing material parameters and the other, F_D, the deposited energy. The derivation of Λ involves a description of the number of recoil atoms that can overcome the surface barrier and escape from the solid:

$$Y = \Lambda F_D, \tag{4.22a}$$

where

$$\Lambda = \frac{4.2 \times 10^{-3}}{N U_0} \text{(nm/eV)}, \tag{4.22b}$$

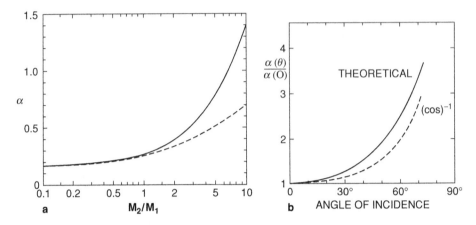

FIGURE 4.6. Factor α occurring in the backward-sputtering yield formula (4.3). (a) Dependence on mass ratio M_2/M_1. The solid line is the theoretical prediction, evaluated for elastic scattering only, with no surface correction applied. The dashed line is interpolated from experimental sputtering yields for 45 keV ions on Si, Cu, Ag, and Au. (The difference is mainly due to the neglect of the surface correction at large mass ratios.) (b) Dependence on angle of incidence. The solid line is the theoretical prediction for Ar ions on Cu. The dashed line is $(\cos\theta)^{-1}$ dependence, valid mainly in the high-velocity limit. [From Sigmund, 1981 with permission from Springer Science+Business Media.]

where N is the atomic density (in nm^{-3}) and U_0 (in eV) is the surface binding energy. The value of U_0 can be estimated from the heat of sublimation ($\cong$ heat of vaporization) and typically has values between 2 and 4 eV. For the deposited energy, Eq. 4.3,

$$F_D = \alpha\, N\, S_n. \qquad (4.22c)$$

The value of α is a function of the mass ratio and ranges between 0.2 and 0.4, as shown in Fig. 4.6. The value of α increases with the angle of incidence because of increased energy deposition near the surface. A reasonable average value for normal incidence sputtering with medium mass ions is $\alpha = 0.25$.

For Ar^+ ions incident on Cu (where $N = 85$ atoms/nm^3), the value of $NS_n = 1240$ eV/nm. The surface binding energy, U_0, is ~ 3 eV based on a heat of vaporization of $\cong 3$ eV. If we rewrite Eq. 4.2, the sputtering yield is

$$
\begin{aligned}
Y &= \frac{4.2}{NU_0}\alpha\, N\, S_n \\
&= \frac{4.2 \times 0.25 \times 1240\ \text{eV/nm}}{84.5/\text{nm}^3 \times 3\ \text{eV}} = 5.1.
\end{aligned}
$$

This result is in reasonable agreement with measured values of about 6. These estimates hold for the ideal case of an amorphous single-element target. The sputtering yields from single-crystal, polycrystalline, or alloy targets may deviate significantly from the simple estimates above. With polyatomic targets the preferential sputtering of one of the elements can lead to changes in composition of the surface layer. These changes will be reflected in the Auger yields, which give the composition of the altered, not original, layer. Another complication is ion beam mixing (redistribution within the collision cascade), which can lead to broadening of the interface during the profiling of layered targets. In many of these cases, it is possible to use Rutherford backscattering to establish layer thicknesses and the concentration of the major constituents. This information will then provide a calibration for the sputter profile.

4.5   Secondary Ion Mass Spectroscopy (SIMS)

Surface layers are eroded by the sputtering process, and hence the relative abundance of the sputtered species provides a direct measure of the composition of the layer that has been removed. Sputtered species are emitted as neutrals in various excited states; as ions both positive and negative, singly and multiply charged; and as clusters of particles. The ratio of ionized to neutral species from the same sample can vary by orders of magnitude, depending on the condition of the surface. Analysis of sputtered species is the most sensitive of the surface-analysis techniques. The common use is the detection and measurement of low concentrations of foreign atoms in solids.

One of the most commonly used sputtering techniques is the collection and analysis of the ionized species—the secondary ions. As shown in Fig. 4.7, the secondary ions enter an energy filter, usually an electrostatic analyzer, and then are collected in a mass spectrometer. This process gives rise to the acronym SIMS for *secondary ion*

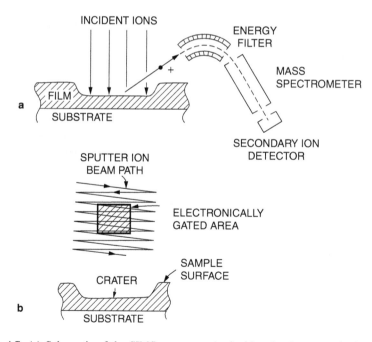

FIGURE 4.7. (a) Schematic of the SIMS apparatus. An incident ion beam results in sputtered ionic species, which are passed through an electrostatic energy filter and a mass spectrometer and finally detected by an ion detector. (b) The beam is usually swept across a large area of the sample and the signal detected from the central portion of the sweep. This setup avoids crater edge effects.

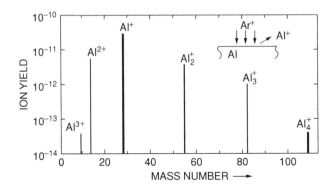

FIGURE 4.8. Secondary ion cluster spectrum from Ar^+ ion bombardment of Al. Note that the ordinate is a log scale. The predominant species is Al^+, but $Al_2{}^+$: and $Al_3{}^+$: are also in abundance. [From Werner, 1978, with permission from Springer Science+Business Media.]

mass spectroscopy. All SIMS instruments possess a capability for surface and elemental depth concentration analysis. In one mode of operation, the sputter ion beam is rastered across the sample, where it erodes a crater in the surface. To insure that ions from the crater walls are not monitored, the detection system is electronically gated for ions from the central portion of the crater. There are also direct imaging instruments—ion microscopes—in which the secondary ions from a defined micro-area of the sample are detected so that an image of the surface composition can be displayed.

The spectra of both positive and negative secondary ions are complex, exhibiting not only singly and multiply charged atomic ions but all ionized clusters of target atoms. As shown in Fig. 4.8, the mass spectrum from Ar^+-bombarded Al shows not only singly ionized atoms but also doubly and triply ionized atoms and two-, three-, and four-atom clusters. In most cases, the yield of singly ionized atoms predominates.

Sputtered particles emerge from the solid with a distribution of energies corresponding to the fluctuations in the many individual events that make up the sputtering process. The sputtered particles have a total yield Y related to the energy spectrum $Y(E)$ such that

$$Y = \int_0^{E_{max}} Y(E)\,dE, \tag{4.23}$$

where E_{max} is the maximum energy of the sputtered particles. The positively ionized secondary ion yield $Y^+(E)$ is related to the sputtering yield $Y(E)$ by

$$Y^+(E) = \alpha^+(E)Y(E), \tag{4.24}$$

and the total secondary positive ion yield is

$$Y^+ = \int_0^{E_{max}} \alpha^+(E)Y(E)\,dE, \tag{4.25}$$

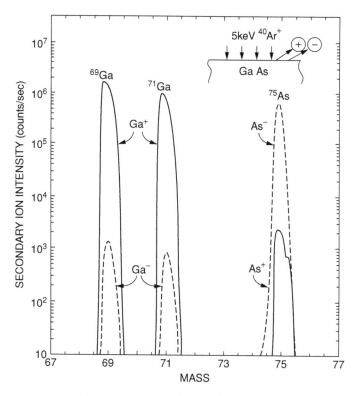

FIGURE 4.9. Positive (solid lines) and negative (dashed lines) ion yields from GaAs bombarded with 5 keV Ar ions. The SIMS spectrum shows the large difference in sensitivity between ionized species for nearly identical sputtering yields of Ga and As from GaAs. [From Magee, *Nucl. Instr. Meth.* 191, 297 (1981), with permission from Elsevier.]

where the ionization probability $\alpha^+(E)$ depends on the particle energy and the nature of the substrate. As shown in Fig. 4.9, the ionization yield can vary by three orders of magnitude between species with nearly identical sputtering yields. The major difficulty in quantitative analysis by SIMS is the determination of $\alpha^+(E)$.

The measured signal I^+, generally given in counts/s, of a mono-isotopic element of mass A at a concentration C_A in the target is given by

$$I_A^+ = C_A i_p \beta T \, \alpha^+(E, \theta) Y(E, \theta) \, \Delta\Omega \, \Delta E, \qquad (4.26)$$

where i_p is the primary beam current (ions/s), θ and E represent the angle and pass energy of the detector system, $\Delta\Omega$ and ΔE are the solid angle and *width* of the energy filter, and β and T are the detector sensitivity and the transmission of the system for the ion species measured. Both α^+ and Y are dependent on the sample composition. The composition dependence can frequently be neglected if concentration profiles of a low-level constituent in a matrix of constant composition are to be determined. A good example of this application is the measurement of the depth profile of ion-implanted

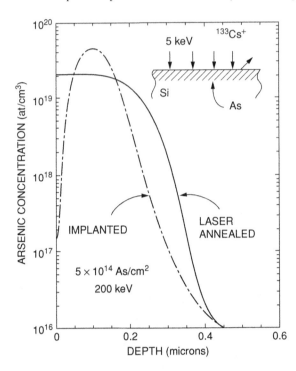

FIGURE 4.10. SIMS concentration profile of As implanted in Si and redistributed by pulsed laser melting of the outer Si layer. The measured concentration profile extends below levels of $10^{18}/cm^3$.

impurities in semiconductors (Fig. 4.10). The maximum impurity concentration is less than 10^{-3}, and hence the presence of the As has minimal effect on α^+. A strong feature of SIMS is the ability to analyze hydrogen over a wide range of concentrations, as shown in Fig. 4.11. In this case, surface contamination by water vapor can influence the dynamic range.

Secondary ion yields are very sensitive to the presence of either electropositive or electronegative ions at the target surface. The picture of neutralization of a positive ion leaving a surface involves the atomic energy levels of the emitted species and the availability of electrons at the solid surface to fill the ionized level. In one view, this process is most efficient when there are electrons in the solid at precisely the same binding energy as the unoccupied level. Under this condition, a resonance tunneling can occur that neutralizes the outgoing species (Fig. 4.12). Thus the probability of neutralization depends on the band structure of the solid and the atomic levels of the sputtered ion. For high yields of ionized particles, one desires to reduce the neutralization probability. This could be accomplished by the formation of a thin oxide layer, which results in a large forbidden gap and a decrease in number of available electrons for neutralization. For example, oxygen adsorption causes an enhancement of secondary ion yield. Figure 4.13 shows secondary ion yields for 3 keV Ar^+ bombardment of clean and oxygen-covered metals. The enhancement in yield covers a wide range of two to three orders of magnitude. The enhancement for Si is shown in Fig. 4.13b as a function of oxygen concentration in the Si. The sensitivity to an oxidized surface can be an

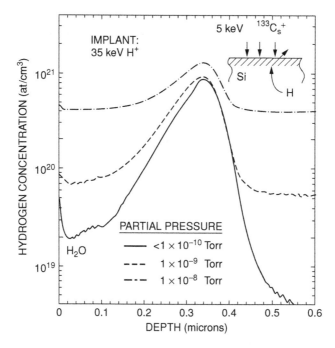

FIGURE 4.11. SIMS measurements of hydrogen depth profiles for 5 keV Cs^+ sputtering of a silicon sample implanted at 35 keV with a dose of 1×10^{18} H ions/cm^2. The effect of the H_2O partial pressure in the analysis chamber upon the H dynamic range is evident. [From Magee and Botnick, 1981.]

advantage; for this reason, SIMS analysis is often carried out with the surface *flooded* with oxygen or bombarded with an oxygen beam.

4.6   Secondary Neutral Mass Spectroscopy (SNMS)

As shown in Fig. 4.13, the secondary ion yield from Si could vary over three orders of magnitude, depending on the oxygen concentration. These matrix effects can be avoided when the sputtered neutral particles are used for composition analysis. The mass analysis system still requires ions for detection, and, in SNMS, the emission (sputtering) and ionization (charge transfer) processes can be decoupled by ionizing the sputtered neutral atoms after (postionization) emission from the sample surface.

An example of an SNMS system is shown in Fig. 4.14, where the major difference from a conventional SIMS system (Fig. 4.7) is the insertion of an ionizing plasma chamber in front of the mass spectrometer. The grids act as an electrical diaphragm between the sample and the chamber, which prevents ions of both signs from entering or leaving the chamber. Thus only neutral species enter the ionizing chamber, and species ionized within the chamber cannot reach the sample.

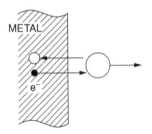

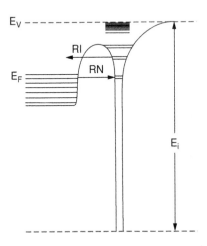

FIGURE 4.12. Model of the electronic structure of an ion or atom close to a metal surface. E_f = Fermi energy; E_i = ionization energy; RI = resonance ionization; and RN = resonance neutralization.

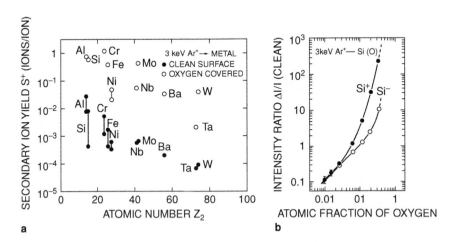

FIGURE 4.13. (a) Comparison of the secondary ion yield of clean and oxygen-covered metals sputtered by 3 keV argon ions. (b) Intensity ratios $\Delta I/I$, for Si$^+$ and Si$^-$ ion yields from oxygen-implanted Si versus oxygen concentration for 3 keV Ar sputtered silicon The oxygen-induced intensity ΔI is given by $\Delta I = I - I_c$, where I is the measured intensity from oxygen-doped Si, and I_c is the ion emission from clean Si. [Both (a) and (b) are from Wittmaack, *Surface Sci.* 112, 168 (1981), with permission from Elsevier.]

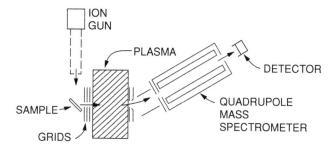

FIGURE 4.14. Apparatus for conducting secondary neutral mass spectroscopy (SNMS), in which neutral sputtered species enter a plasma environment for postionization. The ions are then extracted and detected in the quadruple mass spectrometer system.

Ionization of the neutral species can be achieved in the chamber by use of a low-pressure, high-frequency plasma excited by electron cyclotron wave resonance (Oechsner, 1984). The postionization factor α_A^0 for the sputtered species A depends on the plasma parameters, electron impact ionization of A, and the travel time of A through the ionizer. Values of α_A^0 close to 10^{-1} are achieved for near noble ions and 10^{-2} for transition metal ions like Ta. The postionization factor α^0 is determined by the experimental conditions of the system, and for a particular species A, the factor α_A^0 can be treated as a constant for the apparatus. The measured signal α_A^0 of the neutral species A can be written as

$$I_A^0 = i_p \; Y_A \; \alpha_A^0 \; (1 - \alpha_A^+ - \alpha_A^-) \; \eta_A^0, \tag{4.27}$$

where i_p is the primary beam current, Y_A is the sputtering yield of A, α_A^+ and α_A^- are ionization yields for the formation of secondary ions, and η_A is the instrumental factor. The ionization probabilities α^+ and α^- are usually well below unity, so the factor $(1 - \alpha^+ - \alpha^-)$ can be treated as unity. Since sample matrix effects are small in the postionization factor α^0, calibration can be achieved readily by use of standards. The sensitivity of SNMS to low concentrations of impurities is comparable to that of SIMS, with a detection value of about 1 ppm. In SNMS, however, one does not expect large variations in yield with variation in the properties of the substrate. Instead of plasmas, high-powered lasers can be used to ionize the neutral species.

4.7   Preferential Sputtering and Depth Profiles

In a description of sputtering from a multicomponent system, the influence of preferential sputtering and surface segregation must be included. For a homogeneous sample with two atomic components A and B, the surface concentrations C^s are equal to those in the bulk, C^b, in the absence of segregation to the surface, which might occur due to thermal processes. Then, at the start of sputtering,

$$C_A^s / C_B^s = C_A^b / C_B^b. \tag{4.28}$$

The partial yield of atomic species A and B is defined as in Eq. 4.1 by

$$Y_{A,B} = \frac{\text{Number of elected atoms } A,\ B}{\text{Incident particle}}, \qquad (4.29)$$

where the partial sputtering yield Y_A of species A is proportional to the surface concentration C_A^s, and similarly, Y_B is proportional to C_B^s. The ratio of partial yields is given by

$$\frac{Y_A}{Y_B} = f_{AB}\frac{C_A^s}{C_B^s}, \qquad (4.30)$$

where the sputtering factor f_{AB} takes into account differences in surface binding energies, sputter escape depths, and energy transfers within the cascade. Measured values of f_{AB} generally are in the range between 0.5 and 2.

In the case where f_{AB} is unity, $Y_A/Y_B = C_A^s/C_B^s$, and the yield of sputtered particles is a direct measure of the bulk concentration ratio. In the case where $f_{AB} \neq 1$, the surface concentrations and yields will change from their initial values, $C_A^s(0)$ and $Y_A(0)$, to their values $Y_A^s(\infty)$ and $Y_A(\infty)$ at long times when steady state is achieved.

At the start of sputtering, $t = 0$,

$$\frac{Y_A(0)}{Y_B(0)} = f_{AB}\frac{C_A^s(0)}{C_B^s(0)} = f_{AB}\frac{C_A^b}{C_B^b}. \qquad (4.31)$$

At long times, when steady-state conditions have been achieved, conservation of mass requires that the ratios of partial yields equal the bulk concentration ratio,

$$\frac{Y_A(\infty)}{Y_B(\infty)} = \frac{C_A^b}{C_B^b}. \qquad (4.32)$$

For example, if there is preferential sputtering where $f_{AB} > 1$, the sputtering yield of A is greater than that of B, and the surface will be enriched in B. This enrichment of the surface produces an increase in the sputtering yield of B (more B atoms) and a decrease in the sputtering yield of A (less A atoms). As the process continues with macroscopic amounts (greater than 10 nm) of material removed, the increased concentration of B just balances out the preferential sputtering of A. Therefore, at steady state, the surface concentration ratio will differ from that of the bulk when $f_{AB} \neq 1$:

$$\frac{C_A(\infty)}{C_B(\infty)} = \frac{1}{f_{AB}} \cdot \frac{C_A^b}{C_B^b}. \qquad (4.33)$$

That is, the surface composition is rearranged so that the total sputtering yield gives the bulk composition in spite of differences in yields of the individual atomic species. Analysis of the composition of the remaining surface layer at this point would show a difference from that of the bulk composition.

An example of the change in composition of a silicide layer is shown in Fig. 4.15 for PtSi that was sputtered with 20 keV argon ions and then analyzed with 2 MeV ^{4}He$^+$ ions. The Rutherford backscattering spectrum shows an enrichment of the Pt concentration in the surface region. The ratio of Pt/Si increased from the value of unity associated with that of the bulk values to a value near two in the surface region. The increase in the Pt concentration is due to the fact that the partial sputtering yield of Si is

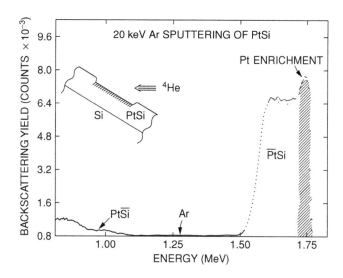

FIGURE 4.15. RBS spectrum of a PtSi film after sputtering with 20 keV Ar ions. The shaded portion in this Pt signal indicates an increase in the concentration of Pt in the near-surface region as a result of the enhanced Si sputtering. [From Liau et al., J. Appl. Phys. 49, 5295 (1978). Copyright 1978 by the American Physical Society.]

greater than that of Pt, $Y_{Si} > Y_{Pt}$. Figure 4.16 shows the partial yields as a function of argon ion dose. As one would expect, at low bombardment doses the sputtering yield of Si is significantly greater than that of Pt. At the onset of sputtering, the yield ratio $Y_{Si}(0)/Y_{Pt}(0) = 2.4$. As the bombardment proceeds, the partial sputtering yields merge into the same value. The equality of the Si and Pt yields merely reflects the fact that the yield ratio, after steady state has been reached, is equal to the bulk concentration ratio, which for PtSi is unity.

4.8   Interface Broadening and Ion Mixing

One of the applications of sputtering is the removal of deposited or grown layers in thin film structures in order to analyze the composition at the interface between the film and substrate. In these applications, the penetration of the ions used in the sputtering beam can induce an intermixing between the film and substrate due to the strong atomic displacements and diffusion that occurs within the collision cascade around the track of the ion used in sputtering. This intermixing leads to an artificial broadening of the concentration depth profiles at the interface.

Sputtering requires bombardment of the surface with primary ions of appreciable energy (typically 1–20 keV) whose range far exceeds the escape depth of the sputtered ions and often exceeds the observation depth in electron spectroscopies. Therefore, due to the ion-induced intermixing in the collision cascade, a zone of altered material precedes the *analytical* zone during layer removal. Ion mixing is illustrated in Fig. 4.17 for SIMS analysis of a Pt film deposited on Si and sputtered by argon ions.

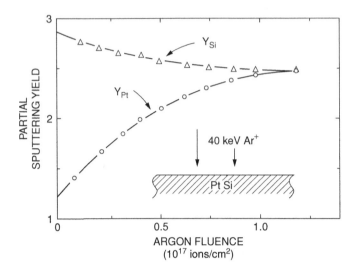

FIGURE 4.16. Dose dependence of the partial sputtering yields of Si and Pt emitted from PtSi for 40 keV Ar bombardment, where fluence denotes dose. [From Liau and Mayer, 1980.]

When the argon ions penetrate through the Pt/Si interface, some of the Si atoms in the substrate will be transported to the top surface of the Pt film, where they can be sputtered. Thus a silicon signal will appear before the Pt film is sputtered away. Platinum is also intermixed with the Si, and, consequently, a Pt signal will persist in the SIMS spectrum at depths well beyond the thickness of the original deposited layer of Pt.

An estimate of the interfacial broadening in such systems can be made by setting the range R of the sputtering ion equal to the half-width of the broadened signal. The ion

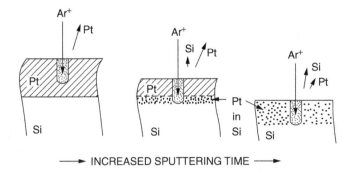

FIGURE 4.17. Schematic diagram of Ar$^+$ ion sputtering of a 100 nm Pt layer on Si at three different times in the sputtering process. When the Ar range is less than the Pt film thickness, only Pt ions are sputtered. When the Ar ion penetrates through the Pt/Si interface, ion-induced intermixing occurs, and a Si signal is found in the sputtering yield. After the initial Pt film has been removed, a Pt signal is still observed due to mixing of Pt into the substrate Si. [From Liau et al., *J. Vac. Sci. Technol.* 16, 121 (1979).]

range is given by

$$R = \int_{E_0}^{0} \frac{1}{dE/dx} dE, \qquad (4.34)$$

which can be approximated for medium mass ions in the keV energy range by assuming that nuclear energy loss dominates and has an energy-independent value,

$$R = E_0/(dE/dx)_n, \qquad (4.35)$$

where $(dE/dx)_n$ is given by Eq. 4.19. The value of $(dE/dx)_n$ for Ar$^+$ ions in Cu is about 1000 eV/nm, which is the basis for the rule of thumb that the altered layer extends 1 nm/keV.

The amount of interface broadening can be minimized by proper choice of ion energies and incident angles during sputter profiling. In many cases, sputter depth profiles can have a better depth resolution than that obtained with backscattering spectrometry.

When possible, it is advantageous to use two or more analytical techniques that provide complementary data. Figure 4.18 shows the analysis of a tungsten silicide film on polycrystalline silicon (poly-Si). Secondary ion mass spectroscopy (Fig. 4.18a) is used to determine the phosphorus concentration in polycrystalline Si and to detect oxygen contamination at the interface. Backscattering spectrometry (Fig. 4.18b) provides the depth scale and the composition of the sputter-deposited silicide—in this case, 183 nm of WSi$_{2.7}$. In the SIMS data, the phosphorus concentration in poly-Si was calibrated from a standard obtained by implanting phosphorus into a silicon sample.

The ratio of W to Si signals in SIMS data (Fig. 4.18a) does not reflect the silicide composition, and there is an order-of-magnitude increase in the Si signal going from the silicide into the Si. These effects are due to the influence of the matrix on the yield of secondary ions. The peak in the W signal at the silicide/Si interface is due to the enhancement of the W ion yield because of oxygen at the interface (note also the enhancement of the phosphorus yield at the Si/SiO$_2$ interface). The RBS spectrum

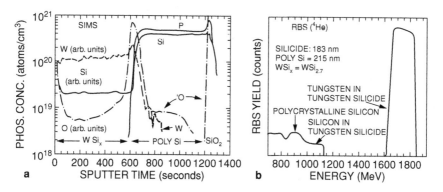

FIGURE 4.18. (a) Secondary ion mass spectroscopy (SIMS) and (b) Rutherford backscattering spectrometry (RBS) analysis of a tungsten silicide layer sputter deposited on a phosphorus-doped polycrystalline silicon layer on a layer of SiO$_2$ on silicon.

(Fig. 4.18b) shows that there is no peak in the W distribution at the interface and that the composition of the silicide is $WSi_{2.7}$. The increase in the Si signal around 950 keV is due to the increase in Si concentration going from silicide to Si, and the decrease around 800 keV is due to the presence of a layer of SiO_2 70 nm thick at the interface between the poly-Si layer and substrate Si. The signal from the 1 atomic % of phosphorus in the poly-Si cannot be detected in the RBS spectrum but is easily detected in the SIMS data. The amount of interface broadening is minimal in the depth profile of the SIMS data, and the concentration of light mass elements (oxygen and phosphorus) can easily be detected. The strong influence of the matrix on the ion yield does not allow an accurate determination of the relative concentration of the major constituents (tungsten and silicon). These quantities are found along with layer thickness from the Rutherford backscattering spectrum. By the use of two complementary analytical techniques (RBS and SIMS), a rather accurate picture of the composition of the sample can be obtained.

4.9   Thomas–Fermi Statistical Model of the Atom

In low-velocity collisions, the impact parameter is sufficiently large such that the nuclear charge is screened by the electrons. This leads to a modification of the scattering potential from that of the unscreened Coulomb potential, $V = Z_1 Z_2 e^2 / r$. The modified potential is found from the Thomas–Fermi description, which treats all atoms as identical aside from scaling factors.

The Thomas–Fermi model assumes that the electrons can then be treated by statistical mechanics, Fermi–Dirac statistics, in which they behave as an ideal gas of particles that fills the potential well around the positively charged core. The density of states, $n(E)$, of a free electron gas is obtained by applying periodic boundary conditions and box normalization to a cell of length L to give

$$n(E) = \frac{L^3}{2\pi^2 \hbar^3} (2m)^{3/2} E^{1/2}. \tag{4.36}$$

The energy of the gas increases as the number of electrons increases. For a collection of electrons, the number at a point r, $Z(r)$, is given by

$$Z(r) = \int_0^{E_F(r)} n(E) \, dE = \frac{L^3 (2m)^{3/2}}{2\pi^2 \hbar^3} \int_0^{E_F(r)} E^{1/2} dE = \frac{L^3 (2m)^{3/2}}{3\pi^2 \hbar^3} E_F(r)^{3/2}, \tag{4.37}$$

where $E_F(r)$ is associated with the maximum energy of the ensemble of electrons at r. The Fermi energy is simply the energy of the highest filled state. In the many-electron atom that we are considering, the total energy E_r of an electron is $E_r = E_K + V(r)$, where E_K is the kinetic energy. For a bound electron, $E_r \leq 0$, which requires that for the maximum kinetic energy electron, $E_F = -V(r)$. From Eq. 4.37,

$$\rho(r) = \frac{Z(r)}{L^3} = \frac{(2m)^{3/2}}{3\pi^2 \hbar^3} [-V(r)]^{3/2}. \tag{4.38}$$

The self-consistency condition is that the potential due to the electron density in Eq. 4.38, as well as that due to the nuclear charge, properly reproduces the potential energy, $-V(r)$. Consequently, the charge density, $-e\rho$, and the electrostatic potential, $-[V(r)/e]$, must satisfy Poisson's equation

$$-\frac{1}{e}\nabla^2 V = -4\pi(-e\rho),$$

or

$$\nabla^2 V = \frac{1}{r^2}\frac{d}{dr}\left(r^2\frac{dV}{dr}\right) = 4\pi e^2\rho = \frac{4e^2[-2mV(r)]^{3/2}}{3\pi\hbar^3}. \tag{4.39}$$

Equations 4.38 and 4.39 may be solved simultaneously for ρ and V, with boundary conditions as follows: As $r \to 0$, the leading term in the potential energy must be due to the nucleus, so that $V(r) \to -Ze^2/r$, and as $r \to \infty$, there must be no net charge inside the sphere of radius r, so that V falls off more rapidly than $1/r$, and $rV(r) \to 0$. Equation 4.39 and the boundary conditions given above are conveniently expressed in a dimensionless form in which Z, E, m, and h appear only in scale factors. We put

$$V(r) = -\frac{Ze^2}{r}\chi, r = ax,$$

and

$$a = \frac{1}{2}\left(\frac{3\pi}{4}\right)^{2/3}\frac{\hbar^2}{me^2Z^{1/3}} = \frac{0.885a_0}{Z^{1/3}}, \tag{4.40}$$

where $a_0 = \hbar^2/me^2$, the Bohr radius. Eq. 4.8 indicates that the scaling parameter to describe the size of an atom is inversely proportional to the cube root of the atomic number. (For electron spectroscopies where transitions are between core levels, the 1s radius is approximated by a_0/Z.) With these substitutions, Eq. 4.39 becomes

$$x^{1/2}\frac{d^2\chi}{dx^2} = \chi^{3/2}. \tag{4.41}$$

In this dimensionless Thomas–Fermi (TF) equation, the potential behaves like a simple Coulomb interaction in the extreme case as $r \to 0$. The accurate solution of Eq. 4.41 is done numerically, and there are also analytical approximations represented in series expansions or exponentials. The Molière approximation to the Thomas–Fermi screening function shown in Fig. 4.4 is most often used in computer simulations and is given by

$$\chi(x) = 0.35e^{-0.3x} + 0.55e^{-1.2x} + 0.10e^{-6.0x}, \tag{4.42}$$

where $x = r/a$.

Problems

4.1. The maximum value of the nuclear energy loss occurs at the reduced energy value of 0.3 for the Thomas–Fermi potential. What energy in keV does $\varepsilon' = 0.3$ correspond to for Ar^+ ion incident on Si?

4.2. Assuming nuclear energy loss dominates and the stopping cross section is energy independent, what is the range of 10 keV Ar ions incident on Cu?

4.3. For a screened Coulomb collision with $\chi = a/2r$, use the impulse approximation to show that b, the impact parameter, is proportional to $(a/E\theta)^{1/2}$, and derive $\sigma(\theta)$, the cross section.

4.4. Calculate the ratio of the unscreened to screened nuclear cross section $d\sigma/dT$ for the following cases: 2.0 MeV He^+ on Au, 0.1 MeV He^+ on Au, and 1 keV Ar^+ on Cu.

4.5. For a scattering potential $V(r) \propto r^{-3}$, what is the energy dependence of the energy loss dE/dx?

4.6. Calculate the sputtering yield for 45 keV Ar ions incident on Si ($U_0 = 4.5$ eV) using a screened potential. Compare your answer with the data given in Figure 4.2.

4.7. If the sputtering yield of species A is twice that of species B in a matrix AB, what is the ratio A to B of the flux of sputtered species at (a) the initial time and (b) the steady-state time, and what is the ratio A to B of surface composition at (c) the initial time and (d) the steady-state time?

4.8. Determine the time in seconds required to sputter 50 nm of Si using a 10 μA/cm^2 beam of 45 keV ions of (a) Ne, (b) Kr, and (c) Xe. (Use data given by the solid line in Fig. 4.2.)

References

1. E. E. Anderson, *Modern Physics and Quantum Mechanics* (W. B. Saunders Co., Philadelphia, 1971).
2. H. H. Anderson and H. L. Bay, "Sputtering Yield Measurements," in *Sputtering by Particle Bombardment I*, R. Behrisch, Ed. (Springer-Verlag, New York, 1981), Chapter 4.
3. R. Behrisch, Ed., *Sputtering by Particle Bombardment I and II* (Springer-Verlag, New York, 1981 and 1983).
4. G. Carter and J. S. Colligon, *Ion Bombardment of Solids* (Elsevier Science Publishing Co., New York, 1968).
5. G. Carter, B. Narvinsek, and J. L. Whitton, "Heavy Ion Sputtering Induced Surface Topography Development," in *Sputtering by Particle Bombardment II*, R. Behrisch, Ed. (Springer-Verlag, New York, 1983).
6. W. K. Chu, "Energy Loss of High Velocity Ions in Matter," in *Atomic Physics*, P. Richard, Ed., *Methods of Experimental Physics*, Vol. 17 (Academic Press, New York, 1980), Chapter 2.
7. Z. L. Liau and J. W. Mayer, "Ion Bombardment Effects on Material Composition," in *Ion Implantation*, J. K. Hirvonen, Ed., *Treatise on Materials Science and Technology*, Vol. 18, N. Herman, Ed. (Academic Press, New York, 1980), Chapter 2.
8. J. Lindhard, M. Scharff, and H. E. Schiott, "Range Concepts and Heavy Ion Ranges (Notes on Atomic Collision, II)," *Mat. Fys. Medd. Dan. Vid. Selsk.* 33(14) (1963).
9. C. Magee and E. M. Botnick, *J. Vac. Sci. Technol.* 19(47) (1981).
10. J. A. McHugh, "Secondary Ion Mass Spectrometry," in *Methods of Surface Analysis*, A. W. Czanderna, Ed. (Elsevier Science Publishing Co., New York, 1975), Chapter 6.
11. J. M. McCrea, "Mass Spectrometry," in *Characterization of Solid Surfaces*, P. F. Kane and G. B. Larrabee, Eds. (Plenum Press, New York, 1974), Chapter 2.
12. H. Oechsner, Ed., *Thin Film and Depth Profile Analysis* (Springer-Verlag, New York, 1984).
13. P. Sigmund, "Sputtering Processes: Collision Cascades and Spikes," in *Inelastic Ion-Surface Collisions*, N. Tolk et al., Eds. (Academic Press, New York, 1977), pp. 121–152.

14. P. Sigmund, "Sputtering by Ion Bombardment; Theoretical Concepts," in *Sputtering by Particle Bombardment I*, R. Behrisch, Ed. (Springer-Verlag, New York, 1981), Chapter 2.
15. I. M. Torrens, *Interatomic Potentials* (Academic Press, New York, 1972).
16. P. D. Townsend, J. C. Kelley, and N. E. W. Hartley, *Ion Implantation, Sputtering and Their Applications* (Academic Press, New York, 1976).
17. H. W. Werner, "Introduction to Secondary Ion Mass Spectrometry (SIMS)," in *Electron and Ion Spectroscopy of Solids*, L. F. Ermans et al., Eds. (Plenum Press, New York, 1978).

5
Ion Channeling

5.1 Introduction

The arrangement of atoms in a solid determines the properties of a material and, in single crystals, determines the magnitude of incident ion–target-atom interactions. The influence of the crystal lattice on the trajectories of ions penetrating into the crystal is known as *channeling*—a term that visualizes the atomic rows and planes as guides that steer energetic ions along the *channels* between rows and planes. The steering action is effective and can lead to hundredfold reductions in the yield of backscattered particles. In this chapter, we describe channeling of high-energy ions in single crystals and show how use of this technique can improve the *depth resolution* of the ion scattering technique and improve its sensitivity to light impurities. The combination of Rutherford backscattering and channeling plays an important role in the materials analysis of a variety of thin film/single-crystal problems. Single crystals are not rare in modern-day technology, as all electronic components are based on single-crystal semiconductors.

5.2 Channeling in Single Crystals

Channeling of energetic ions occurs when the beam is carefully aligned with a major symmetry direction of a single crystal (Lindhard, 1965). By a major symmetry direction, we mean one of the open directions as viewed down a row of atoms in a single crystal. Figure 5.1 shows a side view of this process in which most of the ion beam is steered (channeled) through the channels formed by the string of atoms. Channeled particles cannot get close enough to the atomic nuclei to undergo large-angle Rutherford scattering; hence, scattering from the substrate is drastically reduced by a factor of ~ 100. This improves the ion scattering sensitivity to light impurities on the surface. There is always a full interaction with the first monolayers of the solid. This *surface interaction* results in an improved *depth resolution* in these experiments.

The trajectory of a channeled ion is such that the ion makes a glancing angle impact with the axes (axial channeling) or planes (planar channeling) of the crystal and is steered by small-angle scattering collisions at distances greater than 0.01 nm from the

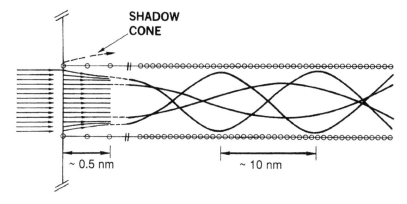

FIGURE 5.1. Schematic of particle trajectories undergoing scattering at the surface and channeling within the crystal. The depth scale is compressed relative to the width of the channel in order to display the trajectories.

atomic cores. Since the steering of the channeled particle involves collisions with many atoms, one may consider in a continuum model that the nuclear charge of the atoms in a row (or plane) is uniformly averaged along the row (or plane). The interaction of a channeled particle with an atomic row is described in terms of a single continuum potential $U_a(r)$, where r is the perpendicular distance from the row. $U_a(r)$ is the value of the atomic potential averaged along the atomic row with atomic spacing d. For the axial case,

$$U_a(r) = \frac{1}{d} \int_{-\infty}^{\infty} V\left(\sqrt{z^2 + r^2}\right) dz, \tag{5.1}$$

where $V(\tilde{r})$ is the screened Coulomb potential and $\tilde{r}$ is the spherical radial coordinate, $\tilde{r}^2 = z^2 + r^2$. Rather than using the Molière potential, we use a more convenient form of the screened Coulomb potential, extensively used in channeling theory because it permits analytical treatments of channeling parameters without significant loss of precision. This *standard potential* is given by

$$V(\tilde{r}) = Z_1 Z_2 e^2 \left(\frac{1}{\tilde{r}} - \frac{1}{\sqrt{\tilde{r}^2 + C^2 a^2}} \right), \tag{5.2}$$

where C^2 is usually taken as equal to 3 and a is the Thomas–Fermi screening distance (Eq. 4.5). Then we obtain for the axial continuum potential

$$U_a(r) = \frac{Z_1 Z_2 e^2}{d} \ln\left[\left(\frac{Ca}{r} \right)^2 + 1 \right], \tag{5.3}$$

where d is the average distance between atoms in the rows. The magnitude of this potential is the order of atomic potentials, that is, 223 eV at $r = 0.01$ nm for He along the $\langle 110 \rangle$ rows of Si ($d = 0.384$ nm).

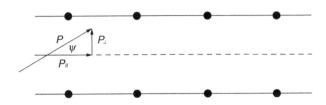

FIGURE 5.2. Components of the initial momentum vector for a particle incident to an atomic string at an angle ψ.

The continuum potential and conservation of energy allow us to find the critical angle for channeling. The total energy E of a particle inside the crystal is

$$E = \frac{p_\parallel^2}{2M} + \frac{p_\perp^2}{2M} + U_a(r),\tag{5.4}$$

where $p_\parallel$ and $p_\perp$ are, respectively, the parallel and perpendicular components of the momentum with respect to the string direction (Fig. 5.2). Then

$$p_\parallel = p \cos \psi, \quad p_\perp = p \sin \psi,$$

and

$$E = \frac{p^2 \cos^2 \psi}{2M} + \frac{p^2 \sin^2 \psi}{2M} + U_a(r).\tag{5.5}$$

Channeling angles are small, and we (1) use a small-angle approximation and (2) equate the last two terms with the transverse energy

$$E_\perp = \frac{p^2 \psi^2}{2M} + U_a(r),\tag{5.6}$$

i.e., a kinetic energy contribution and a potential energy contribution. The total energy is conserved, and, in this approximation, the transverse energy is conserved. Then the critical angle ψ_c is defined by equating the transverse energy at the turning point $U(r_{\min})$ to the transverse energy at the midpoint:

$$E\psi_c^2 = U(r_{\min}),\tag{5.7}$$

or

$$\psi_c = \left(\frac{U(r_{\min})}{E}\right)^{1/2}.\tag{5.8}$$

The thermal smearing of the atom positions sets a lower limit to the minimum distance for which a row can provide the necessary correlated sequence of scatterings required for the channeling condition. The most useful first approximation to the critical angle is obtained by substituting $r_{\min} = \rho$ in Eqs. 5.3 and 5.8, where ρ^2 is two-thirds of the mean square thermal vibration amplitude:

$$\psi_c(\rho) = \frac{\psi_1}{\sqrt{2}} \left| \ln\left[\left(\frac{Ca}{\rho}\right)^2 + 1\right]\right|^{1/2},\tag{5.9}$$

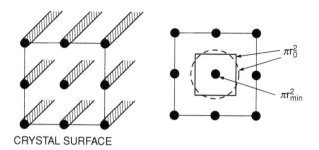

CRYSTAL SURFACE

FIGURE 5.3. View of an ideal crystal showing continuous rows of atoms and defining the area/row, πr_{min}^2.

where

$$\psi_1 = \left(\frac{2Z_1Z_2e^2}{Ed}\right)^{1/2}. \tag{5.10}$$

Thermal vibrations and the definition of ρ are discussed in Section 8.3.

The values of $\psi_c(\rho)$ are within 20% of experimental measurements and follow the measured temperature dependence. For 1.0 MeV He$^+$ incident on Si$\langle 110 \rangle$ at room temperature, $\psi_c(\rho) = 0.65°$, while the experimentally measured value is 0.55°.

The concept of a distance of closest approach, r_{min}, allows a simple geometric derivation of the fraction of channeled particles for incidence parallel to a crystal axis, $\psi = 0$. In Fig. 5.3, we are looking head-on into the crystal. Around each string of atoms is an area πr_{min}^2 in which particles cannot channel; particles incident at $r > r_{min}$ can channel. Then the fraction not channeled is simply $\pi r_{min}^2 / \pi r_0^2$, where r_0 is the radius associated with each string:

$$\pi r_0^2 = \frac{1}{Nd}. \tag{5.11}$$

Here N is the atomic concentration of atoms and d is the atomic spacing along the string. In the literature, the ratio $\pi r_{min}^2 / \pi r_0^2$ is usually referred to as the *minimum yield*, χ_{min}. In a backscattering experiment, for example, χ_{min} is the yield of close-encounter events as a result of the channeling process. Since $r_{min} \sim 0.01$ nm, the minimum yield is on the order of 1% or the fraction of particles channeled is $\sim 99\%$.

The continuum description can be applied to planar channeling as well as to axial channeling. For planes, two-dimensional averaging of the atomic potential results in a sheet of charge, as with the corresponding planar continuum potential $U_p(y)$ being defined as

$$U_p(y) = Nd_p \int V\left(\sqrt{y^2 + r^2}\right) 2\pi r \, dr, \tag{5.12}$$

where Nd_p is the average number of atoms per unit area in the plane, d_p is the spacing between planes, and y is the distance from the plane. For the standard potential, $U_p(y)$ is given by

$$U_p(y) = 2\pi Z_1 Z_2 e^2 a N d_p \left\{ \left[\left(\frac{y}{a}\right)^2 + C^2\right]^{1/2} - \frac{y}{a} \right\}. \tag{5.13}$$

Similarly to axial channeling, a critical angle can be defined as

$$\psi_p = \left(\frac{U_p(y_{min})}{E}\right)^{1/2},$$ (5.14)

where $y_{min} \cong \rho/\sqrt{2}$ is the one-dimensional vibrational amplitude. We define the characteristic angle for planar channeling,

$$\psi_2 = \left(\frac{2\pi Z_1 Z_2 e^2 a N d_p}{E}\right)^{1/2},$$ (5.15)

which is on the order of the critical angle for planar channeling. Experimentally, planar critical angles are a factor of 2–4 smaller than characteristic critical angles for axial channeling.

A geometric picture of planar channeling indicates that the minimum yield, χ_{min}(planar), is approximately given by

$$\chi = \frac{2y_{min}}{d_p},$$ (5.16)

which is substantially larger than the corresponding value for axial channeling. The value of the minimum yield for good planar channeling directions is typically on the order of 10–25%.

5.3   Lattice Location of Impurities in Crystals

One application of channeling is the determination of the lattice site (substitutional or nonsubstitutional) of an impurity in a single crystal. For small concentrations of impurities, <1%, the presence of the impurities does not affect the channeling properties of the host lattice. Therefore, the close-encounter probability of a substitutional impurity follows the same angular dependence as that of the host lattice. Impurities that occupy an ensemble of different sites, such as might exist in an impurity cluster, show no angular dependence in the yield curve. The angular yield curves for both the substitutional case and the *random* cluster of nonsubstitutional impurity atoms are shown schematically in Fig. 5.4.

Rutherford backscattering is typically used to investigate the lattice site location of impurities that have a larger atomic mass than the host atoms. Scattering kinematics separates the signal of the impurity and host. For lighter-mass impurities, the yield from nuclear reactions (Chapter 13) or ion-induced X-ray interactions (Chapter 11) is used to monitor particle-impurity interactions. The angular yield curve, shown in Fig. 5.4, is obtained by monitoring the yield of the impurity and the host lattice. The simultaneous accumulation of both angular yield curves provides a sensitive experimental test of the actual substitutionality of the impurity. For substitutional impurities (i.e., those that replace a host atom on its lattice site), the angular scan and minimum yield for scattering from impurities follow that of the host crystal (Fig. 5.4). Impurities located at interstitial sites show a different angular scan. The interpretation of these angular scans requires knowledge of the flux distribution of channeled particles.

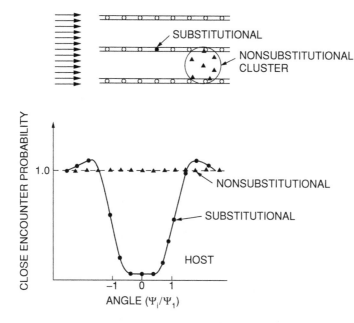

FIGURE 5.4. Schematic of the close-encounter probability curve expected for substitutional impurities and for a nonsubstitutional cluster.

5.4   Channeling Flux Distributions

A channeled particle is confined within equipotential contours, U_T, such as the continuum contours shown in Fig. 5.5. Here $U_T = \sum U_i$ is the sum of the individual potentials U_i of the nearby rows or planes. Thus a particle with a given transverse energy $E_\perp$ must always lie within a region given by $U_T(r) \leq E_\perp$.

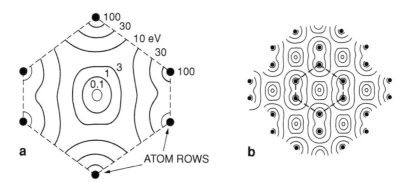

FIGURE 5.5. (a) Equipotential contours of the axial continuum potential for the case of He$^+$ in the ⟨110⟩ direction of Si. Note the change in shape of the potential contours corresponding to the geometry of the channel. (b) The potential contours for an array of the channels of the type shown in Fig. 5.5a.

For silicon the 3 eV potential contour closes within the center of the channel (Fig. 5.5), and hence particles with a transverse energy less than 3 eV will have their trajectories confined to a particular channel. Channeled particles with higher transverse energies, $E_\perp \geq 10$ eV, are not confined within one channel but are guided by the cylindrically symmetric potentials around the axial rows (Fig. 5.5). If we normalize to unit probability of finding a particle somewhere in its allowed area $A(E_\perp)$, the probability to finding a particle of transverse energy $E_\perp$ at any point $\mathbf{r}$ is

$$P(E_\perp, \mathbf{r}) = \begin{cases} \dfrac{1}{A(E_\perp)}, & E_\perp \geq U_T(\mathbf{r}) \\ 0, & E_\perp < U_T(\mathbf{r}) \end{cases}. \tag{5.17}$$

The area $A(E_\perp)$ is defined by an equipotential contour such as that shown in Fig. 5.5. For example, particles with $E_\perp = 10$ eV (i.e., 1 MeV particles entering the center of the channel at an angle $0.18°$) have equal probability of being found at any point within the area defined by the equipotential contour $U_T(r) = 10$ eV.

In the following, the flux distribution of ions for a channeled beam ($\psi = 0$) is calculated:

1. Due to conservation of transverse energy, a particle that enters at r_{in} cannot get closer than r_{in} to the string.
2. For two-dimensional axial channeling, a particle has a uniform probability of being found in its *allowed* area; for cylindrical symmetry, the allowed area is πr_{in}^2, where r_{in} is the initial distance from the string at the crystal surface, i.e.,

$$P(r_{in}, r) = \begin{cases} \dfrac{1}{\pi r_0^2 - \pi r_{in}^2}, & r > r_{in} \\ 0, & r < r_{in} \end{cases}. \tag{5.18}$$

3. We assume cylindrical symmetry for simplicity (Fig. 5.6).

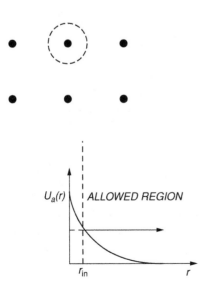

FIGURE 5.6. Geometry associated with the flux distribution calculation showing the cylindrical geometry and the *allowed* area.

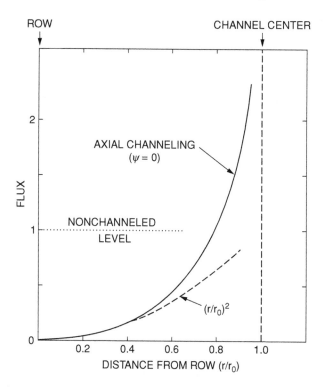

FIGURE 5.7. Flux distribution of the channeled beam as a function of distance from the row for the case of the parallel incidence. $\psi = 0$, and assuming statistical equilibrium and cylindrical symmetry. The small r/r_0 approximation (Eq. 5.20) is shown for comparison.

The flux distribution inside the crystal, $f(r)$, then corresponds to integrating over all initial impact parameters:

$$f(r) = \int_0^{r_0} P(r_{in}, r) 2\pi r_{in} \, dr_{in}$$

$$= \int_0^{r_0} \frac{1}{\pi r_0^2 - \pi r_{in}^2} 2\pi r_{in} \, dr_{in}$$

$$= \ln \frac{r_0^2}{r_0^2 - r^2}. \tag{5.19}$$

The effect of channeling is to transform a spatially uniform distribution to the peaked distribution shown in Fig. 5.7. This flux distribution displays the most prominent feature of channeling, namely, that the flux intensity and hence the close-encounter probability approaches zero near the atomic rows (as $r \to 0$). Expansion $f(r)$ for small r, near the atom rows, gives

$$f(r) \approx \frac{r^2}{r_0^2}, \tag{5.20}$$

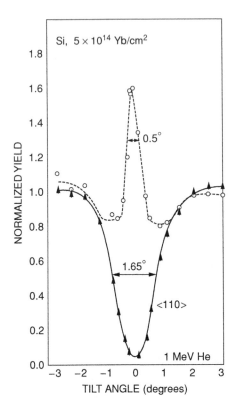

FIGURE 5.8. Close-encounter probability (normalized yield) as a function of the angle between the beam and ⟨110⟩ symmetry direction for the case of 1.0 MeV He$^+$ in Si implanted with 5×10^{14} Yb/cm^2 (60 keV, 450°C). The high yield for Yb (open data points) was the first indication of the flux peaking effects. [Courtesy of John A. Davies.]

where the approximation is shown by the dashed line in Fig. 5.7. This simple flux distribution is useful for estimating the scattering intensity from substitutional impurities.

Examination of Fig. 5.7 indicates another property of the channeling flux distribution: the peaking of the particle density in the center of the channels ($r \cong r_0$). The intensity is far in excess of unity, the value corresponding to the nonchanneling particle density. Thus the yield from interstitial impurities located near the channels ($r \approx r_0$) will be greater than the nonchanneling value.

The *flux peaking* effect at $r = r_0$ was first shown in lattice location experiments of Yb implanted into Si (Fig. 5.8). A substitutional impurity would have an angular scan equivalent to that of the host. Clearly, the Yb is not substitutional but is located at a position where the flux intensity is appreciably greater than that for a nonchanneling direction.

5.5   Surface Interaction via a Two-Atom Model

The two-atom model is a simple and illustrative example of the interaction of the ion beam with the atomic structure at a surface. We calculate the shadow behind a repulsive scattering center, the topmost atom on a surface. Essentially, we calculate the flux distribution $f(r_2)$ at the second atom as a result of scattering interactions with the

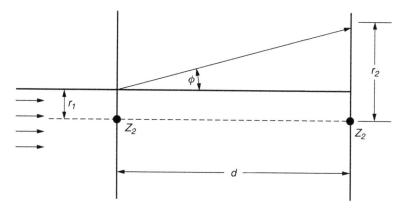

FIGURE 5.9. Geometry for the calculation of the shadow cone and the flux distribution at the second atom.

first atom. The small-angle approximation to pure Coulomb scattering is used so that $\phi = Z_1 Z_2 e^2 / E r_1$, as shown in Fig. 5.9. Then

$$r_2 = r_1 + \phi d$$

$$= r_1 + \frac{Z_1 Z_2 e^2 d}{E r_1}$$

$$= r_1 + \frac{R_C^2 / 4}{r_1}, \tag{5.21}$$

where the Coulomb shadow cone radius R_C is defined as

$$R_C = 2 \left(\frac{Z_1 Z_2 e^2 d}{E} \right)^{1/2}. \tag{5.22}$$

This quantity is the distance of closest approach to the second atom (Fig. 5.10). The flux distribution at the second atom, $f(r_2)$, is given by

$$f(r_2) \, 2\pi r_2 \, dr_2 = f(r_1) 2\pi r_1 dr_1, \tag{5.23}$$

where $f(r_1)$, the incident flux distribution, is uniform and normalized to unity so that

$$f(r_2) = \frac{r_1}{r_2} \left| \frac{dr_1}{dr_2} \right|. \tag{5.24}$$

As shown in Fig. 5.10, r_2 has a minimum, i.e., $dr_2/dr_1 = 0$. It is simplest to evaluate $f(r_2)$ in two steps, for $r_1 < r_{1c}$ and for $r_1 > r_{1c}$; $r_2(r_{1c}) = R_C$. The final result is

$$f(r_2) = \begin{cases} 0, & r_2 < R_C \\ \dfrac{1}{2} \left[\dfrac{1}{\sqrt{1 - R_C^2/r_2^2}} + \sqrt{1 - R_C^2/r_2^2} \right], & r_2 > R_C \end{cases}. \tag{5.25}$$

The function $f(r_2)$ is sketched in Fig. 5.11.

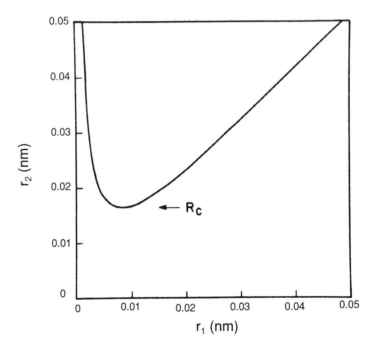

FIGURE 5.10. The dependence of r_2 on the initial impact parameter r_1 for the case of 1.0 MeV He$^+$ ions on $W\langle 100\rangle$; r_2 is calculated for $d = 0.316$ nm corresponding to the atom spacing in the $\langle 001\rangle$ direction. The minimum value of r_2 is the Coulomb shadow cone with radius R_C.

The flux distribution is so sharp that the curvature occurs within a distance that is small compared to the thermal vibration amplitude of atoms in a crystal. Thus we approximate $f(r_2)$ to a delta function, i.e.,

$$f(r_2) = \begin{cases} 0, & r_2 < R_C \\ 1 + \dfrac{R_C^2}{2} \dfrac{\delta(r_2 - R_C)}{r_2}, & r_2 \geq R_C \end{cases} \qquad (5.26)$$

The intensity of scattering from the second atom, I_2, which we treat as a measure of

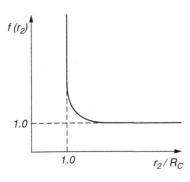

FIGURE 5.11. Flux distribution $f(r_2)$ as a function of r_2/R_C.

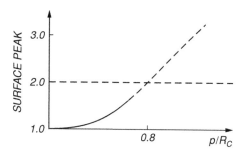

FIGURE 5.12. Surface peak intensity as a function of ρ/R_C from "two-atom" calculation.

the close-encounter probability, is given by the overlap of the flux distribution with the Gaussian position distribution $P(r_2)$ of the second atom:

$$I_2 = \int \tilde{P}(r_2) f(r_2) 2\pi r_2 \, dr_2, \tag{5.27}$$

where

$$\tilde{P}(r) = \frac{1}{\pi \rho^2} e^{-r^2/\rho^2}. \tag{5.28}$$

This treatment assumes that the close-encounter process has a characteristic interaction distance that is very small compared to thermal vibration amplitudes or R_C; such processes are nuclear reactions, Rutherford backscattering, and inner-shell X-ray excitation. If we use Eqs. 5.26 and 5.28,

$$I_2 = \left[1 + \frac{R_C^2}{\rho^2} \right] e^{-R_C^2/\rho^2}, \tag{5.29}$$

and the total surface peak intensity I is given by the unit contribution for the first atom and I_2 (Fig. 5.12). Note that the intensity I_2 is determined by a single parameter ρ/R_C, which defines the ratio of the thermal vibration amplitude to the shadow cone:

$$I = I_2 + 1. \tag{5.30}$$

For values of $\rho < R_C$, the topmost atom does indeed shadow the underlying atoms from direct, close encounters with the analysis beam. This surface shadowing effect is most vividly revealed in the backscattering spectrum from clean, single crystals. The spectrum is dominated by the surface peak corresponding to interactions with the first few monolayers of the solid; at lower energies, there is a continuum of scattering corresponding to scattering by the relatively few nonchanneled particles. As described previously, the intensity of the surface peak is sensitive to the arrangement of surface atoms.

5.6 The Surface Peak

The application of ion beams to surface structure determination depends upon (1) accurate measurement of a surface peak in monolayers and (2) the ability to predict the surface peak for a given surface structure. For example, the aligned spectra in Fig. 5.13

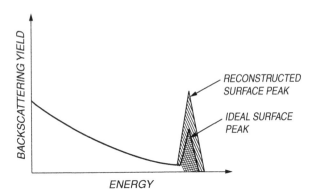

FIGURE 5.13. Channeling spectrum showing the surface peak for an ideal surface and a reconstructed surface.

differ in their surface peak intensity for different surface structures. The reconstructed surface shows a higher yield than the ideal surface due to surface displacements.

More sophisticated treatments of the surface peak take into account that more than two atoms may contribute to the surface peak, and the governing potential is not a pure Coulomb description but a screened potential, for example, the Molière potential. The intensity of the surface peak is established by numerical techniques. The results of many calculations demonstrate that the surface peak scaling still holds; the scaling parameter is ρ/R_M, where R_M is the shadow cone radius associated with the Molière potential, roughly $R_M \cong R_C$.

The values of the surface peak in terms of atoms/row have been calculated for a broad range of cases in which an ideal surface with a *bulklike* structure is assumed. The results are plotted in Fig. 5.14, as a *universal curve* for the intensity of the surface peak as a function of ρ/R_M, where ρ is the two-dimensional vibrational amplitude.

Four simple cases of surface structure are shown with their corresponding surface peak spectra in Fig. 5.15. The dashed spectra represent the scattering yield from a crystal with an ideal surface for the case where the thermal vibration amplitude ρ is much less than the shadow cone radius R_M. This condition assures that the surface peak intensity corresponds to one atom/row in this ideal case. The crystal with a reconstructed surface where the surface atoms are displaced in the plane of the surface (Fig. 5.15b) represents a situation where the second atom is not shadowed. The surface peak intensity in this case is twice that of the ideal crystal. To test for relaxation where the surface atoms are displaced normal to the surface plane (Fig. 5.15c), one must use nonnormal incidence so that the shadow cone established by the surface atoms is not aligned with the atomic rows in the bulk. Here, normal incidence would yield a surface peak intensity equivalent to one monolayer. These two measurements, then, at normal and oblique incidence reveal the presence of relaxation. A surface adsorbate atom can shadow the atoms of the substrate if $R_{adsorbate} > \rho_{substrate}$. The atomic mass sensitivity of ion scattering permits discrimination between substrate and adsorbate. In Fig. 5.15d, the adsorbate is positioned exactly over the surface atoms and hence reduces the substrate surface peak.

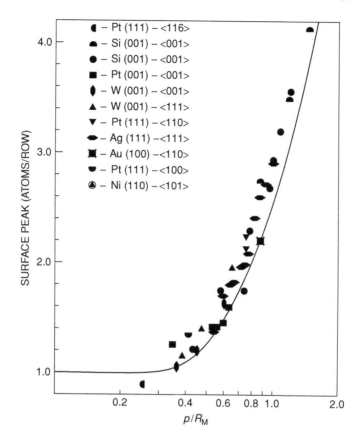

FIGURE 5.14. Comparison of the universal curve with experimental values for a number of different *bulklike* surface. The experimental values were determined from backscattering measurements. The notation Pt (111) ⟨116⟩ indicated a Pt crystal with a (111) surface plane; the backscattering measurement is in the ⟨116⟩ axial direction.

The comparison between a spectrum obtained with the incident beam aligned with the ⟨100⟩ axis of W (an *aligned* spectrum) and with the spectrum obtained with the beam oriented away from a crystallographic direction (a *random* spectrum) is shown in Fig. 5.16. The surface peak is clearly visible and corresponds to about two atoms per row ($\rho/R_M = 0.65$). Scattering from the bulk of the crystal in this aligned geometry is two orders of magnitude smaller than the scattering yield under random incidence due to the bulk channeling effect. It is this suppression of the scattering from the bulk in the aligned spectrum that permits measurement of the surface peak.

5.7 Substrate Shadowing: Epitaxial Au on Ag(111)

An important application of ion scattering is the study of the initial stages of epitaxy. The ability to monitor epitaxial growth from the very first monolayer is indicated by

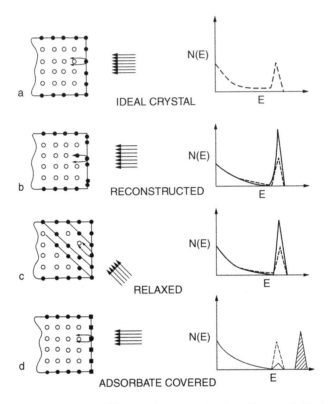

FIGURE 5.15. Representations of different surfaces on a simple cubic crystal. The backscattering spectra shown on the right-hand side represent the expected signal from the different structures. The dashed line represents the signal from the bulklike crystal.

Fig. 5.15d. If the deposited atoms are in perfect registry with the substrate, the shadow cones established by these adsorbed atoms will shield the substrate atoms from the incident beam.

The shielding concept is demonstrated in Fig. 5.17 for the deposition of monolayer coverage of Au on Ag(111). The upper portion of the figure shows a cross-sectional view of the (111) surface. The plane of the figure is the (011) plane, which contains the ⟨111⟩ normal direction and the off-normal ⟨011⟩ direction. In the ⟨111⟩ direction, the ion beam sees the first three monolayers of the uncovered Ag surface; whereas in the ⟨011⟩ direction, only the first monolayer is visible. Figure 5.17 shows the backscattering spectra for 1.0 MeV ^{4}He$^+$ incident along the ⟨011⟩ direction of a clean Ag surface and a surface covered with approximately one, three, and four monolayers of Au. For the covered surface, the decrease in the Ag surface peak is direct evidence that the Au overlayer is registered with respect to the Ag substrate; that is, the Au is epitaxial on Ag. The determination of registry, monolayer by monolayer, is more sensitive for the ⟨011⟩ direction in which only one monolayer of Au can cover all the available Ag sites.

The decrease of the Ag surface peak as a function of Au coverage is shown in Fig. 5.18 for low temperature (140 K) deposition and analysis in the ⟨011⟩ direction. The solid

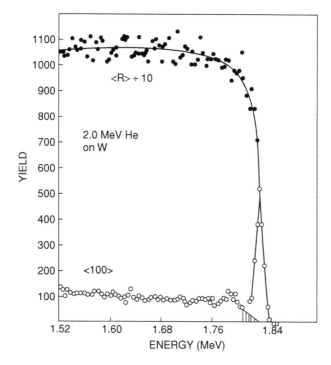

FIGURE 5.16. Backscattering spectra from 2.0 MeV He$^+$ incident on a clean W(001) surface for the beam aligned with the ⟨100⟩ axis (○) and for the beam aligned away from any major crystallographic direction (●). Note that the nonchanneling spectrum ⟨R⟩, termed *random incidence*, has been reduced by a factor of 10.

line is the result of computer simulations, which assume that Au atoms form uniform coverage monolayer by monolayer. The agreement between the data and calculated curve shows that the Au has good epitaxy and uniform coverage.

The registration of the first monolayer of Au with the Ag substrate can be monitored through the reduction of the Ag surface peak. The registration of subsequent Au layers with the initial Au layer can be monitored with the Au–Au shadowing effect. The ratio of the Au signal for ⟨011⟩ incidence to random incidence—χ_{min} (Au)—exhibits a break at one monolayer coverage and a pronounced decrease at higher coverage (Fig. 5.18). The dashed curve is a computer simulation. The agreement between the calculated curve and the data shows that the Au is indeed epitaxial. The decrease of the Ag surface peak with Au coverage measures the registry of the epitaxial layer to the substrate; the decrease of χ_{min} (Au) measures the quality of the epitaxial film.

5.8 Epitaxial Growth

The Au/Ag case is a good example of the growth of a high-quality epitaxial film. This combination of materials satisfies the most important criterion for epitaxy, namely, a

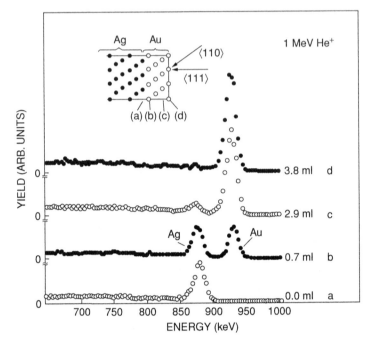

FIGURE 5.17. Backscattering spectra for 1.0 MeV He$^+$ ions incident along the $\langle 110 \rangle$ axis of a clean Ag(111) surface (a) and an Au-covered surface (b), (c), and (d). The Ag surface peak decreases because of the Au absorption.

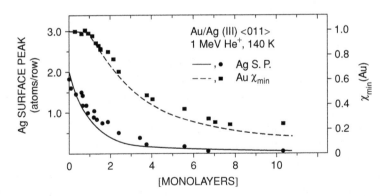

FIGURE 5.18. Intensity of the Ag surface peak as a function of Au coverage in terms of monolayers on a (111) surface for 1.0 MeV He$^+$ incident along the $\langle 110 \rangle$ direction of a Ag(111) surface at 140 K. Also shown (right-hand scale) is the ratio of the Au signal in the aligned and random direction as a function of Au coverage. The solid line and dashed line are calculated assuming pseudomorphic, monolayer-by-monolayer growth. Bulk vibrations are used, and correlations in thermal vibrations are not included. [From Culbertson et al., 1981 ©1981, by The American Physical Society]

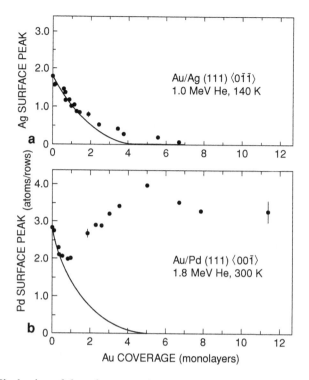

FIGURE 5.19. Shadowing of the substrate surface peak as a function of Au coverage: (a) For Ag(111) along the $\langle 110 \rangle$ direction, with deposition and analysis at 140 K using 1.0 MeV He$^+$; (b) for Pd(111) along the $\langle 100 \rangle$ direction, with deposition and analysis at 300 K using 1.8 MeV He$^+$. The full curve is a theoretical line assuming a pseudomorphic, monolayer-by-monolayer growth and includes correlations in thermal vibration.

good lattice constant match; in this case, the mismatch is $< 0.2\%$. Figure 5.19 compares the reduction of the substrate surface peaks for the case of Au epitaxy on Ag(111) and Pd(111). In this latter case, the mismatch is poor: $\sim 4.7\%$. Note that in both cases there is an immediate decrease in the surface peak corresponding to pseudomorphic growth. However, in Au/Pd this pseudomorphic growth is disrupted after only two layers. This outcome is consistent with epitaxy theory, which considers the strain in a mismatched, epitaxial film, and the eventual onset of misfit dislocations. The calculated thickness corresponding to this onset of dislocations for Au/Pd is ~ 2 monolayers. In a film with dislocations, the overlayer atoms are not registered and hence do not shadow the substrate.

5.9 Thin Film Analysis

An important use of channeling is the suppression of scattering from the single-crystal substrate. Scattering from amorphous overlayers such as oxides are not suppressed. The net result is an increased sensitivity to light impurities as well as structural information

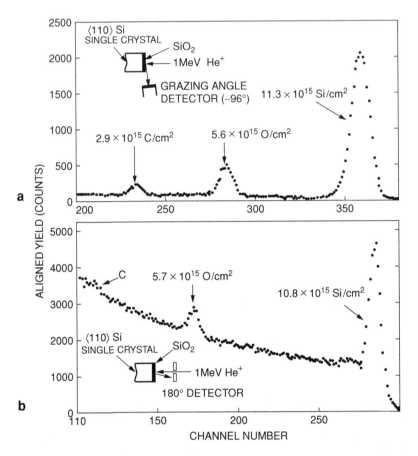

FIGURE 5.20. Energy spectra taken in channeling configuration from an Si(110) single crystal with $a = 0.15$ nm of SiO2. The top figure shows a detector placed at grazing exit angle geometry; the bottom figure shows a detector placed at ~180°.

for the near-surface region. Here we discuss an experiment that reveals the stoichiometry and subsurface strain in the SiO_2/Si system.

The sensitivity of these types of experiments has been improved by use of a grazing exit angle geometry, as shown by the spectra in Fig. 5.20 for a Si crystal with a small overlayer of SiO_2. The placement of the detector has no influence on the close-encounter interactions and channeling of the incident beam, but does influence the relation between detected energy widths and depth intervals. At grazing exit angles (Fig. 5.20a), the outgoing path length of the emergent particles can be five times that for the path length near 180° scattering (Fig. 5.20b). The stretching of the depth scale spreads the total number of detected scattering events in a given thickness over a greater energy interval and hence decreases the number of counts per energy channel in the bulk crystal. In the sample, the surface layer (~1.3 nm of oxide) is thin compared with the depth resolution, so the energy width of the signal is determined by the energy resolution of the detector system.

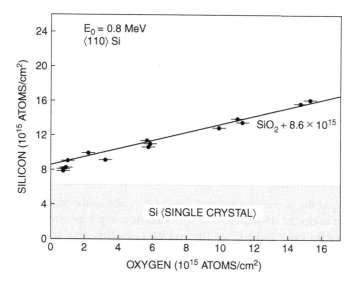

FIGURE 5.21. Si versus oxygen areal densities for a range of oxides up to ∼0.4 nm. The data points are extracted from spectra of the type shown in Fig. 5.30. The shaded portion represents the silicon surface peak intensity expected for clean Si(110) with bulklike surface structure.

The scattering spectrum from the aligned Si crystal with a thin oxide layer consists of (1) a Si peak with scattering contributions from Si in the oxide, nonregistered Si in the surface region, and the intrinsic surface peak from the underlying single crystal and (2) at lower energy, an oxygen peak corresponding to the oxygen coverage. These peak areas can be converted to atoms/cm^2 with approximately 5% accuracy.

The Si versus oxygen intensities are shown in Fig. 5.21 for a range of oxides up to ∼4.0 nm. Over most of the range, the data are well fitted by a line expected for stoichiometric SiO$_2$ plus an additional offset of 8.6×10^{15} Si/cm^2. Most of this offset is the expected contribution from the intrinsic Si surface peak. The results from Fig. 5.21 show that the oxide is primarily stoichiometric SiO$_2$ and that the interface is sharp. The data suggest that the interface consists of either two monolayers of Si not registered to the bulk or a thin region (< 0.5 nm) of nonstoichiometric oxide.

Problems

5.1. Calculate and compare values of the *standard* potential, the Thomas–Fermi potential, and the $1/r^2$ potential at $r/a = 0.01, 0.1$, and 1.0, for He$^+$ ions incident on Si.

5.2. Cu is an FCC metal with a lattice constant of 0.3615 nm. Calculate the axial and planar critical angles and minimum yield for 2 MeV ^{4}He$^+$ ions incident along 100 axial and planar directions of Cu with a thermal vibration amplitude $\rho = 0.012$ nm.

5.3. Calculate the shadow cone radius R, and the surface peak in the two-atom and *universal* model (assume $R_C = R_M$) for 1.0 MeV ^{4}He$^+$ ions incident along the

100 axis of Al (lattice constant $= 0.405$ nm and $\rho = 0.014$ nm). What energy is required for a unity surface peak intensity ($\rho/R_M = 0.4$)?

5.4. Use the r^2/r_0^2 approximation to the flux distribution (Eq. 5.20), and calculate the minimum yield of an impurity displaced $0.01, 0.03$, and 0.05 nm from the atomic row. Take the vibration amplitude as 0.01 nm and use an r_0 value appropriate to the Si $\langle 010 \rangle$ channeling direction.

5.5. Channeling occurs not only along the strings of atoms but also between the sheets of atoms that make up atomic planes. In this case, the potential governing the transverse notion can be described by a parabolic potential of the form

$$V(y) = \frac{1}{2}ky^2, \quad 0 \le |y| \le d_p/2,$$

where d_p is the planar spacing and y is the distance to the planar wall measured from the midpoint. Using the concept of motion in a harmonic potential, derive a formula for the wavelength of the oscillatory motion. Evaluate this wavelength for 1.0 MeV ions channeling along the $W(100)$ planes. The *spring constant k* can be estimated by noting that $V(d_p/2) = U_p(0)$ from Eq. 5.13.

References

1. J. U. Anderson, O. Andreason, J. A. Davies, and E. Uggerhøj, *Rad. Eff.* 7, 25 (1971).
2. B. R. Appelton and G. Foti, "Channeling," in *Ion Beam Handbook for Material Analysis*, J. W. Mayer and E. Rimini, Eds. (Academic Press, New York, 1977).
3. W. K. Chu, J. W. Mayer, and M.-A. Nicolet, *Backscattering Spectrometry* (Academic Press, New York, 1978).
4. G. Dearnaley, J. H. Freeman, R. S. Nelson, and J. Stephen, *Ion Implantation* (North-Holland, Amsterdam, 1973).
5. L. C. Feldman, J. W. Mayer, and S. T. Picraux, *Materials Analysis by Ion Channeling* (Academic Press, New York, 1982).
6. D. S. Gemmell, "Channeling and Related Effects in the Motion of Charged Particles Through Crystals," *Rev. Mod. Phys.* 46(1), 129–227 (1974).
7. J. Lindhard, *Mat. Fys. Medd. Dan. Vid. Selsk.* 34(14), 1 (1965).
8. D. V. Morgan, Ed., *Channeling* (John Wiley and Sons, New York, 1973).
9. R. S. Nelson, *The Observation of Atomic Collisions in Crystalline Solids* (North-Holland, Amsterdam, 1968).
10. P. D. Townsend, J. C. Kelly, and N. E. W. Hartley, *Ion Implantation, Sputtering and Their Applications* (Academic Press, New York, 1976).
11. R. J. Culbertson, L. C. Feldman, and P. J. Silverman, *Phys. Rev. Lett.* 47, 657 (1981).

6
Electron–Electron Interactions and the Depth Sensitivity of Electron Spectroscopies

6.1 Introduction

Detection of an element from the near-surface region of a solid often involves the measurement of an electron energy characteristic of a particular atom. The depth resolution of these techniques is then determined by the thickness of the material that an emitted electron can traverse without undergoing an inelastic event and thus altering the electron energy. We consider these inelastic electron processes in order to obtain a quantitative understanding of the thickness of the analyzed layer. An understanding of these phenomena is particularly useful in the design of surface studies, as an experimenter can often choose the electron energy and thus determine the depth probed. In this chapter, we use the particle-scattering concepts developed in Chapters 2 and 3 to derive classical relations for electron–electron collisions. These provide a guide for useful approximations for electron escape depths and impact ionization cross sections.

6.2 Electron Spectroscopies: Energy Analysis

The surface analytical tools, photoelectron and Auger electron spectroscopy, discussed in the following chapters use photons or electrons to excite electrons that escape from the solid with sharply defined energies. Emitted electrons in the 100 eV range have escape depths on the order of 1.0 nm. By choosing the appropriate incident beam parameters and detection systems, these electron spectroscopies become extremely surface sensitive. Electron spectroscopies have found their widest application in surface analysis for a number of reasons. Electrons are easily focused into beams, are efficiently detected and counted, and may be analyzed with respect to angular and energy distribution using electrostatic lenses and deflection systems.

The electron spectroscopies are based on an analysis of the energy distribution of electrons emitted from the surface. The different features of the emission spectra and the requirements imposed by the analytical technique have led to the development of a variety of analyzers for measuring the electron energy distributions (see also Chapter 10). In the cylindrical mirror analyzer (CMA) (Fig. 6.1), the emitted electrons are focused electrostatically in such a way that only those electrons with energies within

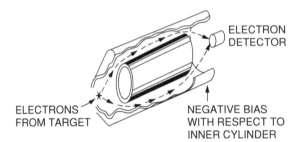

FIGURE 6.1. Schematic diagram of a cylindrical mirror analyzer used for electron energy detection in various electron spectroscopies.

ELECTRON
DETECTOR

ELECTRONS
FROM TARGET

NEGATIVE BIAS
WITH RESPECT TO
INNER CYLINDER

a certain small range pass through the analyzer and arrive at the collector. Focusing is achieved by applying a potential V_a to create a cylindrical electric field between the two coaxial electrodes. The outer cylinder is held at a negative potential with respect to the inner cylinder. Electrons entering the analyzer through the annular entrance are deflected toward the inner cylinder. The analyzer design allows electrons with energy $E = eV_a$ and energy spread ΔE to pass through the exit slit to the collector; the energy resolution $\Delta E/E$ is usually between 0.1% and 1%. The transmission of the analyzer is high, since it accepts electrons over a large solid angle. A modulating AC voltage is superimposed upon the cylinder potential to allow single and double differentiation so that the sharp characteristic energy peaks can be removed from the background.

6.3   Escape Depth and Detected Volume

For quantitative analysis, it is important to determine the escape depth: the distance that electrons of a well-defined energy, E_C, can travel without losing energy (Fig. 6.2). The incident radiation, whether photons or electrons, is sufficiently energetic so that it penetrates deeply into the solid, well beyond the escape region for characteristic-energy electrons. The electrons that undergo inelastic collisions and lose energy δE in transport from the point of excitation to the surface leave the solid with a lower energy

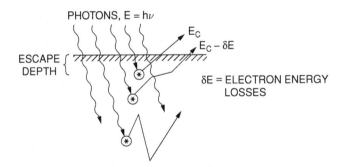

PHOTONS, E = hν

E_C
$E_C - \delta E$

ESCAPE
DEPTH

δE = ELECTRON ENERGY
LOSSES

FIGURE 6.2. Schematic of energetic photons incident on a surface creating characteristic electrons relatively deeply in the solid. Only those electrons created near the surface escape with no loss of energy.

and contribute to a background signal or tail that can extend several hundred eV below the main signal peak. In analogy with the experimental methods used to determine the escape depth, we consider the substrate as a source of a flux I_0 of electrons of well-defined energy E_C and deposit a thin film on the substrate. Any inelastic collisions within the thin film will remove electrons from the group of electrons of energy E_C. Consider that the cross section for the inelastic collision is σ and that there are N' scattering centers/cm^3 in the deposited film. The number dI of electrons removed from the initial group is σI per scattering center, and the number of electrons removed per thickness increment dx is

$$-dI = \sigma I \, N' \, dx,$$

which gives

$$I = I_0 e^{-\sigma N' x}. \tag{6.1}$$

The mean free path is related to the cross section by definition as

$$I/\lambda = N'\sigma, \tag{6.2}$$

so Eq. 6.1 can be written as

$$I = I_0 e^{-x/\lambda}. \tag{6.3}$$

The number of electrons that can escape from the surface of the deposited absorber film then decreases exponentially with film thickness. In this discussion, we treat the mean free path to be synonymous with the escape depth and use the same symbol λ. The yield of electrons from a solid excited uniformly in depth is given by $\int I(x)\,dx = I_0\lambda$, such that a thick substrate appears as a target of thickness λ.

A method used to characterize the attenuation of electrons is to monitor the signal from electrons generated in a substrate as a function of the thickness of a deposited overlayer of different metal. Figure 6.3 shows the relative intensity of Auger electrons (92 eV) from Si as a function of the thickness of an overlayer of Ge. The Ge thickness was determined from Rutherford backscattering analysis (Chapter 3). The data of Fig. 6.3 show that the intensity decreases exponentially with Ge film thickness with a decay length λ equivalent to 2.5×10^{15} Ge atoms/cm^2 (~ 0.5 nm). The attenuation of electrons depends upon the characteristic energy of the outgoing electrons. Figure 6.3b illustrates the energy dependence of the attenuation for two different characteristic energies of electrons emitted from Ge (LMM, 1147 eV; MVV, 52 eV) as a function of Si overlayer thickness. The mean free paths correspond to 9.81×10^{15} atoms/cm^2 and 9.81×10^{15} atoms/cm^2, respectively. The terminology for Auger transitions is discussed in Chapter 12.

These values of mean free paths are in agreement with other measurements of electron mean free paths shown in Fig. 6.4. The data show that the mean free path is energy dependent with a broad minimum centered around 100 eV. The mean free path is relatively insensitive to the material traversed by the electrons. Such curves of mean free path versus energy have come to be called *universal curves*.

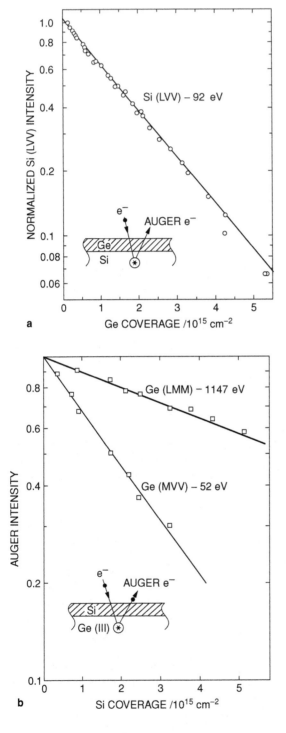

FIGURE 6.3. (a) Attenuation of the Si LVV (92 eV) as a function of Ge coverage. The mean free path corresponds to ∼ 0.5 nm of Ge. In this example, the Ge deposition is at 300 K. (b) Attenuation of Ge LMM (1147 eV) and Ge MVV (52 eV) Auger lines as a function of Si coverage. [Adapted from Feldman and Mayer, 1986]

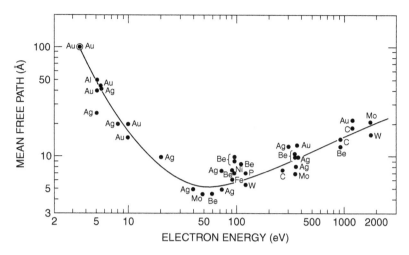

FIGURE 6.4. Universal curve for electron mean free path. [From G. Somerjai, *Chemistry in Two Dimensions: Surfaces* (Cornell University Press, Ithaca, NY, 1981).]

6.4   Inelastic Electron–Electron Collisions

The cross section for an inelastic collision can be derived using the impulse approximation (Chapter 3) for scattering in a central force field. For an electron of velocity v, the amount of momentum transferred to a target electron is

$$\Delta p = \frac{2e^2}{bv},$$ (6.4)

where b is the impact parameter. Here we have taken the small-angle scattering results from Chapter 3 with $Z_1 = Z_2 = 1$ and $M_1 = M_2 = m$ (Fig. 6.5). Let T denote the energy transferred by an electron of energy $E = \frac{1}{2}mv^2$; then

$$T = \frac{(\Delta p)^2}{2m} = \frac{e^4}{E\,b^2}.$$ (6.5)

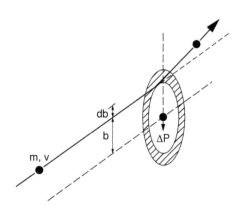

FIGURE 6.5. Schematic of an electron of momentum mv interacting with a free electron at an impact parameter b.

The differential cross section $d\sigma(T)$ for an energy transfer between T and $T + dT$ is

$$d\sigma(T) = -2\pi b \, db. \tag{6.6}$$

From Eq. 6.5, $2b \, db = -(e^4/ET^2) \, dT$, so

$$d\sigma(T) = \frac{\pi e^4}{E} \frac{dT}{T^2}. \tag{6.7}$$

The cross section for an electron to transfer an energy between T_{min} and T_{max} is

$$\sigma_e = \int_{T_{min}}^{T_{max}} d\sigma(T), \tag{6.8}$$

$$\sigma_e \cong \pi \frac{e^4}{E} \left(\frac{1}{T_{min}} - \frac{1}{T_{max}} \right). \tag{6.9}$$

For energetic electrons with energy E of several hundred eV or greater, the maximum energy transfer ($T_{max} = E$ for $M_1 = M_2$) is very much greater than T_{min}; therefore,

$$\sigma_e \cong \frac{\pi e^4}{E} \frac{1}{T_{min}} = \frac{6.5 \times 10^{-14}}{E \, T_{min}} \, \text{cm}^2, \tag{6.10}$$

with E and T_{min} in eV and using the values $e^2 = 1.44$ eV-nm.

6.5   Electron Impact Ionization Cross Section

The value of the cross section can be estimated from Eq. 6.10 with $T_{min} = E_B$,

$$\sigma_e = \frac{\pi e^4}{E E_B} = \frac{\pi e^4}{U E_B^2}, \tag{6.11}$$

where $U = E/E_B$ and E_B is the binding energy of an orbital electron. For incident energies less than E_B, $U < 1$, the cross section must be equal to zero. The actual

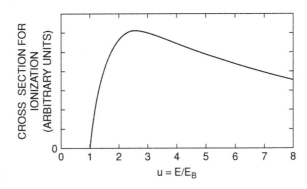

FIGURE 6.6. Ionization cross section versus reduced energy U for interactions within a solid. [From C. C. Chang, in P. F. Kane and G. B. Larrabee, Eds., *Characterization of Solid Surfaces* (Plenum Press, NY, 1974), with permission from Springer Science+Business Media.]

FIGURE 6.7. Maximum electron impact ionization cross section per electron (Near $U = 4$) versus binding energy for various electron shells.
[From J. Kirschner in H. Ibach, 1977, with
permission from Springer Science+Business
Media.]

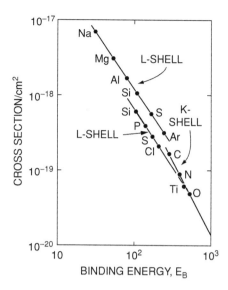

shape of the cross section as a function of U is shown in Fig. 6.6. The value of the cross
section has a maximum near reduced energy value of $U \cong 3$ to 4. For $E_B = 100$ eV and
$U = 4$, the value of the cross section is 1.6×10^{-18} cm^2. The value is in reasonable
agreement with measured values of the maximum electron impact ionization cross
section (measured near $U = 4$) that are shown in Fig. 6.7.

6.6   Plasmons

In solids, the collective excitation of the conduction electron gas leads to discrete peaks
in the energy loss of electrons. The plasmon is a quantum of a plasma oscillation and
has an energy $\hbar\omega_p$ of about 15 eV. From a classical viewpoint, the plasma frequency
is determined by oscillations of the valence electrons in a metal with respect to the
positively charged cores (Fig. 6.8). Consider the fluctuation δr in radial distance r
from a positive core of a free electron gas containing a concentration n of electrons.
If the gas expands from its equilibrium radius δr, the number of electrons in the shell,

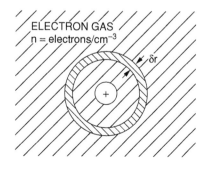

FIGURE 6.8. Electron gas with $4/3\,\pi r^3 n$ electrons
around a positive core and undergoing a radical con-
traction δr.

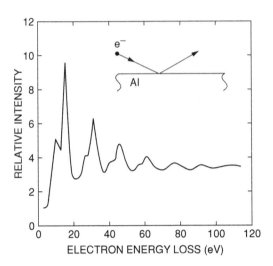

FIGURE 6.9. Energy loss spectrum for electrons reflected from Al for incident primary electron energy of 2 keV. The loss peaks are made up of a combination of surface and bulk plasmon losses.

$\delta n = 4\pi\, nr2\, \delta r$, establishes an electric field

$$\mathscr{E} = \frac{e}{r^2}\delta n = 4\pi e\, \delta r. \tag{6.12}$$

The retarding force F created by the expansion is

$$F = -e\,\mathscr{E} = -4\pi e^2 n\, \delta r. \tag{6.13}$$

The solution for the frequency of a harmonic oscillator with a force given in Eq. 6.13 is

$$\omega_p = \left(\frac{4\pi e^2 n}{m_e}\right)^{1/2}, \tag{6.14}$$

where m_e is the mass of the electron. For metals, the value of $n \cong 10^{23}/cm^2$ gives an oscillator frequency $\omega_p = 1.8 \times 10^{18}$ rad/s and an energy $\hbar\omega_p = 12$ eV. The plasma frequency may be thought of as a *natural* frequency of the electron–ion system that is excited by an incoming charged particle.

The measured value of the plasmon energy for Mg is 10.6 eV and for Al is 15.3 eV. Figure 6.9 shows energy loss spectra for electrons reflected from a film of Al. The loss peaks are made up of combinations of the bulk plasmon $\hbar\omega_p = 15.3$ eV and the surface plasmon at an energy of 10.3 eV. The surface plasmon frequency $\omega_p(s)$ has the following relationship to the bulk plasmon:

$$\omega_p(s) = \frac{1}{\sqrt{2}}\omega_p. \tag{6.15}$$

This equation is found to hold for many metals and semiconductors. The calculated plasmon energies of Si and Ge are 16.0 eV based on four valence electrons per atom with the entire valence electron sea oscillating with respect to the ion cores. The measured values for Si are 16.4–16.9 eV and for Ge are 16.0–16.4 eV.

6.7   The Electron Mean Free Path

The mean free path can be estimated using the general formulation for electron energy loss in a solid of n electrons/unit volume (see Eq. 3.10),

$$-\frac{dE}{dx} = \frac{4\pi e^4 n}{mv^2}\ln B, \qquad (6.16)$$

where B represents a ratio of particle energy to excitation energy. The excitation of plasma oscillations, plasmons, in distant collisions is the dominant mode of energy loss of electrons in solid. The electron energy loss for 80 keV electrons traversing a thin NiAl foil is shown in Fig. 6.10. The dominant feature in the figure is the energy distribution of electrons that have lost 17.8 eV, the bulk plasmon energy, in transmission through the film. This finding suggests that the energy losses occur in quantum jumps of $\hbar\omega_p$, and we set

$$B = \frac{2mv^2}{\hbar\omega_p}, \qquad (6.17)$$

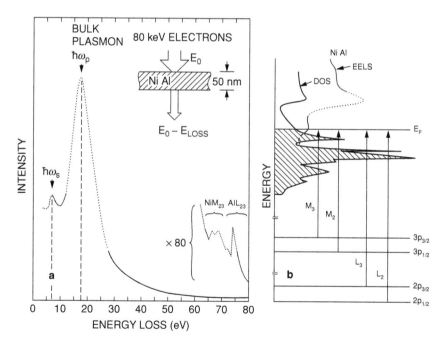

FIGURE 6.10. (a) Electron energy loss spectrum for 80 keV electrons transmitted through a 50 nm NiAl film. The dominant peak in the loss spectrum is the single plasmon losses at $\hbar\omega_p = 17.8$ eV. The surface plasmon ($\hbar\omega_s = 7.1$ eV) and core excitation (Ni M_{23} and Al L_{23}) are indicated. (b) The calculated density of states (DOS) for NiAl (solid line) and Al L_{23} core excitation (data points) from EELS spectra similar to that shown in (a). Core level threshold transitions are shown by arrows. [Adapted from Fledman and Mayer, 1986]

and from $\omega_p = (4\pi e^2 n/m)^{1/2}$ (Eq. 6.14), write the energy loss in terms of ω_p:

$$-\frac{dE}{dx} = \frac{\omega_p^2 e^2}{v^2} \ln \frac{2mv^2}{\hbar\omega_p}, \qquad (6.18)$$

A similar formulation is presented by Ibach (1977). If we treat the plasmon as the major source of energy loss in determining the mean free path λ for electrons, we can write λ as

$$\frac{1}{\lambda} = \left(-\frac{dE}{dx}\right)\frac{1}{\hbar\omega_p}. \qquad (6.19)$$

This gives

$$\frac{1}{\lambda} = -\frac{\omega_p e^2}{\hbar v^2} \ln \frac{2mv^2}{\hbar\omega_p}. \qquad (6.20)$$

For example, we calculate the value of λ to be 0.92 nm for 350 eV electrons in Al ($\hbar\omega_p = 15$ eV, $v^2 = 2E/m = 1.23 \times 10^{18}$ cm^2/s^2, and $\hbar = 6.6 \times 10^{-16}$ eV-s). This value is in reasonable agreement with the data presented in Fig. 6.4.

6.8   Influence of Thin Film Morphology on Electron Attenuation

One of the many prominent uses of electron spectroscopies is in the characterization of different modes of film growth. Questions about the growth mode are particularly important in the creation of uniform layered films in which one desires extreme uniformity in composition for films thinner than 10 nm. Common types of growth include the following:

1. Layer-by-layer, in which the deposited film completes one monolayer of coverage, then the second, etc. This mode is commonly referred to as *Frank–van der Merwe growth*.
2. Layer plus islanding, in which the first layer completely covers the surface of the substrate and subsequent layers form islands of deposited material. This mode is referred to as *Stranski–Krastanov growth*.
3. Complete islanding, in which the material immediately forms islands on the surface. This mode is commonly referred to as *Volmer–Weber growth*.
4. Statistical deposition, in which the growth corresponds to the random occupancy of surface sites according to Poisson statistics.

6.8.1   Layer-by-Layer Growth

In the following discussions, we continually make use of the exponential probability of inelastic scattering for a characteristic electron emerging from a solid. As a first example, consider the attenuation of electrons from a substrate as a result of an overlayer that deposits in a layer-by-layer mode. Although the derivation here is for attenuation of substrate electrons, very similar considerations can be used to derive formulae for the increase of the electron yield from the overlayer.

A typical experiment involves a measurement of the intensity I_s of a characteristic Auger electron from the substrate as a function of overlayer thickness t. (Naturally, the overlayer is a different material.) The mean free path of the substrate Auger electron in the material of the overlayer is λ. Then, for coverages up to one monolayer, the Auger intensity can be written as

$$I_S/I_{S_0} = (1 - x) + xe^{-l/\lambda}, \quad 0 \le x \le 1, \tag{6.21}$$

where x is the fraction of the surface covered by the overlayer, I_{S_0} is the intensity from the clean surface, and I is the thickness of a monolayer. In this regime of up to one monolayer, the dependence of the intensity on coverage is linear in x. Hence, a similar formula can be written for the region from 1 to 2 monolayers as

$$I_S/I_{S_0} = (1 - x)e^{-l/\lambda} + xe^{-2l/\lambda}, \quad 0 \le x \le 1, \tag{6.22}$$

where x is now the fraction of the surface covered with two layers and $(1 - x)$ is the fraction covered with one layer. Again, in the region of coverage of one to two layers, the attenuation is linear in x.

In general, a similar formula can be written for the transition from the n to the $(n + 1)$ layer as

$$I_S/I_{S_0} = (1 - x)e^{-nl/\lambda} + xe^{-(n+1)l/\lambda}, \quad 0 \le x \le 1, n = 0, 1 \ldots, \tag{6.23}$$

where $(1 - x)$ is the fraction of the surface covered with n layers and x is the fraction covered with $(n + 1)$ layers. The characteristic shape of this curve is a series of straight lines (on a linear plot) with breaks at coverages corresponding to an integral number of monolayers. The envelope of points corresponding to integral coverages describes an exponential decay of the form $e^{-nt/\lambda}$ (Fig. 6.11).

6.8.2   Single Layer plus Islanding

A second type of growth mode corresponds to the deposition of a single uniform layer that is then followed by islanding. From the discussion in the previous section, we can write the attenuation of up to one monolayer of coverage as

$$I_S/I_{S_0} = (1 - x) + xe^{-l/\lambda}, \quad 0 \le x \le 1, \tag{6.24}$$

where x represents the fraction of the covered surface. In the second stage of growth, the deposited material forms islands of unspecified dimensions, and we cannot write a simple analytical formula, as the actual attenuation will depend on the fraction of covered surface. As an example, however, we take as a very simple case that the islands cover 50% of the surface. Then, for the second monolayer's worth of material, the coverage formula can be written as

$$I_S/I_{S_0} = (1 - x)e^{-l/\lambda} + xe^{-3l/\lambda}, \quad 0 \le x \le 0.5, \tag{6.25}$$

where x is the fraction of the surface covered with two-monolayer-high islands. A similar formula can be written for any coverage regime, but there will always be a term

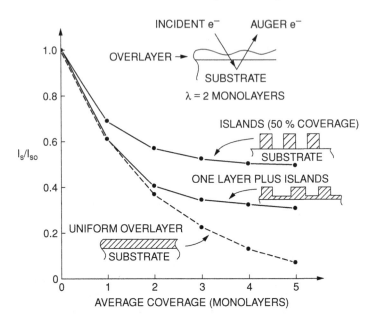

FIGURE 6.11. Extinction curves for characteristic substrate electrons as a function of average coverage of an overlayer. In these examples, the mean free path, λ is taken as two monolayers, close to the minimum mean free path achievable. In the case of the uniform overlayer, the growth is assumed to be layer by layer and the extinction curve is a series of straight lines with the envelope of points at integral coverage corresponding to an exponential decay. The other curves correspond to a single-layer-plus-islanding growth mode and a case of pure islanding. For the cases involving islanding, it is assumed that 50% of the surface is covered.

of magnitude $0.5\,e^{-1/\lambda}$ in the formula, plus additional positive contributions. The net result is that the Auger yield from the substrate is always finite and does not approach the zero yield associated with layer-by-layer growth. The decay curve for this type of growth is shown schematically in Fig. 6.11. An actual case is shown in Fig. 6.12 for the growth of Pb on Cu(100). Note the almost constant level of the Cu Auger yield beyond one monolayer of coverage. Such curves can be fitted with simple models of islands as described above. The Auger signal from the Pb overlayer increases with coverage but saturates again consistent with islanding.

6.8.3   Islanding

As described in the preceding section, the substrate yield curve as a function of coverage does not show a simple exponential decay in cases of islanding. In this type of growth, the substrate yield remains high, since some fraction of the substrate is not covered by the overlayer. Figure 6.11 shows the expected curve for the simple case of islands covering 50% of the surface. Obviously, the yield can never have values below 0.5 in this scheme. The differences in the extinction curves (Fig. 6.11) reveal the different modes of growth. The interpretation of these curves requires an accurate measurement of the absolute coverage and knowledge of the mean free path.

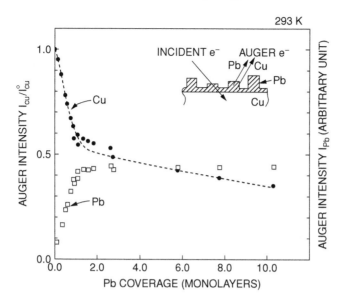

FIGURE 6.12. Auger yield curve for the deposition of Pb on Cu(100) at room temperature. The solid circles represent the attenuation of Cu substrate signal and the open squares the growth of the Pb signal. This type of the growth corresponds to the "one layer plus islanding" mode. In this example, the (average) Pb coverage is measured by RBS. [From R.J. Culbertson, unpublished data.]

6.8.4 Distribution of Deposited Atoms

By a statistical distribution we mean that deposited atoms simply reside on the surface in a Poisson distribution—as if they impinged randomly and simply stuck at the landing site. For an average coverage θ, the probability of finding a structure that is k atom layers high is

$$P_k = \frac{\theta^k e^{-\theta}}{k!}, \tag{6.26}$$

so

$$P_0 = e^{-\theta},$$

the usual coverage formula for the fraction of the surface not covered by the adsorbate. Then the Auger intensity factor is given for layers of thickness I by

$$I_S/I_{S_0} = \sum_k P_k e^{-kl/\lambda}, \tag{6.27}$$

or

$$I_S/I_{S_0} = e^{-\theta} \sum_k \frac{\theta^k}{k!} e^{-kl/\lambda}. \tag{6.28}$$

This expression can be rewritten as

$$I_S/I_{S_0} = e^{-\theta} \sum_{k=0} \frac{(\theta e^{-l/\lambda})^k}{k!}, \tag{6.29}$$

or

$$I_S/I_{S_0} = e^{-\theta}(1 - e^{-l/\lambda}). \tag{6.30}$$

Note that this attenuation curve is a pure exponential for all coverages. The effective decay length is now $(1 - e^{-l}/\lambda)^{-1}$ rather than λ.

6.9   Range of Electrons in Solids

In materials analysis, energetic electrons are used to generate inner-shell vacancies that decay by Auger emission or X-ray emission. In measurements of the emitted characteristic X-rays in the electron microprobe, the concern is with the depth over which the X-rays are generated. The situation is more complex for incident electrons with energies between 1 and 50 keV than for heavy ions whose path is relatively straight over most of the range. For electrons, significant deviations from the incident direction occur due to elastic scattering. Monte Carlo calculations of the paths of electrons in Fe ($E_0 = 20$ keV) are shown schematically in Fig. 6.13. The elastic scattering is composed of both large-scale scattering events and multiple small-angle scattering events that can

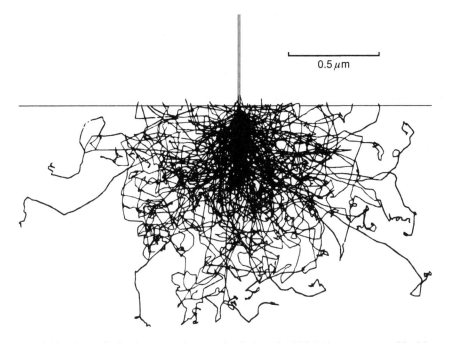

FIGURE 6.13. Monte Carlo electron trajectory simulation of a 20 keV beam at normal incidence on Fe. The density of trajectory gives a visual impression of the interaction volume. [From Goldstein et al., 1981, with permission from Springer Science+Business Media.]

also lead to a large change in the direction of the electrons. The electron range R is defined as the total distance that an electron travels in the sample along a trajectory and can be written as

$$R = \int_{E_0}^{0} \frac{dE}{dE/dx}, \tag{6.31}$$

where the energy loss expression has been discussed previously. The energy loss formula, dE/dx, is of the form

$$\frac{dE}{dx} \propto \frac{NZ_2}{E} \ln \frac{E}{I}, \tag{6.32}$$

or

$$\frac{dE}{dx} \propto \frac{\rho}{E} \ln \frac{E}{I}, \tag{6.33}$$

where N is the atomic density so that NZ_2 is the proportional to ρ and I is the average ionization energy, $I - 10Z_2$ (eV). From experimental results, the dependence of the range on incident energy has the form

$$R = \frac{K}{\rho} E_0^{\gamma}, \tag{6.34}$$

where ρ is the density (g/cm^3), K is a material-independent constant, and γ varies from 1.2 to 1.7. It is convenient to use the mass range ρR, since, to a first approximation, ρR is independent of target material. The electron range R as a function of energy is given in Fig. 6.14 for $K = 0.064$ and $\gamma = 1.68$.

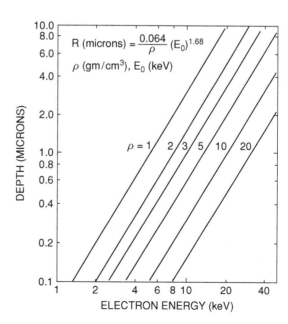

FIGURE 6.14. The electron range R (μm) versus incident electron energy for different density ρ materials. The lines are calculated from Eq. 6.24, with $K = 0.064$ and $\gamma = 1.68$.

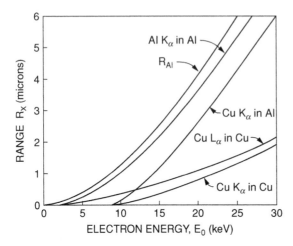

FIGURE 6.15. The effective electron ranges in Cu and Al for X-ray production. The ranges denote the end point of penetration of electrons for generation of characteristics K_α and L_α X-rays. [From Goldstein et al., 1981, with permission from Springer Science+Business Media.]

The mass range ρR_X for characteristic X-ray production is smaller than the range of electrons, since characteristic X-rays can only be produced at energies above the critical excitation energy, or binding energy E_B, for a given element. The mass range for characteristic X-ray production is given by

$$\rho R_X = K(E_0^\gamma - E_B^\gamma). \tag{6.35}$$

As discussed by Goldstein et al. (1981), the mass range equation for X-ray production can be expressed as

$$\rho R_X = 0.064 \left(E_0^{1.68} - E_B^{1.68} \right), \tag{6.36}$$

where E_0 and E_B are in keV, ρ is in g/cm^3, and R_X is in microns (μm). Figure 6.15 shows the electron range R in Al and R_X for the Al K_α and Cu K_α lines generated in Cu-doped Al and the Cu K_α and Cu L_α lines in pure Cu. The ranges for X-ray production depend, of course, on the density of the matrix (Al = 2.7 g/cm^3, Cu = 8.9 g/cm^3) and on the value of E_B (Cu K_α, $E_B = 8.98$ keV; Cu L_α, $E_B = 0.93$ keV).

6.10   Electron Energy Loss Spectroscopy (EELS)

The characteristic energy losses of electron beams penetrating through a film or reflected from a surface can give important information about the nature of the solid and the relevant binding energies. Electron energy loss spectroscopy (EELS) is carried out from ≤ 1 eV to ~ 100 keV. The choice is based on a variety of experimental considerations and the energy range of interest. The low-energy regime is used primarily in surface studies, where the investigation centers on the energies of vibration in states associated with absorbed molecules. The energy loss spectrum contains discrete peaks corresponding to the vibrational states of absorbed molecules.

At higher energies, as shown in Section 6.7, the dominant peak corresponds to a plasmon loss or losses. A detailed examination of the energy loss spectrum would also show discrete edges in the spectra corresponding to excitation and ionization of atomic core levels. These features represent a means of element identification, particularly useful in cases where the spatial resolution of the electron microscope is required. The features tend to be broad, since the incident electron can transfer a continuum of energies to the bound electron. For example, a core electron may be excited to the unoccupied states within the solid (excitation) or actually ejected from the solid (ionization). The cross section tends to strongly favor small energy transfers, thus making excitation dominant. An examination of the loss features under high resolution can then yield information on the unoccupied density of states. In the following, we show EELS spectra of Ni_xSi_y films using the inelastic spectrum from an $\sim$100 keV electron beam. High-energy electrons, approximately 100 keV, are used because the long distance between collisions, about 50.0–100.0 nm, permits examination of self-supporting films that can be mounted on conventional electron microscopy sample grids. Electrostatic analyzers of 0.1–0.5 eV energy resolution are generally used so that changes in the density of states can be monitored.

The energy loss of 80 keV electrons transmitted through 50.0 nm films of crystalline NiAl was given earlier in Fig. 6.10. The dominant feature in the spectrum is the large bulk plasmon peak, labeled $\hbar\omega_p$, that is centered at 17.8 eV. This resonance involves all the valence electrons and is displaced to higher energy loss in NiAl than the bulk plasmon found at 15.0 eV in Al. The sensitivity of the bulk plasmon to the composition of the sample is shown in Fig. 6.16, which gives spectra for samples of Si, $NiSi_2$ (the

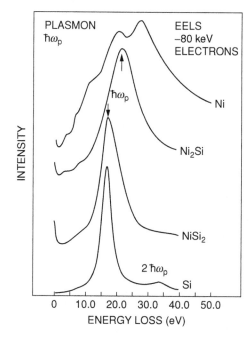

FIGURE 6.16. Electron energy loss spectra for 80 keV electrons incident on thin ($\sim$40 nm) self-supporting films of Si, $NiSi_2$, Ni_2Si, and Ni. The positions of the bulk plasmon peaks $\hbar\omega_p$ are shown for two nickel silicides. [From J. C. Barbour et al., in *Thin Films and Interfaces II*, J. Baglin, J. Campbell, and W.-K. Chu, Eds. (North-Holland, Amsterdam, 1984), with permission from Elsevier]

most Si-rich nickel silicide), Ni_2Si, and Ni. The bulk plasmon peaks become broader and are centered at increasingly higher energies with increasing Ni concentration. Silicon, $NiSi_2$, and Ni_2Si have bulk plasmon energies, $\hbar\omega_p$, of 16.7, 17.2, and 21.8 eV, respectively. The spectra in Fig. 6.16 were scaled such that the plasmon peak heights are equivalent, even though the absolute intensity of the Si plasmon peak is much greater than the corresponding peaks in the Ni spectrum.

The bulk plasmon may also be used to estimate the relative amount of plural scattering. Scattering of the incident electron beam by two sequential bulk plasmon events produces a peak in the loss function at twice the bulk plasmon energy, as can be seen in the peak in the Si spectrum at 33.4 eV in Fig. 6.16. From the figure we can see that the ratio of the double to single plasmon intensity is quite small, which indicates that the sample thickness is less than the mean free path for excitation of bulk plasmons.

In the EELS spectrum of NiAl (Fig. 6.10), weak but sharp peaks appear at high energy loss values of $\sim$70 eV. These correspond to excitation of individual deeply bound core electrons to the unfilled conduction band states. The Al L_{23} transition of about 75 eV in the spectrum of Figure 6.10 corresponds to excitations of the Al 2p core electrons to unfilled states above the Fermi level. In this figure, the measured Al L_{23} data points are aligned with calculated values (solid line) of the density of states (DOS). The experimental points mirror the DOS shape, indicating that EELS measurements can be used to determine the density of states above the conduction band.

The energy loss spectra for nickel silicides also display features due to core excitations. Figure 6.17a is an electron energy loss spectrum of a 40 nm thick $NiSi_2$ self-supporting film taken over the energy region from 0 to 138 eV. The largest peak is the bulk plasmon ($\hbar\omega_p$), with the Ni M_{23} and Si L_{23} core excitation peaks at higher energies, magnified by 100 and 350 times, respectively. No multiple scattering events are discernible in the EELS spectrum (no higher-order plasmon loss peaks are present), indicating that the background on the low energy loss side of the Ni before the Ni M edge is primarily due to the tail of the plasmon peak. The heights of the steps at the Ni M and Si L edges can be used to determine the composition.

At energies below the bulk plasmon peak (energy loss values 0–15 eV in Fig. 6.17a), there are peaks in the spectrum that correspond to interband transitions. The interband transitions involve a convolution of the valence and conduction band densities of states and hence are more difficult to interpret than core level spectra, where the initial states are sharp.

The differences in the spectrum step heights at the Ni and Si edges for different Ni to Si concentration ratios can be seen in Fig. 6.17b, which gives spectra taken from samples of Ni_2Si and $NiSi_2$. The yield Y_A for detection of an incident electron that loses energy E_A when passing through a material of thickness t containing a concentration N_A of atoms is given by

$$Y_A = Q\, N_A\, t\, \sigma_A\, \eta\, \Omega, \tag{6.37}$$

where Q is the integrated incident electron current density, $N_A t$ is the number of A atoms contributing to the inelastic scattering events, σ_A is the cross section for excitation of an electron in a given core energy level of atom A, η is a collection efficiency, and Ω is the detector collection angle. Equation 6.37 assumes the collected electrons experience only single inelastic scattering events. Provided that the collection efficiency

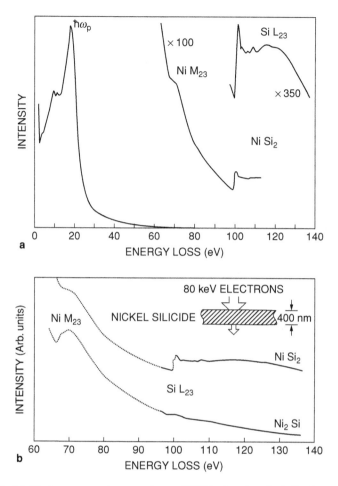

FIGURE 6.17. (a) Electron energy loss spectrum of NiSi$_2$ showing the bulk plasmon peak and the characteristic Ni M_{23} and Si L_{23} core level excitations. (b) A comparison of EELS spectra in the energy region of the characteristic Ni and Si level excitations. [From J. C. Barbour et al., *Ultramicroscopy* 14, 79 (1984).]

in scattering from A atoms is equal to that in scattering from B atoms, then the atomic ratio of A atoms to B atoms is

$$(N_A/N_B) = (Y_A/Y_B)(\sigma_B/\sigma_A),\qquad(6.38)$$

where Y_A and Y_B can be experimentally measured as the areas above background in an energy window above the edge. Therefore, the accuracy of the atomic ratio is sensitive to the cross section calculations and the accuracy in determining the relative Ni M to Si L areas by fitting the background after the edges.

Electron energy loss spectroscopy is not the most straightforward or sensitive method of detecting average composition or trace impurities. Its major advantage is analysis of small areas (<100 nm) for detecting microprecipitates and composition variations.

6.11   Bremsstrahlung

The energy loss of electrons passing through matter contains an additional significant component corresponding to the radiation loss. Classical physics tells us that an accelerated charge emits radiation. This acceleration is caused, for example, by the deflection of an electron in the field of an atom as the charged particle penetrates matter. Since the acceleration is essentially the ratio of the electrostatic force to the mass, this radiation component is significantly more important for electrons than heavy projectiles. As we will show, this *bremsstrahlung* (German for *braking radiation*) yields a continuum of photons up to the energy of the incident electron. It is of interest to the materials analyst, since bremsstrahlung may provide a significant background in analysis techniques that use incident electrons and detect characteristic photons. It is most obvious in electron-μ-probe analysis as a background underlying the characteristic X-ray spectrum.

The elastic scattering cross section of a charged particle, Z_1, by a nucleus of charge Z_2 is given by Eq. 2.17:

$$\frac{d\sigma}{d\Omega} = \left(\frac{Z_1 Z_2 e^2}{4E}\right)^2 \frac{1}{\sin^4 \theta/2}, \tag{6.39}$$

where θ is the scattering angle. It is convenient to express this cross section in terms of the associated momentum transfer Δp (Chapter 2).

$$\Delta p = 2p \sin \theta/2, \tag{6.40}$$

and

$$d\Omega = 2\pi \sin \theta d\theta = \frac{2\pi \Delta p \, d\Delta p}{p^2}.$$

Then the cross section for a momentum transfer Δp is given by

$$\frac{d\sigma}{d\Delta p} = 8\pi \left(\frac{Z_1 Z_2 e^2}{v}\right) \frac{1}{(\Delta p)^3}. \tag{6.41}$$

Classical electromagnetic theory shows that the total energy radiated per unit frequency interval per collision (Jackson, 1975) is

$$\frac{dI}{d\omega} = \frac{2}{3\pi} \frac{(Z_1 e)^2 \Delta p^2}{m^2 c^3}, \tag{6.42}$$

where m is the mass of the deflected particle. This formula is derived for the nonrelativistic case and in the limit of low ω; both conditions are of interest here. Then the differential radiation cross section is

$$\frac{d^2 \chi}{d\omega \, d\Delta p} = \frac{dI}{d\omega} \frac{d\sigma}{d\Delta p},$$

that is, the probability of emitting a photon of energy $\hbar\omega$ associated with a momentum transfer Δp times the probability of a momentum transfer, Δp. Explicitly,

$$\frac{d^2 \chi}{d\omega \, d\Delta p} = \frac{16}{3} \frac{Z_2^2 e^2 \left(Z_1^2 e^2\right)^2}{m^2 v^2 c^3} \frac{1}{\Delta p}. \tag{6.43}$$

We integrate over all possible energy transfers Δp_{min} to Δp_{max} to find the frequency spectrum,

$$\frac{d\chi}{d\omega} = \frac{16}{3} \frac{Z_2^2 e^2 \left(Z_1^2 e^2\right)^2}{m^2 v^2 c^3} \ln \frac{\Delta p_{max}}{\Delta p_{min}}.$$

To determine the ratio, $\Delta p_{max}/\Delta p_{min}$, we consider in detail the kinematics of the process. Energy conservation and momentum conservation can be written as

$$E = E' + \hbar\omega,$$

and

$$(\Delta p)^2 = (\mathbf{p} - \mathbf{p}' - \mathbf{k})^2 = (\mathbf{p} - \mathbf{p}')^2, \tag{6.44}$$

where E, p and E', p' refer to the energy and momentum of the particle before and after the collision, and $\hbar\mathbf{k}$ is the momentum of the emitted bremsstrahlung photon. Here we have neglected the momentum associated with the photon. Then

$$\frac{\Delta p_{max}}{\Delta p_{min}} = \frac{p + p'}{p - p'} = \frac{\left(\sqrt{E} - \sqrt{E - \hbar\omega}\right)^2}{\hbar\omega},$$

so

$$\frac{d\chi}{d\omega} = \frac{16}{3} \frac{Z_2^2 e^2 (Z_1^2 e^2)^2}{m^2 v^2 c^3} \ln \left(\frac{2 - \hbar\omega/E + 2\sqrt{1 - \hbar\omega/E}}{\hbar\omega/E}\right). \tag{6.45}$$

This function falls off as $(\hbar\omega/E)$ for small $\hbar\omega/E$ and flattens out with a sharp cut-off at $E = \hbar\omega$ (Fig. 6.18). This form of the bremsstrahlung spectrum was first derived by Bethe and Heftier in 1934.

The radiation cross section, $d\chi/d\omega$, is proportional to $Z_2^2 Z_1^4/m^2$, showing that the emission is most important for light particles (electrons) in materials of high atomic number. The total energy lost by radiation for a particle traversing a material of N nuclei per unit volume is

$$\frac{dE_{rad}}{dx} = N \int_0^{\omega_{max}} \frac{d\chi(\omega)}{d\omega} \, d\omega.$$

Letting $x = \hbar\omega/E$ and noting that

$$\left(\frac{1 + \sqrt{1 - x}}{\sqrt{x}}\right)^2 = \left(\frac{2 - x + 2\sqrt{1 - x}}{x}\right),$$

we have

$$\frac{dE_{rad}}{dx} = \frac{16}{3} \frac{N Z_2^2 e^2 \left(Z_1^2 e^2\right)^2}{c^3 \hbar} \int_0^1 \ln \left(\frac{1 + \sqrt{1 - x}}{\sqrt{x}}\right) dx. \tag{6.46}$$

The dimensionless integral has the value of unity. The radiative loss is essentially energy independent. The ratio of radiative loss (for electrons) to nonradiative loss,

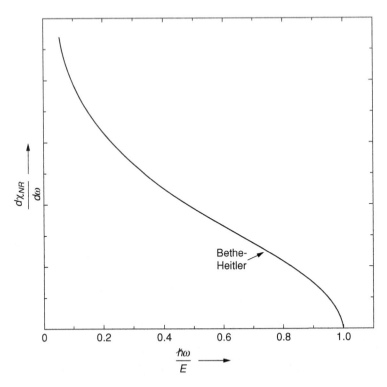

FIGURE 6.18. Radiation cross section (energy × area / unit frequency) for nonrelativistic Coulomb collisions as a function $\hbar\omega_p/E$ (Eq. 6.45).

dE_{rad}/dE_{nr}, is approximately given by

$$\frac{dE_{rad}}{dE_{nr}} \cong \frac{4}{3\pi}\frac{Z_2}{137}\left(\frac{v}{c}\right)^2, \tag{6.47}$$

and is small for $v < c$. For electron probe analysis, we are concerned that the bremsstrahlung photon may obscure the characteristic energy of interest. As shown by Evans, the ratio of the ionization cross section to the radiative cross section is roughly $Z_2(v/c)^2/137$, or -0.01 for $Z_2/137 = 4$ and 100 keV electrons. This is what sets the limit for detecting approximately 1% of an impurity in a solid matrix.

Problems

6.1. Draw the curve of Auger yield versus overlayer thickness for the signal from the overlayer atoms in the case of
   (a) uniform growth in a layer-by-layer fashion and
   (b) island growth with 50% coverage.
   Assume that the mean free path λ is equivalent to two monolayers.

6.2. Typically, electron impact ionization cross sections have a maximum at $E/E_B = 3$. Calculate the value of E at σ_{max} for the K shell and L shell of Si for incident electrons. As discussed in the text, ionization cross sections depend primarily on the velocity of the incident charged particles; calculate the energies of incident protons where the protons have the same velocities as those of the electrons in the examples for the ionization of K- and L-shell electrons.

6.3. You deposit an Al film on a copper substrate, and use 3 keV electrons to excite Cu L_α X-rays for the determination of Cu penetration into Al. What thickness of the deposited Al film is required so that the electrons do not excite Cu L_α X-rays in the Cu substrate? Use values of $K = 0.064$ and $\gamma = 1.68$, and Eq. 6.34.

6.4. You are carrying out a transmission electron energy loss experiment with 100 keV electrons incident on an Al film where the bulk plasmon energy is 15.3 eV. Estimate the value of λ for 100 keV electrons. Compare the values of the range of 100 keV electrons using Eq. 6.34 with that of Eq. 6.31 where you assume that dE/dx is a constant evaluated at 50 keV.

6.5. The Si LVV Auger electron is emitted with 92 eV. Assuming that the only mechanism for energy loss is ionization of Si atoms (Appendix 4), calculate the mean free path for these electrons
   (a) with the relations given in Section 6.7 and
   (b) with $\lambda = 1/N\sigma$, where σ is the cross section given in Eq. 6.11.

6.6. In a vacuum system, the flux of gas atoms impinging on a surface is $Nv/4$, where N is the density of atoms per cm^3 and v is the thermal velocity of the gas atoms. Assuming every gas atom sticks, calculate the value of N such that an absorbed layer of oxygen atoms is thinner than two monolayers after one hour. Express N in torr, noting that one atmosphere (760 torr) is approximately 2×10^{19} atoms/cm^3. It is these basic requirements that determine the need for good vacuum in thin film analysis.

6.7. Consider a semiconductor structure consisting of 2.5 nm of Si, one monolayer of Ge, 2.5 nm of Si, and one monolayer of Ge on a thick Si substrate. Imagine an Auger/sputter profiling analysis of this structure in which one detects the Ge MVV Auger line ($\lambda = 1.0$ nm in Si). Assume that the sputtering process removes one monolayer of material at a time with no interface broadening or other mixing effects. Write an equation for the yield of Ge as a function of material removed, and plot the Ge profile expected in this kind of analysis. (Ignore the effect of the single monolayers of Ge in the extinction path.)

References

1. C. R. Brundle and A. D. Baker, Eds., *Electron Spectroscopy: Theory, Techniques and Applications* (Academic Press, New York, 1981).
2. M. Cardona and L. Ley, Eds., *Photoemission in Solids, I, II, Topics in Applied Physics*, Vols. 26 and 27 (Springer-Verlag, New York, 1978 and 1979).
3. T. A. Carlson, *Photoelectron and Auger Spectroscopy* (Plenum Press, New York, 1975).
4. W. Czanderna, Ed., *Methods of Surface Analysis* (Elsevier Science Publishing Co., New York, 1975).

5. G. Ertl and J. Kuppers, *Low Energy Electrons and Surface Chemistry* (Verlag Chemie International, Weinheim, 1974).

6. R. D. Evans, *The Atomic Nucleus* (McGraw-Hill, New York, 1955).

7. J. I. Goldstein, D. E. Newbury, P. Echlin, D. C. Joy, C. Fiori, and E. Lifshin, *Scanning Electron Microscopy and X-ray Microanalysis* (Plenum Press, New York, 1981).

8. H. Ibach, Ed., *Electron Spectroscopy for Surface Analysis, Topics in Current Physics*, Vol. 4 (Springer-Verlag, New York, 1977).

9. J. D. Jackson, *Classical Electrodynamics*, 2nd ed. (John Wiley and Sons, New York, 1975).

10. L. C. Fledman and J. W. Mayer, *Fundamentals of Surface and Thin Film Analysis* (Prentice Hall, New Jersey, 1986).

7
X-ray Diffraction

7.1 Introduction

Crystalline structures are deduced by the diffraction of known radiation incident from solid materials. The three-dimensional regularity of unit cells in crystalline materials results in coherent scattering of the radiation. The directions of the scattered beams are a function of the wavelength of the radiation and the specific interatomic spacing (d_{hkl}) of the plane from which the radiation scatters. The intensity of the scattered beam depends on the position of each atom in the unit and on the orientation of the crystal relative to the direction of the incident X-ray beam. Those beams that scatter in a constructive manner result in *allowed* reflections, i.e., the intensity is a nonzero value. Those beams that scatter in a destructive manner result in *unallowed* reflections, i.e., the intensity has a minimal value. Each of the constructively scattered beams depends directly on the wavelength of the incident radiation. Having said this, we must address the issue: Why is the wavelength important? One of the requirements for diffraction is that the wavelength of the incident radiation λ must be smaller than the distance between scattering sites. Typically, in crystalline solids, we desire to measure atomic spacing on the order of the lattice constants, e.g., 0.2–0.4 nm. Hence, we must use radiation with wavelengths less than 0.2 nm. This range of wavelength includes those of X-rays and high-energy electrons. Another criteria is that the scattering occurs in a coherent manner, i.e., the energy of the incident radiation equals the energy of the scattered radiation.

W.L. Bragg derived a simplistic description of coherent scattering from an array of periodic scattering sites, such as atoms in a crystalline solid. The scalar description of diffraction considers the case of monochromatic radiation impinging onto sheets of atoms spaced at d_{hkl} within the crystal. The wavelength λ of the radiation is smaller than the interatomic spacing d_{hkl} of the specific (hkl) plane. Based on this description of diffraction, we can conduct experiments to determine the distance between reflecting planes, crystallographic structure, coefficient of thermal expansion, texture, stress, and composition of thin films.

X-rays typically used in diffraction analysis penetrate to a depth of 1–2 μm, depending on the atomic number and destiny of the sample (see Section 8.8). This limits to approximately one micron the depth at which useful information can be derived.

7.2   Bragg's Law in Real Space

We now know that when a solid is irradiated under specific conditions, unique reflections are allowed. However, the question posed is:

In a straightforward manner, can we predict the angle of specific reflections?

Historical records show evidence of Laue's diffraction experiments prior to the publication of W.L. Bragg's explanation of diffraction of radiation by crystalline solids. However, it was the inquisitive attempts by W.L. Bragg to resolve an intellectual debate between his father, W.H. Bragg, and Laue that resulted in the formulation of the famous Bragg's Law.

Laue argued that only optical rules determine the occurrences of diffraction (i.e., wave interactions only). Radiation acts as waves, and scattering depends on the optical laws of scattering (e.g., Law of Reflectivity) and of constructive interference (i.e., waves in phase add constructively and waves out of phase add destructively). These published rules even gained the name *the Laue Equations*. (We will review these equations when we develop Bragg's Law vectorally in Section 8.4). Laue's description of diffraction in three-dimensional space required solving three equations with twelve unknowns. Even then, the derivation is arduous. W.H. Bragg's rebuttal stated that any symmetrical elements present in the diffraction pattern reduces the number of unknowns and that only particle–particle interactions induce diffraction. Kinetic theory required that the particles scatter from atoms coherently, i.e., the incident particle does not lose any energy during the scattering event. Even the earliest Laue experiment supported this account. W.L. Bragg utilized all the above precepts to derive a simplistic description of coherent scattering. The actual development first appeared in vector form. However, the most recognized form appears in scalar form and is represented in Fig. 7.1.

The scalar description of diffraction considers the case of monochromatic radiation impinging on two sheets of atoms in the crystal, spaced at a distance of d_{hkl} between the reflecting planes. Invoking the Law of Reflectivity (or Reflections), $\theta_{in} = \theta_{out}$, gives

$$180° = \psi + 90° + \alpha$$
$$= (90° - \theta) + 90° + \alpha$$
$$\theta = \alpha.$$

Constructive inference occurs if and only if the wave scattered by the atoms results in a total path difference $(2\Delta P)$ for the reflected waves that is equal to integer (n) multiples of λ:

$$n\lambda = 2\Delta P = 2d_{hkl} \sin \theta. \tag{7.1}$$

Hence, the scalar form of Bragg's Law, $n\lambda = 2d_{hkl} \sin \theta$, defines the condition for diffraction.

7.2.1   X-ray Powder Analysis

The simplest of all modern X-ray analyses is powder analysis using an X-ray diffractometer. The technique can be used to characterize powders as well as polycrystalline materials. The material of interest is ground to produce a fine, randomly orientated

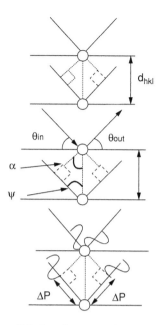

FIGURE 7.1. Development of Bragg's Law.

powder, with each particle in the powder consisting of small single crystals or an aggregate of crystals 10 μm or less. The powder is placed into a recess of a plastic sample holder and leveled flat in the sample holder with a straight edge.

The key components of a modern diffractometer include a monochromatic radiation source, sample stage (goniometer), radiation detection system, enclosure, and safety features. One of the most common configurations is the θ–θ upright. This type of diffractometer has a movable detector and X-ray source, rotating about the circumference of a circle centered on the surface of a flat powder specimen. Figure 7.2 shows the beam path schematically. The intensity of a diffracted beam is measured directly by an electronic solid-state detection system. The scattered X-rays dissipate energy by generating electron–hole pairs in the detector. The electronic system converts the collected charge into voltage pulses. The electronics counts the number of pulses per

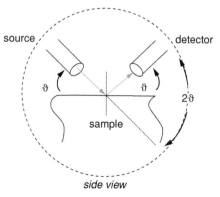

FIGURE 7.2. Schematic of a θ–θ upright shows how the source and detector each move at a constant rate of θ/s relative to the sample. Hence, the source and detector move at constant rate of 2θ/s relative to one another. This tool is ideal for loose powders and large samples.

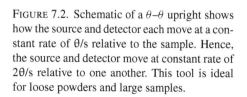

TABLE 7.1. Normalized $h^2 + k^2 + l^2$ values for the BCC and FCC structures.

	BCC			FCC	
(hkl)	$h^2 + k^2 + l^2$	nomalized	(hkl)	$h^2 + k^2 + l^2$	nomalized
(0 1 1)	2	1.00	(1 1 1)	3	1.00
(2 0 0)	4	2.00	(2 0 0)	4	1.33
(1 1 2)	6	3.00	(2 2 0)	8	2.67
(0 2 2)	8	4.00	(3 1 1)	11	3.67
(0 1 3)	10	5.00	(2 2 2)	12	4.00
(2 2 2)	12	6.00	(4 0 0)	16	5.33
(1 2 3)	14	7.00	(3 3 1)	19	6.33
(4 0 0)	16	8.00	(4 2 0)	20	6.67

unit of time, and this number is directly proportional to the intensity of the X-ray beam entering the detector.

7.2.2 Data Analysis

Software for peak location and intensity determination allows for significant time savings during the identification of an unknown specimen. Even with this luxury, experienced operators still examine each peak of every spectrum, inspecting peak positions and shapes, the presence or absence of peaks, and relative background levels. The use of software packages without understanding the algorithms that are employed can lead to serious misinterpretation. Hence, we will index a diffraction pattern from first principles and will start with the simplest case: the analysis of FCC and BCC structures.

Indexing a diffraction pattern (also called a *diffractogram* or *spectrum*) involves determining the lattice constant and structure and labeling each peak with its appropriate *hkl* designation. Starting with Bragg's Law, we rewrite d_{hkl} in terms of the lattice parameter a_o and then rearrange the equation and square both sides:

$$n\lambda = 2d_{hkl_1} \sin\theta = 2\frac{a_{o_1}}{\sqrt{h^2 + k^2 + l^2}} \sin\theta,$$

$$\sqrt{h_1^2 + k_1^2 + l_1^2} = 2\frac{a_{o_1}}{n\lambda} \sin\theta = \kappa \sin\theta, \tag{7.2}$$

$$h_1^2 + k_1^2 + l_1^2 = \kappa^2 \sin^2\theta.$$

The next step is to normalize all the equations relative to the equation for the first reflection:

$$\frac{(h_i^2 + k_i^2 + l_i^2)_i}{(h_1^2 + k_1^2 + l_1^2)_1} = \frac{\kappa^2 \sin^2\theta}{\kappa^2 \sin^2\theta} = \frac{\sin^2\theta}{\sin^2\theta}. \tag{7.3}$$

In a later section, the structure factor calculations give the relationship for allowed reflections for the cases of FCC and BCC structures:

FCC—h, k, l unmixed (all odd or all even numbers)

BCC—$h + k + l$ equals an even number

Table 7.1 lists the normalized values for both BCC and FCC structures based on Eq. 7.3.

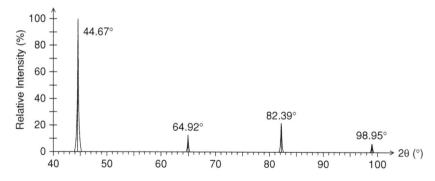

FIGURE 7.3. X-ray spectrum from an unknown metallic specimen.

Figure 7.3 displays a diffractogram obtained from an unknown FCC or BCC metal using an incident λ equal to 0.154056 nm. The data are typically presented in the form of intensity as a function of 2θ, θ, or d-spacing. Maxima in the intensity correspond to the peak positions. Using Table 7.1 and Eq. 7.3, the results presented in Table 7.2 confirm that the material has a BCC structure. Next we calculate a lattice constant from each peak and report the average value as the lattice constant. The structure and lattice parameter (BCC and $a_o = 0.2864$ nm) match that of α-Fe. The final step is to label each peak with the appropriate hkl.

TABLE 7.2. Confirmation that the material has a BCC structure.

line	2θ	$\sin^2(\theta)$	$\dfrac{\sin^2(\theta)}{\sin^2(\theta)_1}$	(hkl)	a_o (nm)
1	44.67	0.1444	1.00	(0 1 1)	0.2867
2	64.92	0.2881	1.99	(2 0 0)	0.2870
3	82.39	0.4338	3.00	(1 1 2)	0.2865
4	98.95	0.5778	4.00	(0 2 2)	0.2866
				$<a_0>=$	0.2867

7.3   Coefficient of Thermal Expansion Measurements

Thermal expansion is a phenomenon that influences the adhesion and mechanical properties of solids. For example, if layers of electronic materials are made of two dissimilar materials, an increase in temperature during device processing or operation results in each material expanding by different amounts. This outcome is a result of differences in the coefficients of thermal expansion (CTE, α). A classical example is the attachment of integrated circuit chips of silicon to alumina substrates using Cu or Sn–Pb solder bumps; note that each of these materials has a different CTE. For the case of good adherence between the layer and the substrate, the expansion of the layer is constrained by the expansion of the substrate and vice versa. Even small dimensional changes result in stresses. When these stresses exceed a critical value, the material fails by either delamination or facture. By measuring the change in the lattice constant and knowing the elastic constant, one can calculate the thermal stress and determine the safe processing and operating conditions. This calculation, however, requires accurate measurement of the CTEs.

In-situ heating during X-ray diffraction analysis is a means by which to measure the thermal expansion of crystalline solids. As the temperature of the specimen increases, the lattice constant increases in each unit cell. The net effect is an increase in the overall length. The thermal expansion coefficient can be determined by measuring the increase in the lattice constant (the decrease of the 2θ angle for a given diffraction peak) with increasing temperature. This technique can provide the coefficients of linear thermal expansion along different crystallographic axes for single or polycrystalline materials, and can also provide information regarding volume changes.

In-situ heating during X-ray diffraction requires the use of specially equipped diffractometers that have the capability to controllably heat and cool the sample during analysis in an open or closed ambient. Analysis is performed at moderate temperatures, e.g., $T < 0.5T_{melt}$ for metals. Typically, at temperatures higher than $0.5T_{melt}$, metals substantially increase their contribution of vacancies to the overall lattice parameter. However, in-situ heating can also be used to monitor phase transformation, as well as solid solution formation as a function of heating. Some modern tools can heat up to 1200 °C in vacuum or inert ambient. The design of these tools allows the sample to remain on the goniometer during heating without any damage to the diffractometer or the operator.

The analysis consists of initially measuring the lattice spacing for a specific reflection to determine the value of a_0 at room temperature. The sample is then heated to an elevated temperature. Once the temperature equilibrates, the lattice constant a_i is again measured. This step is repeated several times. The CTE (α) is the fractional change in length (Δl) per unit length (l) per unit change in temperature (ΔT):

$$\Delta l/l = \alpha \Delta T,$$
$$\text{CTE}(\alpha) = (\Delta l/l)(\Delta T^{-1}). \tag{7.4}$$

In the simplest case, the cubic case, CTE is also equal to the fractional change in lattice constant per unit change in temperature:

$$\text{CTE}(\alpha) = (\Delta a/a_i)(\Delta T^{-1}). \tag{7.5}$$

Starting with Bragg's Law, a relationship for the relative change in lattice spacing with respect to the change in θ is developed:

$$\lambda = 2d\sin\theta$$

$$d = \frac{\lambda}{2}\frac{1}{\sin\theta}$$

$$\Delta d = -\frac{\lambda}{2}\frac{\cos\theta}{\sin^2\theta}\Delta\theta$$

$$\frac{\Delta d}{d} = -\frac{\lambda}{2}\frac{\cos\theta}{\sin^2\theta}\Delta\theta\left(\frac{2}{\lambda}\frac{\sin\theta}{1}\right) = -\Delta\theta\cot\theta. \tag{7.6}$$

For the case of cubic materials,

$$\frac{\Delta d}{d} = -\Delta\theta\cot\theta = \frac{\Delta a}{a_0}. \tag{7.7}$$

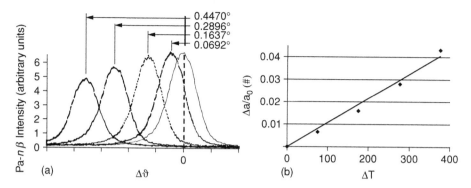

FIGURE 7.4. a) X-ray spectra of the (*hkl*) reflection of β-parylene-*n* at room temperature, 100, 200, 300, and 375 °C; b) a/a_0 is calculated and plotted against the change in temperature.

Combining Eqs. 7.7 and 7.8 yields Eq. 7.9. This technique requires letting a_0 equal the theoretical value (i.e., tabulated value) for that specific material and replacing θ with the theoretical value of θ_B, where θ_B is calculated using again the theoretical value of a_0:

$$-\Delta\theta \cot\theta_B = \frac{\Delta a}{a_0} = \frac{\Delta l}{l} = \alpha\,\Delta T. \qquad (7.8)$$

With the extracted value of $\Delta\theta$, the CTE is determined:

$$\alpha = \frac{-\Delta\theta \cot\theta_B}{\Delta T}. \qquad (7.9)$$

Figure 7.4a shows a series of scans taken from an *n*-parylene sample. The diffractometer is equipped with a hot stage capable of heating the sample to temperatures of 100, 200, 300, and 400 °C. For each temperature, values of ΔT and $\Delta\theta$ are extracted from the plot. Using Eq. 7.8, $\Delta a/a_0$ is calculated and plotted against the change in temperature (Fig. 7.4b). The straight line fit yields a slope and CTE of 1.13×10^{-4}.

7.4   Texture Measurements in Polycrystalline Thin Films

The texture of metallic thin films influences the electrical and metallic properties that are important to the reliability of semiconductor devices. Texture is characterized through the use of X-ray diffraction pole-figure analysis. In order to obtain a pole-figure, the detector and sample geometry are set such that the incident and diffracted X-rays make a specific angle with the sample surface. This angle is the same as the angle necessary to satisfy the Bragg condition for a specific set of {*hkl*} planes of the thin film (Fig. 7.5a). One-dimensional pole-figures (Fig. 7.5b) measure the intensity of X-rays diffracted from the sample as a function of tilt angle (ψ in this text), without rotation about

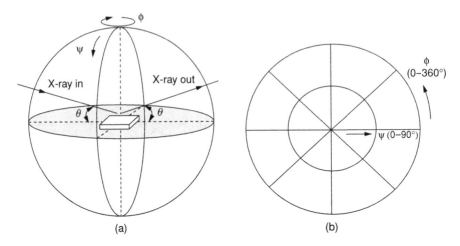

FIGURE 7.5. Schematic drawing (a) showing the geometry of pole-figure measurements with an X-ray diffractometer, and (b) a diagram showing the corresponding stereographic projection.

an axis perpendicular to the sample surface. The results of one-dimensional pole-figure measurements are shown in the form of intensity versus tilt-angle (ψ) plots (Fig. 7.6). The {200} texture of Ag films on amorphous SiO_2 is determined for samples annealed at different temperatures. The Ag films display preferred {111} orientation

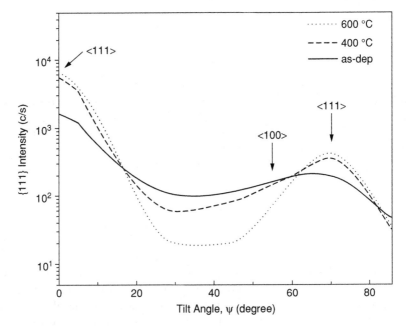

FIGURE 7.6. {111} pole-figures obtained from Ag thin films on SiO_2 substrates, as-deposited (solid line) and annealed at 400 °C (dashed line) and 600 °C (dotted line) for 1 hour in vacuum.

(texture) when annealed at different temperatures, respectively. At ψ (tilt angle) $= 0°$, the $\{111\}$ intensity is the highest, indicating that the Ag films are highly textured, with $< 111 >$ directions of grains in the films being normal to the surface of the substrate. Theoretically, given a $< 111 >$ texture orientation at $\psi = 0°$, another $\{111\}$-type set of planes can be expected to be present at $\psi = 70.5°$ in cubic crystal systems. Consistent with this expectation, high $\{111\}$ intensities are detected at $\psi = 70.5°$. Note that the texturing increases with increased annealing.

A two-dimensional pole-figure is obtained by first setting the tilt angle, and then measuring intensity as a function of sample rotation (ϕ) about an axis perpendicular to the surface of the sample (see Fig. 7.5a,b). The sample is rotated from $0°$ to $360°(\phi)$. After completing an azimuthal rotation, the sample is tilted (ψ) by a single tilt-angle step (typically $1°$). Azimuthal rotation of the sample is repeated, while the intensity of the diffracted beam is recorded. This process is repeated for the entire range of desired rotation and tilt angles. Three-dimensional contour plots are then generated to show texture. The two-dimensional pole-figure measurements and three-dimensional contour maps are obtained from a 150 nm layer of Ag on SiO_2 substrates, annealed in vacuum. Two-dimensional $\{200\}$ pole-figures are plotted in Fig. 7.7a,b, and the corresponding contour maps are displayed in Fig. 7.7c,d. Silver exhibits fiber-texture for both as-deposited and annealed samples. In addition, the fiber-textured grains ($\{111\}$ and $\{200\}$) are distributed randomly about the normal to the substrate, i.e., the film possesses *mosaic structure*. This is indicated by the contour line-shapes being circular at each tilt-angle position. Due to the mosaic structure of the Ag, the $\{111\}$ intensity at $\psi = 70.5°$ is lower than the $\{111\}$ intensity at $\psi = 0°$. Upon annealing, the $\{200\}$ intensity is reduced and the $\{111\}$ increases with temperature.

7.5 Strain Measurements in Epitaxial Layers

The strain in an epitaxial layer is determined by comparing the perpendicular and parallel lattice spacings (a) of the film to that of the underlying substrate a_{sub} and determining whether they are larger or smaller and by how much. The film is pseudomorphic when lattice planes of the epilayer align with the lattice planes of the substrate. The resulting strain values are expressed as $[\Delta a/a]_\perp$ and $[\Delta a/a]_\parallel$, and are based on the difference in the perpendicular and parallel lattice constants, respectively. Since these differences are measured relative to the substrate, the substrate lattice spacing must be known. From these parameters, a strain value is calculated by determining the difference in the lattice spacing of the film from its bulk relaxed state.

The application of X-ray diffraction in thin film samples is illustrated by diffraction measurement of epitaxial layers of SiGeC layers on silicon. Initially, an asymmetrical scan is taken about a plane (hkl) not parallel to the surface plane, where that plane makes an angle ψ relative to the surface (see Fig. 7.8). Since, in this case, $\theta_{in} + \psi$ and $\theta_{out} - \psi$ are not identical, the asymmetrical scan provides information about both perpendicular spacing ($a_\perp$) and parallel spacing ($a_\parallel$). For analysis of Si-based structures, both $(\bar{2}\,\bar{2}\,4)$ and $(2\,2\,4)$ reflections are typically used (Fig. 7.9). In one case, the spectrum is a glancing incidence $(2\,2\,4)$ reflection and the other is a $(\bar{2}\,\bar{2}\,4)$ glancing exiting reflection. Note the presence of a peak from the Si substrate and from the SiGeC film in

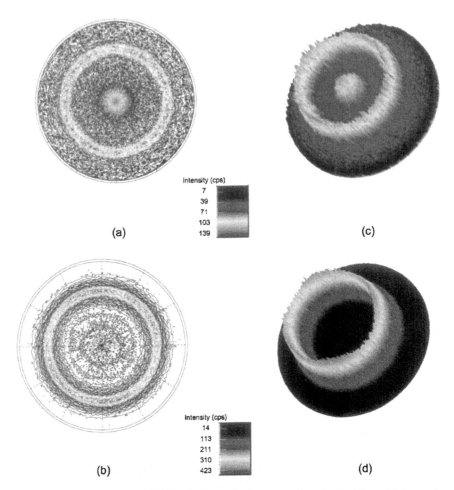

FIGURE 7.7. Two-dimensional {200} pole-figures and contour plots obtained from (a) Ag, as-dep; (b) Ag, annealed at 600 °C in vacuum for 1 hour, and corresponding three-dimensional {200} pole-figure (c) as-dep, and (d) 600 °C, respectively.

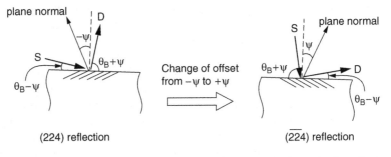

FIGURE 7.8. Schematic showing symmetrical scan and asymmetrical scan is taken about a plane (*hkl*) not parallel to the surface plane, where that plane makes an angle ψ relative to the surface.

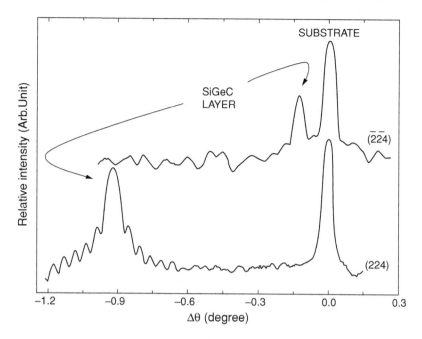

FIGURE 7.9. Asymmetrical scan (2 2 4) and ($\bar{2}\,\bar{2}$ 4) reflection of a 180 nm thick Ge–Si alloy on a Si(0 0 1) single-crystal substrate.

both scans. Satellite peaks above and below the substrate reflection are due to thickness interference fringes. The absence of satellites for the ($\bar{2}\,\bar{2}$ 4) peaks is due to dispersion of the diffracted beam.

For epilayer strain calculations, the substrate is typically considered to be unstrained, and hence all measurements are made relative to the substrate's lattice constant. For this reason, it is useful to plot rocking curve measurements of intensity as a function of $\Delta\theta$ from the substrate peak. The perpendicular and parallel lattice constants are calculated by first measuring the angular peak separation from glancing incident (2 2 4) reflections, ω_1, and the glancing exit ($\bar{2}\,\bar{2}$ 4) reflections, ω_2. The deviation of the Bragg angle between the substrate and layer, $\Delta\theta$, is calculated from

$$\Delta\theta = 0.5(\omega_1 + \omega_2). \tag{7.10}$$

Due to tetragonal distortion in the epitaxial layer, the angle between the {224} planes and the surface in the substrate will not be the same as this angle in the film. This difference $\Delta\psi$ is calculated from

$$\Delta\psi = 0.5(\omega_1 - \omega_2). \tag{7.11}$$

The perpendicular lattice constant of the film is calculated by

$$[\Delta a/a]_\perp = -\Delta\theta \cot(\theta_B) + \Delta\psi \tan(\psi), \tag{7.12}$$

where θ_B is the Bragg angle for the reflection measured relative to the substrate. The

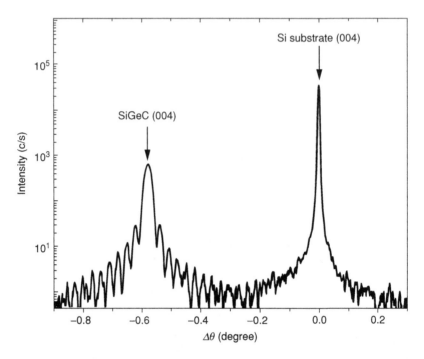

FIGURE 7.10. Symmetrical scan (0 0 4) reflection of a 180 nm thick Ge–Si alloy on a Si(0 0 1) single crystal substrate.

parallel lattice constant of the film is calculated from

$$[\Delta a/a]_{\parallel} = -\Delta\theta \cot(\theta_B) + \Delta\psi \cot(\psi). \tag{7.13}$$

Using Eqs. 7.12 and 7.13, values of $\theta_B = 44.015°$, $\psi = 35.26°$, $\Delta\theta = -0.529$, and $\Delta\psi = 0.3963$ yield values of $[\Delta a/a]_{\perp}$ and $[\Delta a/a]_{\parallel}$ as $+0.0144$ and -0.0002, respectively. Note that negative values for Δa indicate that the dimension in the film is smaller than that of the substrate; hence, the film is in tension. When the perpendicular lattice constant for a pseudomorphic film is larger than the substrate, the film has a larger unit cell and is in compression in the plane of the film. As an independent check of the procedure above, a symmetrical scan is taken from an allowed reflection parallel to the (0 0 1) surface. The (0 0 4) reflection is an example of such an allowed reflection. If we use the (0 0 4) reflection in Fig. 7.10, Eq. 7.24, and $\Delta\psi = 0$, a value of $[\Delta a/a]_{\perp} = 0.0146$ is obtained and is consistent with values obtained from the asymmetrical scans.

High-resolution X-ray diffraction can also be used to generate reciprocal space maps. The significance of reciprocal space maps is that they can directly distinguish between mosaic-related defects (which result in diffuse scattering along the ω axis) and from alloying-induced d_{hkl} spacing variations (which cause diffuse scattering along the $\omega/2\theta$ axis). With a single asymmetric reflection, a map can also quickly reveal whether the film is pseudomorphic or relaxed. Reciprocal lattice points for a pseudomorphic (in plane) and tetragonally distorted (out of plane) layer lie directly along the same [0 0 1] vector. For relaxed layers, the lattice point lies parallel to the [0 0 1] vector;

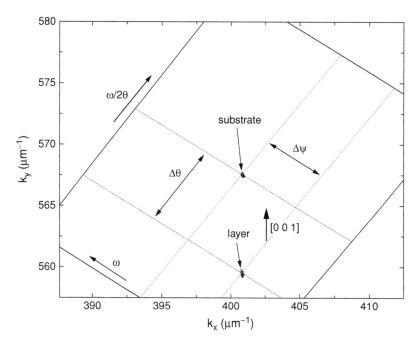

FIGURE 7.11. A reciprocal lattice map of the same sample previously shown in Figs. 7.9 and 7.10.

however, the point has a finite ω separation. This ω separation of the maxima is used to quantify both the parallel and perpendicular lattice constants. Figure 7.11 shows a reciprocal space map for a glancing incidence (2 2 4) reflection from the same SiGeC sample described above. The reciprocal space maps are oriented with the [0 0 1] growth direction on the vertical axis and the [1 1 0] direction along the x-axis. The area of reciprocal space scanned during the X-ray measurement axes is labeled as ω and $\omega/2\theta$. The values needed for Eqs. 7.12 and 7.13 can be measured directly from the map. Such measurement yields values of $[\Delta a/a]_\perp$ and $[\Delta a/a]_\parallel$ as $+0.0147$ and -0.0001, respectively. In addition, a layer reciprocal lattice point lies directly below the substrate lattice point on the [0 0 1] axis. This indicates that the layers are pseudomorphic, since all the deflection is in the [0 0 1] growth direction. Film relaxation would result in an increase in the parallel lattice parameter, which would be reflected in a change in the [1 1 0] dimension. This shift would cause the layer reciprocal lattice point to move out of this [0 0 1] axis toward the diffraction vector. The reciprocal space maps also show a high structural quality of the layers by the lack of diffuse scattering of the layer lattice points. The low values of the full width at half maximum in the $\omega/2\theta$ direction indicates a lack of variability in the d_{hkl} spacing in the layer.

7.6   Crystalline Structure

Crystallographers describe each of the over 40,000 known crystalline structures through the use of one of the 14-point lattices. At first inspection, one would think, *"How is*

FIGURE 7.12. Schematic of a structure consisting of one point lattice and a unique basis.

this possible?" Crystallographers use a specific and yet simple nomenclature to define crystal structures. The term *structure* refers to a composite of a point *lattice* and a unique *basis* (Fig. 7.12). The basis refers to a motif (i.e., repeating theme) placed on each lattice point. In some structures, the motif consists of a single atom; in other structures, it consists of a group of atoms placed on each lattice point.

Any introductory materials engineering or solid-state physics book will address the uniqueness of the 14 Bravais point lattices. The uniqueness of structures is also based on lattice point positions in addition to the atom positions in the basis. The position of the jth atom in the basis is described by the *repeat* vector $\mathbf{r}_j$,

$$\mathbf{r}_j = a_j\,\mathbf{x} + b_j\,\mathbf{y} + c_j\,\mathbf{z}, \qquad (7.14)$$

where $0 \le a_j, b_j, c_j \le 1$, as measured relative to a lattice point (typically the origin of the specific lattice). The origin is defined as the position 0,0,0. We then measure atomic positions relative to this origin, using vector notation. For this example, let the basis consist of a molecule *AX* (given in Figure 7.13a). For the two-atom basis, the smaller atom is placed at a position vector [0 0 0] relative to the origin, and the second atom is placed at position $[\tfrac{1}{2}\tfrac{1}{2}\tfrac{1}{2}]$ relative to the origin. Hence the positions of the atoms in the basis are given by the relationship atom *A* at [0 0 0] and atom *X* at $[\tfrac{1}{2}\tfrac{1}{2}\tfrac{1}{2}]$. In short-hand notation, this relationship is expressed as *A* 000, $X\tfrac{1}{2}\tfrac{1}{2}\tfrac{1}{2}$. As indicated above, the lattice is built by repetitive displacements of these basic units cells. For example, consider the tetragonal lattice in Fig. 7.13b. Now place the motif on every lattice point. The formal description of the structure is a primitive tetragonal plus a

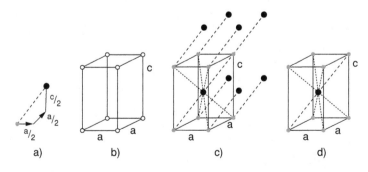

FIGURE 7.13. (a) Schematic of the spatial relationship between atoms in a basis, and the formation of a crystal structure: (b) take a point lattice, (c) place a common basis on each lattice point, and (d) the final structure.

two-atom basis 000, $\frac{1}{2}\frac{1}{2}\frac{1}{2}$. Later we will use the formal notation to determine the allowed reflections during diffraction analysis. Note that Fig. 7.13c and Fig. 7.13d maintain the same symmetry and hence they each describe the same structure.

Let us now examine the crystal structures of some widely used materials. The metal Fe has a BCC structure that consists of a BCC lattice and a one-atom basis of Fe on each lattice site. We describe its structure as BCC lattice + Fe 000. The atomic arrangement of the diamond structure consists of two interpenetrating FCC sublattices displaced from each other by $a\sqrt{3}/4$. Both sublattices are occupied by identical atoms. Silicon, diamond, and germanium crystallize in this structure: FCC lattice + Si 000, $\frac{1}{4}\frac{1}{4}\frac{1}{4}$. The zinc-blend structure also consists of two interpenetrating FCC sublattices displaced from each other by $\sqrt{3}\,{}^{a}/4$, where a is the lattice parameter. In this case, one of the sublattices is occupied by one type of atom and the other sublattice is occupied by another type of atom. GaAs, InP, and GaP crystallize in this structure: FCC lattice + Ga 000, As $\frac{1}{4}\frac{1}{4}\frac{1}{4}$. In the NaCl structure, Na and Cl ions reside on their respective FCC sublattices. Furthermore, the nearest neighbors of Na^+ ions are Cl^- ions. This structure occurs to avoid the electrostatic repulsion between the positively or negatively charged ions. Based on the symmetry of the basis, the NaCl structure is described as FCC lattice + Na^+ 000, Cl^- $\frac{1}{2}\frac{1}{2}\frac{1}{2}$ or FCC lattice + Cl^- 000, Na^+ $\frac{1}{2}\frac{1}{2}\frac{1}{2}$. In the CsCl structure, Cs and Cl ions reside on their respective simple cubic sublattice, and the structure is SC lattice + Cs^+ 000, Cl^- $\frac{1}{2}\frac{1}{2}\frac{1}{2}$ or SC lattice + Cl^- 000, Cs^+ $\frac{1}{2}\frac{1}{2}\frac{1}{2}$.

7.7 Allowed Reflections and Relative Intensities

In X-ray diffraction, we view the incident and emergent radiation both as discrete particles with a specific energy and as electromagnetic radiation with λ on the order of 0.1 nm. Engineers and scientists use diffraction analyses to quantitatively determine crystal structure, lattice parameter, orientation, defects, and mechanical properties. The analysis involves interrogating the sample with X-rays or electrons and interpreting the resulting diffraction pattern. The pattern consists of a series of intensity peaks that arise from coherent scattering from specific planes. Each peak corresponds to a specific *hkl* reflection and conveys information about crystal structure and symmetry. Note that planes are (*hkl*) and reflections are *hkl*. This leads us to the following question:

Can we predict the allowed reflections and their relative intensities?

Structure Factor Calculations (*SFC*) determine the resultant scattering power of the whole crystal structure. Given the fact that the whole structure consists of a large number of unit cells (all of which scatter in phase under Bragg diffraction conditions), the resultant scattering power need only be calculated for the contents of one unit cell. Structure factor calculations start by initially determining the relative scattering intensity of the atom relative to the scattering by a single electron. Secondly, we predict the relative scattering intensity of the unit cell relative to the scattering by a single electron. Those reflections that result in nonzero intensities are allowed. Those reflections that result in a zero intensity are not allowed.

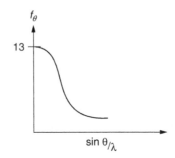

FIGURE 7.14. Atomic scattering factor f_ϑ for Be as a function of $^{\sin\theta}/\lambda$.

7.7.1   Scattering by the Atom—Atomic Scattering Factor

The efficiency of the atom in scattering waves is quantified relative to the scattering by a single electron. The atomic scattering factor f_ϑ is the amplitude of the wave scattered by the atom relative to the amplitude of a wave scattered by an isolated electron.

f_ϑ = amplitude scattered by the atom/amplitude scattered by a single electron

$$= A_n/A_e \qquad (7.15)$$

The f_θ depends on both scattering angle θ and λ of the incident radiation. Figure 7.14 displays a schematic diagram showing the variation of the atomic scattering factor f_θ with $^{\sin\theta}/\lambda$ for Be. Note that these two species have an identical number of electrons. At a zero scattering angle, the scattered waves are in phase and the scattered amplitude is the sum from all the electrons, i.e., f_θ equals Z (atomic number) in the atom or ion. As λ increases, the intensity decreases because the efficiency for scattering through large angles is lower. From Appendix 5, the values can be determined for atoms and ions. For example, the f_θ value can be determined for the case when Mo $K\alpha$ (λ of Mo $K\alpha$ is 0.07107 nm) is incident on a Be atom at an angle of 16.52°. First calculate the value of $^{\sin\theta}/\lambda = {}^{\sin(16.52°)}/_{0.07107\,nm} = 4.0\,nm^{-1}$. The value of f_θ can be obtained from Appendix 5, and is found to be equal to 1.6. In a similar manner, Mo $K\alpha$ incident on N^{3+} at the same angle results in a f_θ value of 2.0. Upon inspection, it becomes evident that atoms with greater numbers of electrons scatter radiation more effectively.

7.7.2   Scattering by Unit Cell and Structure Factor Calculations

A monochromatic beam incident upon a regularly spaced three-dimensional array of atoms can result in an interference pattern. In this particular case, the path difference between scattered beams results in a phase difference. A phase difference of zero results in constructive inference. A phase difference of 180° results in destructive inference. The superimposition of all the scattered waves from the single unit cell produces the resultant intensity from the whole crystal. This resultant wave and the phase difference are a function of the number of electrons and the positions of each of the atoms in the unit cell. The structure factor F_{hkl} therefore corresponds to the summation of amplitude A of scattering of all the electrons of one unit cell relative to the amplitude of scattering

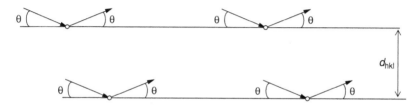

FIGURE 7.15. Schematic of coherent scattering from a lattice with a basis of one atom.

from an isolated electron:

$$F_{hkl} = \Sigma A_n / A_e$$
$$= \text{amplitude of all atoms of the unit cell/amplitude of a single electron.}$$

Structure factor calculations establish the allowed hkl reflections. In addition, SFC also conveys the amplitude and the phase for each reflection. The intensity of the reflection is influenced by the direction and wavelength of the incident radiation and by the structure of the crystalline solid. Note that the presence of the reflection is not dependent on the method used to obtain the reflection or the type of radiation used.

Starting with the simplest case of a primitive unit cell with a single atom basis at each lattice point, Fig. 7.15 displays a schematic of the incident and scattered waves for a crystalline lattice. For a specific set of (hkl) planes, each atom has an atomic scattering factor f_0 at the correct Bragg angle. For all the atoms residing on the successive planes, the path difference is zero for the case of constructive interference. Hence, the total scattered amplitude relative to a single electron is the sum of the contribution from all atoms [($1^{\text{atoms}}/8_{\text{corners}}$)*8 corners = 1 atom] in the unit cell, F_{hkl}, and this equals f_0.

Next we consider the case for a two-atom basis (Fig. 7.16). Consider an atom placed at the origin with atomic scattering factor f_0 and another atom placed at a position defined by $\mathbf{r}_1$, the position vector, and with atomic scattering factor f_1. As previously stated, $\mathbf{r}_1$ is expressed in terms of fractional coordinates [uvw] along the axes $\mathbf{x}$, $\mathbf{y}$, and $\mathbf{z}$, respectively. The incident wave and scattered waves can be described by the vectors $\mathbf{s}_0$ and $\mathbf{s}$, respectively. For a wave front scattering from points A and D, constructive interference will occur if the path difference $\overline{PD} = \overline{AB} - \overline{CD}$ is equal to an integral

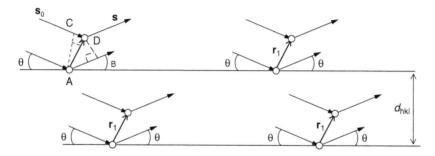

FIGURE 7.16. Schematic of coherent scattering from a lattice with a basis of two atoms.

multiple of λ:

$$\overline{PD} = \overline{AB} - \overline{CD}$$
$$= \mathbf{r}_1 \cdot \mathbf{s}_0 - \mathbf{r}_1 \cdot \mathbf{s} \tag{7.16}$$
$$= \mathbf{r}_1 \cdot (\mathbf{s}_0 - \mathbf{s}).$$

At this point, we need to solve Bragg's Law vectorally in reciprocal space. We shall do this later in Section 8.4. For the time being, we will accept that

$$\overline{PD} = \mathbf{r}_1 \cdot (\mathbf{s}_0 - \mathbf{s}) = \lambda(hu_1 + kv_1 + lw_1), \tag{7.17}$$

where h, k, and l are Miller indices of the reflecting planes. Eq. 7.17 expresses the scattered wave's path difference from different atoms in the unit cell (and from the basis) conveniently as an angular phase difference (ϕ). This value is measured relative to an atom at the origin of the unit cell (i.e., generally the origin of the basis as well). The resultant is calculated by the superimposition of waves, one from each atom in the unit cell (and from the basis). The amplitude of each wave depends on the number of electrons in the atom (atomic scattering factor, $f_{\theta i}$, and scattering angle θ) and the phase, which depends on the position of the atom in the unit cell (and from the basis):

$$\phi_n = (2\pi/\lambda) \cdot \overline{PD}$$
$$= (2\pi/\lambda) \cdot \lambda(hu_n + kv_n + lw_n)$$
$$= 2\pi(hu_n + kv_n + lw_n). \tag{7.18}$$

Equation 7.18 gives the phase difference for any unit cell of any shape. In general, to add combinations of the atoms, we need only to use a vector-phase relationship in which the vectors are proportional to the atom's atomic scattering.

7.7.3 Allowed Reflections

Now we only need to sum the scattering of all the atoms in the unit cell to determine the relative intensity of allowed reflections. This sum is called the structure factor F_{hkl} for the crystal structure. For a primitive cell, the scattering from the whole unit cell (given by the structure factor F_{hkl}) is the same as the scattering from one single atom (given by the atomic scattering factor f_θ). The scattered amplitude (A) from each atom in the unit cell can be described as

$$A = f_\theta \exp[i\phi], \tag{7.19}$$

where f_θ is the scattering factor for a single atom and ϕ is the phase of the wave with respect to the wave scattered from the origin of the unit cell. If diffraction occurs from the (hkl) plane, then the phases are such that scattering from the atom n at cell position uvw can be given by

$$A_n = f_n \exp[i\phi_n] = f_n \exp[2\pi i(hu_n + kv_n + lw_n)]. \tag{7.20}$$

The scattering from each atom and its corresponding phase difference are summed over the entire unit cell:

$$F_{hkl} = {}^1\!/_{A_e} \Sigma A_n = \Sigma \, f_n \exp[2\pi i(hu_n + kv_n + lw_n)]. \tag{7.21}$$

TABLE 7.3. Useful relationships.

$e^{ix} = \cos(x) + i\sin(x)$	$e^{in\pi/2} = e^{in5\pi 2} \ldots = i$
$e^{ix} + e^{-ix} = 2\cos(x)$	$e^{n\pi i} = e^{3\pi ni} \ldots = -1$
$e^{2n\pi i} = e^{4n\pi i} \ldots = 1$	$e^{i3\pi n/2} = e^{in7\pi/2} \ldots = -i$

Typically, XRD analyses only involve measurements of relative intensities; hence, only the nonzero and zero values of intensity (I_{hkl}) determine the allowed and unallowed reflections, respectively. The structure factor F_{hkl} for a given reflection is related to the intensity (I_{hkl}) by the relationship ($I_{hkl}) \propto (F_{hkl})^2$. If F_{hkl} equals a complex number ($x + iy$), then we multiply the complex number by its complex conjugate ($x - iy$) to determine the intensity values of ($F_{hkl})^2$. Several useful relationships are given in Table 7.3.

We will review two approaches for determination of allowed reflections. The first is the *long approach*. Initially, we identify each atom in the unit cell by its position vector [uvw] and by its contribution to the unit cell (corners $1/8$ atom, face $1/2$ atom, and body 1 atom). By using Eq. 7.18, the sum all the atoms is obtained. We simplify the equation and develop relationships that generate integer sums of the exponents. It is then useful to summarize results. Below are a few examples.

In the first example, we determine the allowed reflections for the case of α-Fe (BCC structure). Eight Fe atoms reside on the corners and one in the body position. Each corner atom contributes $1/8$ atom to the unit cell, and the body atom contributes one atom to the unit cell. Using Eq. 7.21, we sum over all atom positions:

$$F_{hkl} = \Sigma f_n \exp[2\pi i(hu + kv + lw)]$$

$$= f_{Fe}[1/8 e^{(0*2\pi i)} + 1/8 e^{(h2\pi i)} + 1/8 e^{(k2\pi i)} + 1/8 e^{(l2\pi i)} + 1/8 e^{(h+k)2\pi i}$$

$$+ 1/8 e^{(h+l)2\pi i} + 1/8 e^{(k+l)2\pi i} + 1/8 e^{(h+k+l)2\pi i} + (1)e^{(h/2+k/2+l/2)2\pi i}].$$

Note that all hkl are Miller indices, and hence they must be integers:

$$F_{hkl} = f_{Fe}[1/8(1) + 1/8(1) + 1/8(1) + 1/8(1) + 1/8(1) + 1/8(1) + 1/8(1)$$

$$+ 1/8(1) + (1)e^{(h+k+l)\pi i}]$$

$$= f_{Fe}[1/8(8) + (1)e^{(h+k+l)\pi i}] = f_{Fe}[1 + e^{(h+k+l)\pi i}].$$

At this point, we develop relationships that generate integer sums of the exponents:

If $h + k + l$ equals an even integer $F_{hkl} = f_{Fe}[1 + e^{(2n)\pi i}] = f_{Fe}[1 + 1] = 2f_{Fe}$

If $h + k + l$ equals an odd integer $F_{hkl} = f_{Fe}[1 + e^{(2n+1)\pi i}] = f_{Fe}[1 - 1] = 0$

where $n = 1, 2, 3. \ldots$

Finally, multiply the SFC by the complex conjugate to determine the relative intensity and summarize results:

If $h + k + l$ equals an even integer $I_{hkl} = (F_{hkl})^2 = 4(f_{Fe})^2$ allowed reflections;

If $h + k + l$ equals an odd integer $I_{hkl} = (F_{hkl})^2 = 0$ reflections absent.

The second example requires identifying the allowed reflections for CsCl. Eight anions reside on the corners and one cation on the body position. Due to the symmetry of the structure, it can also be stated that eight cations reside on the corners and one anion on the body position. Again, each corner atom contributes $1/8$ atom to the unit cell, and the body atom contributes one atom to the unit cell. For this example, we will assume Cs^+ on the corners and Cl^- in the body. Using Eq. 7.18, we sum over all atom positions:

$$F_{hkl} = \Sigma f_n \exp\left[2\pi i(hu + kv + lw)\right]$$

$$= f_{Cs}[^1/_8 e^{(0^*2\pi i)} + {}^1/_8 e^{(h2\pi i)} + {}^1/_8 e^{(k2\pi i)} + {}^1/_8 e^{(l2\pi i)} + {}^1/_8 e^{(h+k)2\pi i}$$

$$+{}^1/_8 e^{(h+l)2\pi i} + {}^1/_8 e^{(k+l)2\pi i} + {}^1/_8 e^{(h+k+l)2\pi i}] + f_{Cl}\, e^{(h/2+k/2+l/2)2\pi i}.$$

Note that all hkl are Miller indices, and hence they must be integers:

$$= [f_{Cs}{}^1/_8(8) + f_{Cl}e^{(h+k+l)\pi i}] = f_{Cs} + f_{Cl}e^{(h+k+l)\pi i}].$$

In this case, we multiply the SFC by the complex conjugate and summarize the intensity results:

If $h + k + l$ equals an even integer,
$$I_{hkl} = (F_{hkl})^2 = (f_{Cs})^2 + 2(f_{Cl})^2 + 2(f_{Cs})(f_{Cl}) \text{ allowed reflection;}$$

If $h + k + l$ equals an odd integer,
$$I_{hkl} = (F_{hkl})^2 = (f_{Cs})^2 + 2(f_{Cl})^2 - 2(f_{Cs})(f_{Cl}) \text{ allowed reflection.}$$

The structure of any crystalline solid is *best described* by its lattice and basis (Section 7.4). In a similar manner, the structure factor is also *best described* by the relationship between the structure factor of the lattice and the structure factor of the basis:

$$F_{hkl} = F_{hkl}{}^{\text{Lattice}} \cdot F_{hkl}{}^{\text{Basis}} = \Sigma\, f_n \exp[2\pi i(hu_n + kv_n + lw_n)]. \qquad (7.22)$$

The relationship above, considered a *short approach*, often reduces the algebra by a considerable amount. Using this advancement, one initially identifies the structure (*lattice* and *basis*). This requires distinguishing each atom in the basis by its position vector $[uvw]$. Each atom of the basis contributes one atomic scattering factor. Set up the relationship for the lattice times the basis equation, and solve for the allowed reflections for the lattice first. The allowed reflections will contribute the number of lattice points per unit cell. Use any known relationships for allowed and absent reflections for that lattice. Simplify the equation and develop relationships that generate integer sums of the exponents. Finally, summarize results.

If we repeat the previous example for CsCl, the structure is defined as a simple cubic lattice plus a basis of Cs 000 and Cl $1/2\ 1/2\ 1/2$. Given the structure's symmetry, a basis of Cl 000 and Cs $1/2\ 1/2\ 1/2$ can also be used. A convenient relationship is that all reflections are allowed for any simple lattice (regardless of the crystal system). Using Eq. 7.12, we multiply the lattice times the basis and sum over all atom positions:

$$F_{hkl} = F_{hkl}{}^{\text{Lattice}} \cdot F_{hkl}{}^{\text{Basis}} = \Sigma\, f_n \exp[2\pi i(hu_n + kv_n + lw_n)]$$

$$= \overset{\text{All } h,k,l \text{ allowed}}{[1]} \cdot \left[f_{Cs}\, e^{(0^*2\pi i)} + f_{Cl}e^{(h/2+k/2+l/2)2\pi i} \right].$$

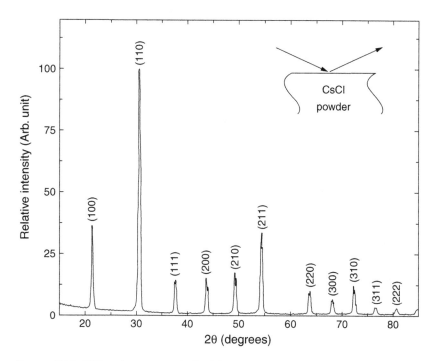

FIGURE 7.17. Diffraction pattern obtained from a CsCl sample using Cu$K\alpha$ radiation.

At this point, the only remaining steps are as those above—multiply the SFC by the complex conjugate and summarize the intensity results. Figure 7.17 shows an actual indexed X-ray diffraction pattern obtained from a CsCl sample using Cu $K\alpha$ radiation.

Problems

7.1. Use Fig. 7.17 to show that the lattice parameter of CsCl is 0.412 nm.
7.2. X-ray diffraction analysis is conducted on a pure iron sample using Mo $K\alpha$. radiation. The peak positions are listed below. Prove that the metal is indeed BCC and calculate the index pattern and determine the lattice parameter.

Peak	2θ (deg)
1	20.16
2	28.66
3	35.29
4	40.95
5	46.10

7.3. Show that the allowed FCC structure reflections correspond to $h\,k\,l$ being unmixed (i.e., all odd or all even) using structure factor calculations. Using the information obtained and Appendix 5, calculate the expected peak positions and quantitatively

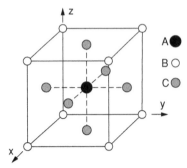

FIGURE 7.18. Problem 7.5.

approximate relative intensities when a piece of copper is analyzed using Cr $K\alpha$ radiation (see Appendix 6 for the wavelength values).

7.4. Use the allowed FCC structure reflections (the short notation) to determine the allowed reflections for diamond. (Hint: start with the lattice and basis.)

7.5. For the structure shown in Fig. 7.18:

a) For the crystal, determine the structure (i.e., lattice and basis) and give the chemical equation.

b) What are the symmetry elements along the $[\bar{1}\ 0\ \bar{1}]$, $[0\ 0\ 1]$, and $[1\ 1\ 1]$ directions?

c) Using structure factor calculations, derive simplified expressions for the relative intensity of each allowed reflection.

7.6. Determine the allowed reflections for HCP Zinc.

7.7. For the case when the atoms in the basis are positioned to the incident and reflected beams as shown in Fig. 7.19, using simple geometry to show that Bragg's Law is satisfied when $(AB + BC) = 2d_{hkl} \sin \theta$.

7.8. A single crystal of Si of Pd is heated during XRD analysis. The 111 reflection is monitored. From the data below, calculate the coefficient of thermal expansion and compare your data to the tabulated values.

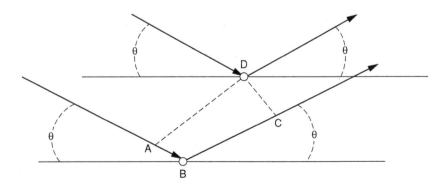

FIGURE 7.19. Problem 7.7.

C°	Δ2θ (deg)
25	0.000
125	−0.024
225	−0.048
325	−0.072

7.9. The data below are taken from an unknown BCC or FCC metal using Mo Ka radiation. Identify the materials and index the pattern.

line	2θ
1	19.57
2	22.63
3	32.22
4	37.98
5	39.73
6	46.21

7.10. Pole-figure analysis is conducted on a single-crystal Pd(0 0 1) layer on a Si(0 0 1) substrate. The layer's normal vector is parallel to the z-axis. Predicate the number of $\{2\,0\,0\}$ poles and calculate the angles between the each pole and the layer's normal. Also calculate the angle between the poles. How would these values differ if a Si(0 0 1) substrate is analyzed?

References

1. B. D. Cullity and S. R. Stocks, *Elements of X-ray Diffraction*, 3rd ed. (Prentice-Hall, New Jersey, 2001).
2. W. Borchardt-Ott, *Crystallography*, 2nd ed. (Springer-Verlag, Berlin, 1995).
3. B. E. Warren, *X-Ray Diffraction* (Dover Publications, Mineola, New York, 1990).
4. D. D. L. Chung, P. W. De Haven, H. Arnold, and D. Ghosh, *X-Ray Diffraction at Elevated Temperatures* (VCH Publishers, New York, 1993).

8
Electron Diffraction

8.1 Introduction

In this chapter, we consider thin film analysis using high-energy electron diffraction with an electron microscope. The most common form of the technique is known as *transmission electron microscopy* (TEM). As discussed in Chapter 7, diffraction reveals detailed information about the crystalline nature of a thin film. Electron microscopy has the critical advantage of providing a nanometer-scale electron beam spot as the radiation source. Thus crystallinity can be examined nanometer by nanometer in order to address questions such as crystal uniformity and to identify individual large defects and a variety of other structures. A simple example illustrates the importance of this parameter. Most metal films, when deposited, form as polycrystalline structures composed of nanoscale crystallites. Clearly, it is of interest to directly characterize this structure, to examine the crystallinity of individual crystallites, and to understand their connectivity. Such issues are addressed by transmission electron microscopy and the accompanying electron diffraction. This chapter presents the fundamental aspects of electron diffraction and its implementation in the transmission electron microscope.

In the case of crystalline solids, we typically desire to measure atomic spacings on the order of the lattice constants, e.g., 0.2–0.4 nm. Hence, we must use radiation with wavelengths less than 0.2 nm. This range of wavelengths corresponds to high-energy electrons. Here we recall from earlier sections that photons are massless particles. The photon's wavelength and energy are given by the relationship λ (μm) = 1.24 (eV-μm)/E (eV). Electrons are not massless particles; hence, the equation above does not yield the correct wavelength for high-energy electrons traveling at relativistic speeds. For such cases, the de Broglie (*particle–wave*) relation $mv = h/\lambda$ and the kinetic energy of the electrons, $\frac{1}{2} mv^2 = eV$, are used to correct the inverse relationship between applied voltage and wavelength:

$$\lambda(\text{nm}) = \left[\frac{0.139 \text{ volts-nm}^2}{V(\text{volts})} \right]^{1/2}, \tag{8.1}$$

where V is the applied voltage in volts and wavelength is in units of nm. Using

Eq. 8.1, we can calculate the wavelength of a 100 keV electron:

$$\lambda(nm) = \left[\frac{1.39 \text{ volts-nm}^2}{100,000 \text{ volts}} \right]^{1/2} = 3.2 \times 10^{-3} \text{ nm.}$$

Hence, the 100 keV electrons typically used in transmission electron microscopy have wavelengths that are appropriate for diffraction analysis of crystalline solids. Given that these wavelengths are two orders of magnitude smaller than typical X-ray wavelengths used for diffraction studies, we will use a different approach in the development of Bragg's Law.

Chapter 7 presents Bragg's Law as a simple law that has application in describing diffraction. This chapter develops Braggs Law in terms of vectors in reciprocal lattice space. Initial encounters with reciprocal space may seen abstract. However, reciprocal space is a powerful vehicle that also describes many other phenomena in solid-state science.

8.2 Reciprocal Space

Real (R) space point (Bravais) lattices are mathematical constructs that are described by specific translation vectors. In a similar manner, this section develops the concept of the reciprocal ($^1/_R$) lattice as a mathematical construct that is best understood in terms of vector algebra. At first glance, the notation and units may appear difficult; however, $^1/_R$ space is an extremely useful means to describe the behavior of electrons in crystals. In this case, the reciprocal space is called *k-space* or *momentum space*. Here in this chapter, we will use $^1/_R$-space relationships to describe the coherent scattering of particles and photons by crystalline solids. Our understanding of the $^1/_R$-lattice in combination with the Ewald sphere construction will simplify the description of electron diffraction in the transmission electron microscope.

Normal vectors can represent families of crystal planes in R-space. Consider a family of planes in a crystal (Fig. 8.1) in R-space. Normal vectors to these planes are drawn from a common origin, and these specify the orientation and spacings of the planes. For example, the (001) and (010) planes of the cubic system have normal vectors along the [001] and [010], respectively. The interplanar spacing d_{001} and d_{010} have a length

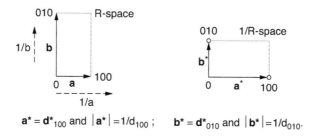

$$\mathbf{a}^* = \mathbf{d}^*_{100} \text{ and } |\mathbf{a}^*| = 1/d_{100} \text{ ; } \quad \mathbf{b}^* = \mathbf{d}^*_{010} \text{ and } |\mathbf{b}^*| = 1/d_{010}.$$

FIGURE 8.1. Schematic of the relationship between R-lattices and $^1/_R$-lattices.

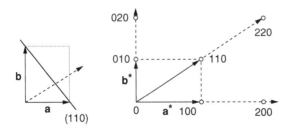

FIGURE 8.2. Schematic of the relationship between R-lattice planes and $^1/_R$-lattice points.

of a_0. In comparison, the (004) plane has a normal vector of [004] and its interplanar spacing d_{004} has a length of $^1/_4\, a_0$. Hence, for a specific plane, the interplanar spacing corresponds to the reciprocal of the length of the normal vector.

In reciprocal space, the reciprocal lattice vector $\mathbf{d}^*$ has the inverse magnitude of the corresponding interplanar spacing d_{hkl} for the specific plane. Reciprocal lattice vectors have dimensions of 1/length (nm^{-1}):

$$|\mathbf{d}^*_{hkl}| = \frac{1}{d_{hkl}}. \tag{8.2}$$

Using Eq. 8.2, we can calculate the magnitudes of the $\mathbf{d}^*_{004}$ and the $\mathbf{d}^*_{\bar{2}11}$ vectors for Si. Given that silicon's lattice parameter a_0 is 0.543 nm, and d_{hkl} is equal to $[h^2 + k^2 + l^2]^{1/2}a_0^{-1}$,

$$|\mathbf{d}^*_{004}| = 1/d_{004} = [0^2 + 0^2 + 4^2]^{1/2}\ (0.543\,\text{nm})^{-1} = 7.37\,\text{nm}^{-1},$$
$$|\mathbf{d}^*_{\bar{2}11}| = 1/d_{\bar{2}11} = [2^2 + 1^2 + 1^2]^{1/2}\ (0.543\,\text{nm})^{-1} = 4.51\,\text{nm}^{-1}.$$

The first axiom used to understand reciprocal lattices in this: *the $^1/_R$-space lattice is directly related to the R-space lattice.* For example, the $^1/_R$-space vector $\mathbf{d}^*$ is composed of reciprocal unit cell vectors $\mathbf{a}^*$, $\mathbf{b}^*$, and $\mathbf{c}^*$ and may be defined in terms of the real-space lattice unit cell vectors $\mathbf{a}$, $\mathbf{b}$, and $\mathbf{c}$. In Fig. 8.2, we draw the reciprocal lattice vectors in a section perpendicular to the z-axis (i.e., containing the $\mathbf{a}$ and $\mathbf{b}$ lattice vectors), from which we will define the reciprocal lattice unit cell vectors $\mathbf{a}^*$ and $\mathbf{b}^*$. This then enables us to express reciprocal lattice vectors in terms of their $^1/_R$-*space lattice* unit vectors $\mathbf{a}^*$, $\mathbf{b}^*$, and $\mathbf{c}^*$ or real lattice unit vectors. For the schematic in Fig. 8.2, the section of the lattice is taken from about the origin and perpendicular to the z-axis. A key point to note is that in the case of the cubic system, the R-space units vectors are parallel to the complementary $^1/_R$-space units vectors, i.e.,

$$\mathbf{a}\|\mathbf{a}^*,\ \mathbf{b}\|\mathbf{b}^*,\ \text{and}\ \mathbf{c}\|\mathbf{c}^*. \tag{8.3}$$

Figure 8-2 shows the (110) plane in R-space and the corresponding $^1/_R$-lattice vector $\mathbf{d}^*_{110}$, with the reciprocal lattice point 110 labeled. Notice the reciprocal relationships between the $\mathbf{d}^*$ lengths and the d_{hkl} spacings—the (002) plane has half the d-spacing of the (001) plane. The corresponding reciprocal lattice point 002 is twice the distance from the origin as the reciprocal lattice point 001. Likewise, the lattice point 003 is three times the lattice point 001 distance from the origin. Hence, the $^1/_R$-lattice points can be propagated throughout $^1/_R$-space.

The reciprocal lattice vectors can now be expressed in terms of their components of the reciprocal unit cell vectors $\mathbf{a}^*$, $\mathbf{b}^*$, and $\mathbf{c}^*$:

$$\mathbf{d}^*_{hkl} = h\mathbf{a}^* + k\mathbf{b}^* + l\mathbf{c}^*. \tag{8.4}$$

Note that the Miller indices hkl are the scalar components of a reciprocal lattice vector. Hence, a vector $\mathbf{d}^*_{hkl}$ drawn from the origin of the $^1/_R$-lattice to any point with coordinates hkl is perpendicular to the R-space plane (hkl); the normal vector $\mathbf{n}_{hkl} \| \mathbf{d}^*_{hkl}$. In the cubic system, any vector in $^1/_R$-space that is drawn from the origin to the point hkl is also perpendicular to the (hkl) plane in the real crystal. Each point in $^1/_R$-space represents a given set of parallel planes in R-space.

8.2.1 The Addition Rule

Just like vectors in R-space, $^1/_R$-space vectors add in a *tail-to-head* manner to generate the resultant vector. The addition rule simply states that the resultant vector is the sum of the indices of the two reciprocal lattice vectors:

$$\mathbf{d}^*_{h_1 k_1 l_1} + \mathbf{d}^*_{h_2 k_2 l_2} = \mathbf{d}^*_{h_1 + h_2\, k_1 + k_2\, l_1 + l_2}. \tag{8.5}$$

Using Eq. 8.5, we can develop several relationships for the reciprocal lattice vector $\mathbf{d}^*_{hkl} = [220]$:

$$\begin{aligned}
\mathbf{d}^*_{220} &= \mathbf{d}^*_{110} + \mathbf{d}^*_{110} = 2\mathbf{d}^*_{110} \\
&= \mathbf{d}^*_{200} + \mathbf{d}^*_{020} = 2\mathbf{d}^*_{100} + 2\mathbf{d}^*_{010} = 2(\mathbf{d}^*_{100} + \mathbf{d}^*_{010}) \\
&= \mathbf{d}^*_{110} + \mathbf{d}^*_{010} + \mathbf{d}^*_{100} \\
&= \mathbf{d}^*_{330} - \mathbf{d}^*_{010} - \mathbf{d}^*_{100}.
\end{aligned}$$

For the case of orthogonal axes (in orthorhombic, tetragonal, and cubic systems), a simple expression is obtained for $\mathbf{d}^*_{hkl} \cdot \mathbf{d}^*_{hkl}$:

$$\begin{aligned}
\mathbf{d}^*_{hkl} \cdot \mathbf{d}^*_{hkl} &= (h\mathbf{a}^* + k\mathbf{b}^* + l\mathbf{c}^*) \cdot (h\mathbf{a}^* + k\mathbf{b}^* + l\mathbf{c}^*) \\
&= h\mathbf{a}^* \cdot h\mathbf{a}^* + k\mathbf{b}^* \cdot k\mathbf{b}^* + l\mathbf{c}^* \cdot l\mathbf{c}^*. \tag{8.6}
\end{aligned}$$

For crystals where $\mathbf{a}^* \cdot \mathbf{b}^* = 0$ and $\mathbf{a}^* \cdot \mathbf{a}^* = 1$, etc.,

$$\begin{aligned}
\mathbf{d}^*_{hkl} \cdot \mathbf{d}^*_{hkl} &= \frac{h^2}{a^2} + \frac{k^2}{b^2} + \frac{l^2}{c^2} \\
&= \frac{1}{d^2_{hkl}}. \tag{8.7}
\end{aligned}$$

The angle γ between plane normal vectors from $(h_1 k_1 l_1)$ and $(h_2 k_2 l_2)$ planes or the angle γ between two vectors $\mathbf{a}$ and $\mathbf{b}$ is given by

$$\cos \gamma = \frac{\overline{\mathbf{a}} \cdot \overline{\mathbf{b}}}{|\overline{\mathbf{a}}||\overline{\mathbf{b}}|}. \tag{8.9a}$$

In a similar manner, the angle γ between two $1/R$-space points $h_1k_1l_1$ and $h_2k_2l_2$ is given by

$$\cos \gamma = \frac{\mathbf{d}^*_{h_1k_1l_1} \cdot \mathbf{d}^*_{h_2k_2l_2}}{|\mathbf{d}^*_{h_1k_1l_1}||\mathbf{d}^*_{h_2k_2l_2}|}. \tag{8.9b}$$

The volume (V) of the unit cell is given by $\mathbf{a} \cdot (\mathbf{b} \times \mathbf{c})$. The cross product ($\mathbf{b} \times \mathbf{c}$) yields a vector parallel to $\mathbf{a}^*$ and with a magnitude equal to the area of the face of the unit cell defined by $\mathbf{b}$ and $\mathbf{c}$; then

$$\mathbf{a}^* = \frac{\mathbf{b} \times \mathbf{c}}{\mathbf{a} \cdot (\mathbf{b} \times \mathbf{c})}, \quad \mathbf{b}^* = \frac{\mathbf{c} \times \mathbf{a}}{V}, \quad \mathbf{c}^* = \frac{\mathbf{a} \times \mathbf{b}}{V}, \tag{8.10}$$

In a similar manner, real space can be defined in terms of the reciprocal space unit vectors:

$$\mathbf{a} = V \cdot (\mathbf{b}^* \times \mathbf{a}^*), \quad \mathbf{b} = V \cdot (\mathbf{c}^* \times \mathbf{a}^*), \quad \mathbf{c} = V \cdot (\mathbf{a}^* \times \mathbf{b}^*). \tag{8.11}$$

8.2.2 Proof of the Zone Law

The intersection of planes $(h_1k_1l_1)$ and $(h_2k_2l_2)$ is a vector. Figure 8.3 depicts the zone axis (**ZA**) as the common intersection of several planes. For example, consider the case of planes $A(h_1k_1l_1)$ and $B(h_2k_2l_2)$ in the cubic system, having normal vectors $[h_1k_1l_1]$ and $[h_2k_2l_2]$, respectively. The vector **ZA** is defined by the relationship

$$\mathbf{ZA} = \mathbf{n_A} \times \mathbf{n_B} = [h_1k_1l_1] \times [h_2k_2l_2] \tag{8.12}$$

Given that the zone axis is $[u\ v\ w]$, then for any plane (hkl) that belongs to that **ZA** (or zone), the following relationship is satisfied:

$$hu + kv + lw = 0 \tag{8.13}$$

In order to belong to a zone, the normal of the plane must be perpendicular to the zone axis; hence, the Zone Law is given as

$$(u\ \mathbf{a} + v\ \mathbf{b} + w\ \mathbf{c}) \cdot \mathbf{n}_{hkl} = 0 \tag{8.14}$$

For example, the zone axis for the (011) and (101) planes is calculated by taking the cross product $[h_1k_1l_1] \times [h_2k_2l_2] = (+1)\mathbf{a} - (-1)\mathbf{b} + (-1)\mathbf{c} = [1\ 1\ \bar{1}]$. To verify that the calculated **ZA** is indeed correct, one uses the zone law ($\mathbf{ZA} \cdot \mathbf{n}_{hkl} = 0$) for validation. Here, $[0\ 1\ 1] \cdot [1\ 1\ \bar{1}] = 0$ and $[1\ 0\ 1] \cdot [1\ 1\ \bar{1}] = 0$. Hence, the calculated **ZA** is correct.

To address the zone law in $1/R$ space, we consider the condition for a $1/R$-lattice point with coordinates hkl with position vector $\mathbf{d}^*_{hkl}$ passing through the origin. For a **ZA**,

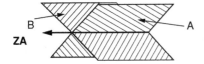

FIGURE 8.3. Schematic showing two intersecting planes and the corresponding zone axis (**ZA**).

vector $[uvw]$ in R-space is perpendicular to $\mathbf{d}^*_{hkl}$, and hence,

$$\mathbf{d}^*_{hkl} \cdot \mathbf{ZA} = 0,$$
$$(h\mathbf{a}^* + k\mathbf{b}^* + l\mathbf{c}^*) \cdot (u\,\overline{\mathbf{a}} + v\,\overline{\mathbf{b}} + w\,\overline{\mathbf{c}}) = 0. \tag{8.15}$$

Since $\mathbf{c}^*$ is perpendicular to both $\mathbf{a}$ and $\mathbf{b}$, the scalar (or dot) products are zero, i.e., $\mathbf{c}^* \cdot \mathbf{a} = 0$, $\mathbf{c}^* \cdot \mathbf{b} = 0$, and similarly for $\mathbf{a}^*$ and $\mathbf{b}^*$, i.e., $\mathbf{a}^* \cdot \mathbf{b} = 0$, $\mathbf{a}^* \cdot \mathbf{c} = 0$, $\mathbf{b}^* \cdot \mathbf{a} = 0$, $\mathbf{b}^* \cdot \mathbf{c} = 0$, so

$$hu + kv + lw = 0. \tag{8.16}$$

8.3 Laue Equations

The Laue Equations represents Laue's use of optical relationships to describe the diffraction of photons by crystalline structures. For simplicity, we consider a simple array of atoms in a crystalline structure with a one-atom basis. (From our previous discussion of Bragg's Law in R-space, the number of atoms in basis does not really matter as long as the basis resides on lattice sites (see Section 7). From this 3-D structure, let us only consider a row of atoms along the x-direction such that the incident radiation $\overline{s}_0$ and the scattered radiation $\overline{s}$ are at angles α_0 and α_n to the row. Note that Fig. 8.4 is somewhat misleading in that it only shows the diffracted beam at angle α_n *below* the atom row—but the same path difference is obtained if the diffracted beam lies in the plane of the paper at angle α_n above the atom row, as well as out of the plane of the paper at angle α_n to the atom row. The condition for constructive interference requires that the path difference $(\overline{AB} - \overline{CD})$ must be an integer multiple of the wavelengths (Fig. 8.4). The path difference between the incident beam and scattered beam:

$$\overline{AB} - \overline{CD} = a(\cos \alpha_n - \cos \alpha_0) = n_x \lambda. \tag{8.17}$$

Equation 8.17 can also be presented in vector form, where $\overline{\mathbf{a}}$ is the translation vector in the x-direction. The path difference of the projection along the x-direction may be represented by the scalar product $\mathbf{a} \cdot \mathbf{s} - \mathbf{a} \cdot \mathbf{s}_0 = \mathbf{a} \cdot (\mathbf{s} - \mathbf{s}_0)$:

$$\mathbf{s}_0 \cdot \mathbf{a} - \mathbf{s} \cdot \mathbf{a}(\cos \alpha_n - \cos \alpha_0) = n_x \lambda. \tag{8.18}$$

For a 3-D solid, we can develop similar relationships for the y- and z-directions:

$$b(\cos \beta_n - \cos \beta_0) = n_y \lambda,$$
$$c(\cos \gamma_n - \cos \gamma_0) = n_z \lambda. \tag{8.19}$$

The scalar forms of Eqs. 8.18 and 8.19 are referred to as the *Laue Equations* and are based solely on optical scattering theory (like that for the scattering of optical

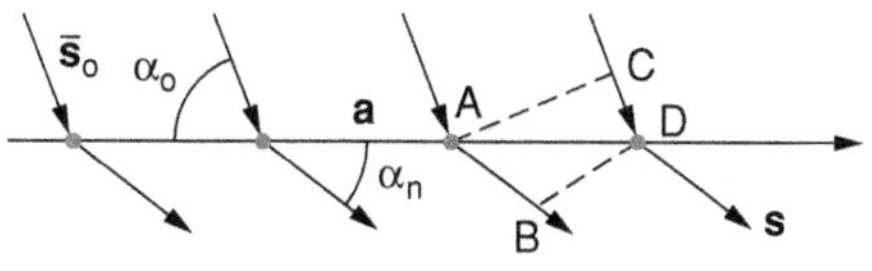

FIGURE 8.4. Schematic of scattering by an array of atoms.

light by slits). For constructive interference to occur simultaneously from all atoms in all three directions, all three Laue equations must be satisfied simultaneously. Each diffracted beam may be identified by three integers n_x, n_y, and n_z, which represent the order of diffraction from each of the atom rows. Hence, to determine or predict the condition for which constructive interference occurs simultaneously from all three rows, we need to determine all the constants. This would mean that for each diffraction peak obtained from our spectrum, we must solve three simultaneous equations with a total of twelve unknowns ($\alpha_n, \alpha_o, a, n_x, \beta_n, \beta_o, b, n_y, \gamma_n, \gamma_o, c, n_z$). Hence, Laue's optical approach has practical disadvantages when used to calculate the angles of the diffracted beams.

8.4   Bragg's Law

W. L. Bragg envisaged diffraction in terms of reflections from crystal planes initially using vector notation. Bragg invoked all the criteria for coherent scattering as prescribed by the senior Bragg (his father) and Laue. Let $\mathbf{s}_0/\lambda$ and $\mathbf{s}/\lambda$ be unit reciprocal space vectors along the directions of the incident and diffracted beams, respectively. Figure 8.5 shows geometrically the development of the Bragg's vector relationship. Initially, we establish that the vector $(\mathbf{s} - \mathbf{s}_0)/\lambda$ is parallel to the reciprocal lattice vector $\mathbf{d}^*_{hkl}$ for the specific (hkl) reflecting planes:

$$(\mathbf{s} - \mathbf{s}_0)/\lambda \parallel \mathbf{d}^*_{hkl} \parallel \overline{\mathbf{n}}_{hkl}.$$

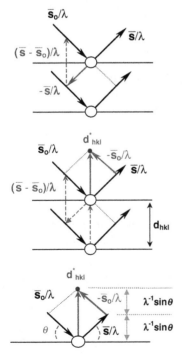

FIGURE 8.5. Development of Bragg's Law in vector form.

Comparing these vectors, $|\mathbf{s} - \mathbf{s}_0|/\lambda = (2/\lambda)\sin\vartheta$ and $|\mathbf{d}^*_{hkl}| = 1/d_{hkl}$. Hence, Bragg's Law may be written as

$$(\mathbf{s} - \mathbf{s}_0)/\lambda = \mathbf{d}^*_{hkl}, \qquad (8.20)$$
$$2(1/\lambda)\sin\vartheta = \left|\mathbf{d}^*_{hkl}\right| = 1/d_{hkl},$$
$$2\,d_{hkl}\sin\vartheta = \lambda.$$

Hence, constructive interference occurs, or Bragg's law is satisfied when the vector $|\mathbf{s} - \mathbf{s}_0|/\lambda$ coincides with the reciprocal lattice vector $\mathbf{d}^*_{hkl}$ of the reflecting planes. Another way of saying this is that when the vector $|\mathbf{s} - \mathbf{s}_0|/\lambda$ coincides with the reciprocal lattice point hkl, Bragg's Law is satisfied:

Bragg's Law: $\quad (\mathbf{s} - \mathbf{s}_0)/\lambda = \mathbf{d}^*_{hkl} = h\mathbf{a}^* + k\mathbf{b}^* + l\mathbf{c}^*.$ $\qquad (8.21)$

It is important to note that Bragg's law applies irrespective of the positions of the atoms in the planes; it is solely the spacing between the planes and the periodicity of the lattice that applies.

Bragg's development must be consistent with the Laue Equations to be meaningful. Upon inspection of Bragg's development, one realizes that the number of variables needed to calculate the directions of the diffracted beams is much less when compared with Laue's approach. Now let us apply Bragg's Law to the Laue Equations (Eqs. 8.18 and 8.19). Consider the case for scattering along the x-direction:

$$n_x\lambda = \mathbf{a}\cdot(\mathbf{s} - \mathbf{s}_0) = \mathbf{a}\cdot(d^*_{hkl}\,\lambda) = \mathbf{a}\cdot(h\mathbf{a}^* + k\mathbf{b}^* + l\mathbf{c}^*)\lambda = h\lambda. \qquad (8.22)$$

Similarly, we can consider the scattering along the y- and z- directions. Hence, the n's in the Laue Equations equate to hkl of the scattering plane: $n_x = h, n_y = k$, and $n_x = l$. Finally, application of relevant symmetry elements, the law of reflections, and structure factor calculations can further reduce the number of variables in the Laue Equations. For example, if the material is cubic, $\alpha_n, \alpha_0, \beta_n, \beta_0, \gamma_n, \gamma_0 = \theta; a, b, c, = a_0$; and $n_x, n_y, n_z = h, k, l$, respectively. Hence, Bragg's Law is satisfied when the vector $(\mathbf{s} - \mathbf{s}_0)/\lambda$ coincides with the $^1/_R$-lattice point hkl where the hkl are allowed reflections.

8.5 Ewald Sphere Synthesis

A useful means to understand the occurrence of diffraction is through use of a geometrical representation of Bragg's Law in reciprocal space. Ewald sphere (ES) synthesis begins by drawing a sphere of radius $1/\lambda$, where λ is the wavelength of the radiation used for coherent scattering. We imagine the real crystal placed at the center of the ES. Figure 8.6a shows the case for a crystal with the (hkl) planes at the correct Bragg angle. The center of the reflecting sphere is at a distance $1/\lambda$ from the origin of the reciprocal lattice and resides along the line of the incident beam. Note again that the origin of the reciprocal lattice is not at the center of the sphere but is at the point where the direct beam exits from the sphere. Bragg's law is satisfied when a reciprocal lattice point hkl lies exactly on the Ewald sphere.

In reciprocal space, the Ewald sphere (Fig. 8.6b) is drawn with radius $1/\lambda$ with the crystal at the center. The origin of the reciprocal lattice is fixed at O, and the reciprocal

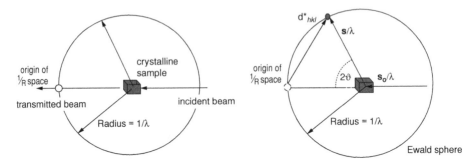

FIGURE 8.6. (a) Scattering conditions used for Ewald sphere synthesis. (b) Schematic of one reflecting plane and one reciprocal lattice point.

lattice point *hkl* intersects the sphere at the exit point of the diffracted beam. The incident radiation is directed toward the sample and is parallel to the incident radiation unit vector, s_0/λ. The scattered vector is along s/λ. The reciprocal lattice vector d^*_{hkl} is drawn from the origin to the point *hkl* in reciprocal space. Since Bragg's Law is satisfied when d^*_{hkl} equals $(s - s_0)/\lambda$, Bragg's Law is equivalent to the statement that the reciprocal lattice point for the reflecting planes (*hkl*) intersects the sphere. Conversely, if the reciprocal lattice point does not intersect the sphere, then Bragg's Law is not satisfied and no diffracted beams occur. It is a simple matter to extend Bragg's Law to all the reciprocal lattice points in a crystal.

Note the Ewald sphere development applies to any diffraction phenomena regardless of the wavelength of the radiation used. For the case of X-ray diffraction, the Ewald sphere is much larger compared with the size of the unit cell in reciprocal space and much smaller than the ES using typical electron wavelengths. This situation typically results in only one reciprocal lattice point satisfying Bragg's Law for a given value of 2θ. For the case of electron diffraction, the ES has a large radius of curvature near the origin of reciprocal space, and this allows the ES to pass through the origin and several other reciprocal lattice points. Consequently, several reciprocal lattice points satisfy Bragg's Law simultaneously.

8.6   The Electron Microscope

This section will acquaint the reader with the basic components and typical operation of a transmission electron microscope (TEM). The TEM allows the user to determine the internal structure of materials. Samples for the TEM must be specially prepared to thicknesses that allow electrons to be transmitted through the sample, as light is transmitted through materials in a conventional transmission optical microscope. Because the wavelength of electrons is much smaller than that of light, the optimal resolution attainable for TEM images is many orders of magnitude better than that from a light microscope. Thus, TEMs can reveal the finest details of internal structure—in some cases as small as individual atoms. Magnifications greater than 300K times are routinely obtained for many materials. Under optimal conditions, even individual atoms can be imaged. Because of the high spatial resolution obtained, TEMs are often employed to determine the detailed crystallography of fine-grained or thin film materials. The biological sciences often use TEM as a complementary tool to conventional crystallographic methods such as X-ray diffraction.

The size of the TEM and number of controls can make the microscope quite daunting. However, if one thinks of an electron microscope as a large light microscope, the TEM becomes less intimidating. For example, the light from a bulb shines (transmits) through the specimen on a glass slide. As the radiation passes through the specimen, it is influenced by the structure and composition of the specimen. This results in the incident light beam being transmitted through only certain parts of the slide, while being scattered or absorbed by other parts. The lenses project an enlarged image of the specimen onto a florescent viewing screen or photographic plate or digital camera.

Passage of electrons through the TEM column requires an ultra-high vacuum in the column. This vacuum level also minimizes the contamination that occurs when the beam interacts with the specimen. The sample and photographic plates are introduced into the high vacuum through separate evacuated load locks without breaking the vacuum.

The energy of the electrons in the TEM determines the relative degree of electron penetration into a specific sample, which in turn directly influences the thickness of material from which useful information may be obtained. At Cornell University, a 1000 kV TEM not only provides close to the highest resolution available but also allows for the observation of relatively thick samples (e.g., $\sim$200 nm to 1000 nm) when compared with the more conventional 100 kV or 200 kV instruments.

Figure 8.7 shows a cross-section schematic of the components and beam path of the typical TEM. The four basic components that are required to produce a magnified image are (1) an electron gun, which emits a beam of monochromatic electrons as the illumination source, (2) a set of condenser lenses to focus the illumination onto the specimen, (3) an objective lens used to form the first image of the specimen, and (4) a series of magnifying lenses to create the final magnified image.

The electron gun produces a beam of monochromatic electrons along the optical axis of the microscope. An intense electron beam is required for imaging at high magnification. The amount of electronic energy deposited in the beam spot causes sample damage, and ultimately limits the resolution. Electrons are emitted by heating a filament (*thermionic emission*, tungsten or LaB_6 filament) or from an unheated filament that has an extremely high potential gradient placed across the filament (*field emission*, fine-tipped single-crystal tungsten). The electron beam is focused into a thin, coherent beam with the use of the first and second condenser lenses. The first lens controls the beam-spot size and dictates the general size range of the final spot that illuminates the sample. The second lens controls the intensity and the size of the spot on the sample. A user-selectable, condenser aperture removes high-angle electrons (those far from the optic axis) and allows for a collimated beam down the optical axis.

The beam strikes the specimen, and a portion of the beam is transmitted through the sample and other parts of the beam are diffracted. The sample stage must allow for easy access to the specimen, mechanical stability, and reproducible translations. Specialty stages allow for excess cooling (liquid nitrogen temperatures) or heating ($\sim$500 °C) or increased tilting ($\pm$60°). State-of-the-art stages allow for in-situ nanoindentation, in-situ electrical probing, or scanning tunneling microscopy.

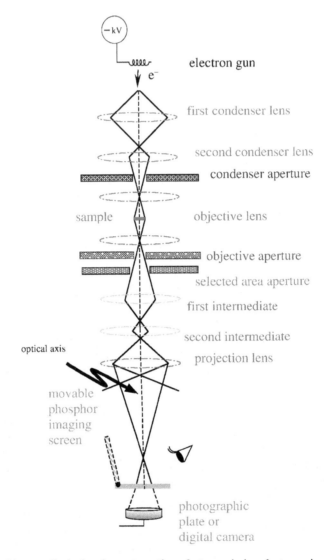

FIGURE 8.7. Diagram displaying the cross section of a transmission electron microscope and the path of the electron beam down the column of the TEM.

The transmitted portion of the beam is focused by the objective lens into an image. The electron image is focused by adjusting the objective focal length. Any defects associated with the objective lens are subsequently magnified onto the final image. Chromatic aberration results from variation in the electron velocity entering the specimen. Spherical aberration is due to variations in focal length as a function of radial position of the electrons in the beam. Such aberration is influenced by the effective aperture of the microscope. Lens-current fluctuations produce lens instabilities, which give unstable images.

The objective aperture and selected area diffraction aperture can both restrict the beam. The objective aperture enhances image contrast by blocking out high-angle diffracted electrons. Use of a selected-area aperture enables the user to image the electron diffraction pattern from a specific portion of the TEM sample. Accurate alignment of the various apertures is important for good imaging. The image is passed down the column through the intermediate and projector lenses. There the image of the sample is magnified onto the viewing screen or onto the recording media. Upon selection of the appropriate control, the projector system magnifies the image of the electron diffraction pattern from the rear focal plane of the objective lens. It is the projection lens setting that defines the camera length.

The electrons that form the image strike a fluorescent screen (typically coated with a fine-grain ZnS). When the electrons strike the screen, fluorescence occurs and results in the formation of a visible image, allowing the user to see the image. The darker areas of the image represent those areas of the sample where fewer electrons were transmitted (they are thicker or denser). The lighter areas of the image represent those areas of the sample where more electrons were transmitted (they are thinner or less dense). Note that X-rays are also generated when the electrons strike the viewing screen. The leaded viewing glass absorbs the X-rays. A photographic plate or digital camera resides beneath the moveable viewing screen.

A potential operational hazard can occur when the electron beam strikes the apertures or part of the column and produces X-rays. Hence, the column is shielded to protect the operator. On the other hand, the column must be shielded from external stray radiation, vibrations, and magnetic fields, if one desires a stable image.

8.6.1   Imaging Modes

In the bright field (BF) mode of the TEM, an aperture is placed in the back focal plane of the objective lens, allowing only the transmitted or direct beam to pass (see Fig. 8.8a). The diffracted beams are blocked. In this case, diffraction contrast contributes to image formation. Crystalline regions that are oriented so as to strongly diffract intensity away from the transmitted beam will appear dark. In addition, Z-contrast occurs when regions with heavy atoms scatter more of the incident beam and appear darker. Useful information is obtainable from BF images. However, interpretation of contrast should be done with care, since the phenomena mentioned above occur simultaneously.

In dark field (DF) images, a selected diffracted beam is allowed to pass through the objective aperture (see Fig. 8.8b). This is accomplished by electromagnetic lenses that effectively tilt the incident beam from the optical axis by an amount $2\theta_{hkl}$ such that the hkl diffracted beam exits the sample parallel to the TEM optical axis. The aperture blocks the direct beam. In contrast to the direct beam, the diffracted beam has interacted strongly with the specimen, and very useful information is often present in DF images, e.g., about planar defects, stacking faults, or particles. Figure 8.9 shows a bright field image and the corresponding dark field image for a TiAl(N,O) layer on an oxidized silicon substrate. In many analyses, correlation between DF and BF images provides useful information about structure and morphology, including layer thickness, grain size, grain orientation, and defect orientations. In general, one or both of these imaging modes are used in conjunction with electron diffraction analysis.

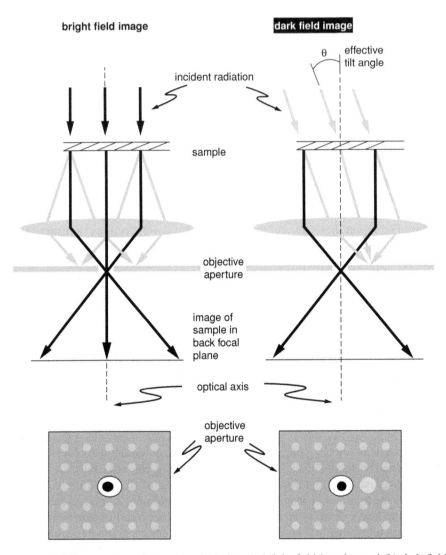

FIGURE 8.8. Diagram of an electron's path during (a) bright field imaging and (b) dark field imaging.

8.6.2   Selected Area Diffraction

Diffraction imaging is similar to BF and DF imaging—an image of the diffracted beams is brought into focus at the back focal plane of the objective lens. Once this real image is formed, it can be projected onto the viewing screen by the intermediate and projection lens system. In many cases, one desires to obtain crystallography information about a specific area of the sample or about a secondary phase present in the sample. For such cases, selected area diffraction (SAD) requires that an intermediate aperture is placed at the first intermediate image focal plane to specify the area from which the diffraction

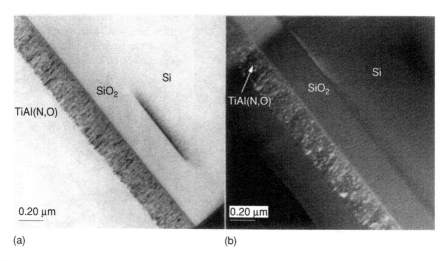

(a)   (b)

FIGURE 8.9. (a) Bright field and (b) dark field image of a TiAl(N,O) layer on an oxidized silicon substrate. (Courtesy N.D. Theodore.)

image is acquired. Through use of this method, crystallographic information is obtained from small features and from small volumes of materials such as precipitates.

Under diffraction conditions, a portion of the incident radiation makes the appropriate angle with a specific set of (*hkl*) planes such that Bragg's Law is satisfied (Fig. 8.10a). Given that the typical operation wavelength can vary between 0.37 and 0.87 pm, the diameter of the Ewald sphere (ES) is very large when compared with the size of the unit cell in reciprocal space. Hence, near the origin of reciprocal space, the curvature of the ES is so small that it is essentially a plane perpendicular to the direction of the incident

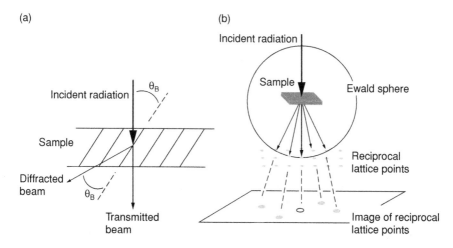

FIGURE 8.10. (a) View of the incident radiation making an angle of θ_B to a set of (*hkl*) planes and the corresponding diffracted radiation. (b) View of the same sample and under the same diffraction conditions showing the Ewald sphere and reciprocal lattice. Note that the diffraction pattern is the projection of the ES surface onto a flat surface.

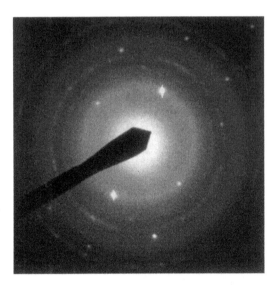

FIGURE 8.11. Selected area electron diffraction pattern from a TiAl(N,O) layer on an oxidized silicon substrate. (Courtesy N.D. Theodore.)

radiation. Hence, the ES passes through several points in reciprocal space that satisfy Bragg's Law. The function of the magnification lenses is to project the image that is in the back focal plane, and that image is a set of points on the surface of the Ewald sphere. The image in fact is a planar section of the reciprocal lattice, perpendicular to the incident radiation. A single-crystal sample produces *spot patterns* associated with a specific zone axis for that crystal (Fig. 8.10b). Under diffraction conditions, polycrystalline samples give rise to *ring patterns*, which are actually the superposition of many single-crystal patterns. An example of this is the SAD pattern from a TiAl(N,O) layer on an oxidized silicon substrate (Fig. 8.11). In general, correlation between DF and BF images can provide extensive information about structure and morphology, including layer thickness, grain size, grain orientation, and defect orientation. Typically, however, one or both of these imaging modes are used in conjunction with electron diffraction analysis.

8.7   Indexing Diffraction Patterns

8.7.1   Single-Crystal Diffraction Patterns

Single-crystal diffraction patterns consist of a series of bright spots, with the transmitted beam being the brightest and the others bright spots being the diffracted beams. The first step in the analysis of electron diffraction patterns is to measure the d_{hkl} values. Figure 8.12 shows a set of reflecting planes at the Bragg angle ϑ (much exaggerated) to the electron beam. The diffracted beam makes an angle of 2ϑ relative to the direct beam and falls on the screen at a distance R from the transmitted spot. Figure 8.12 does not represent the *actual* ray paths in the electron microscope, which are determined and controlled by the lens settings. However, the net effect of the lens settings is encompassed in the camera length (L). This variable is not the actual distance between

FIGURE 8.12. Schematic showing the camera constant relationship.

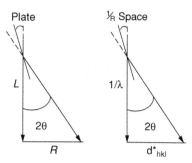

the specimen and screen, but is a projection length (or effective distance) between the sample and the photographic plate (and/or viewing screen). When diffraction spots appear, the Ewald sphere intersects the center spot (i.e., the origin of $^1/_R$ space) and also intersects the $^1/_R$-lattice vector $\mathbf{d}^*_{hkl}$, so Bragg's Law is satisfied.

When Bragg's Law (Eq. 7.1) is satisfied during ED, the wavelength is small ($\sim$ tens of picometers) and the Bragg angle is also small ($0.5°$). From Fig. 8.12, $\tan(2\theta) = \,^R/_L$. Since θ is small, $\tan(2\theta) \approx 2\theta$ (in radians); this gives us the relationship $2\theta = \,^R/_L$. Similarly for Bragg's Law, $\sin(2\theta) \approx \theta$ and hence $\lambda = 2\,d_{hkl}\,\theta$ or $\lambda/d_{hkl} = 2\theta$.

$$^R/_L = 2\theta = \,^\lambda/d_{hkl},$$

$$R\,d_{hkl} = L\,\lambda, \tag{8.23}$$

where λL is known as the *camera constant* and (like L) varies with the lens settings in the microscope. The units of the camera constant are typically expressed in units of mm-nm or cm-nm or m-nm, where R is the distance measured on the screen or photographic plate (e.g., mm or cm or m) and d_{hkl} (e.g., nm) is the interatomic spacing of the planes (*hkl*). The camera constant is a function of the lens settings. It is these lens settings that actually control the optical system's magnification, and the electron energy determines the value of the wavelength.

The accuracy of the calculated d_{hkl} is limited by uncertainty in the diffraction spot positions and of the camera constant. The diffraction spots are often diffuse, and hence the determination of the center position is an approximation. The camera constant is used to determine a known standard sample for a given electron energy setting. The resulting diffraction pattern is indexed, and the calculated camera constant is loaded into the microscope's memory.

Even with this calibration of the camera constant, small variations in lens settings from specimen to specimen and from day to day limit the accuracy of the camera constant to at best four significant figures. The camera constant will have the effect of changing the magnification of the ring patterns; however, a given pattern will have the symmetry of the **ZA** and will only differ slightly in scale with respect to the d_{hkl} values of the diffraction spots. Therefore, the one constant element from microscope to microscope is the symmetry of the resulting patterns for a single-crystal sample. Figure 8.13a is a diffraction pattern obtained from a piece of single-crystal Si using 100 keV electrons. To calibrate the camera constant and the camera length settings on the electron microscope, first draw diagonals from common indices (Fig. 8.13b). Use

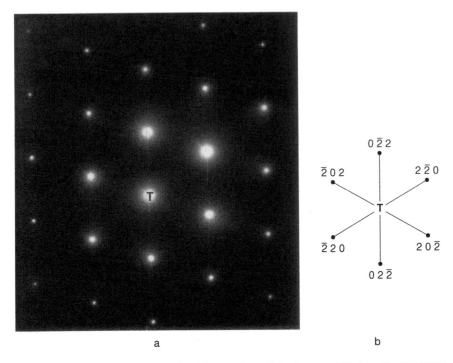

FIGURE 8.13. Diffraction pattern obtained from a piece of single-crystal Si along the [1 1 1] ZA; b) schematic of the indexed diffraction pattern. (Courtesy N.D. Theodore)

as many reflections as are available to increase resolution. For each set of reflections, calculate the average diameter (diagonal) value, and from this, determine the radius length. Next, calculate the d_{hkl} for the common reflection. Using Eq. 8.23, calculate the camera constant:

hkl	d_{hkl} (nm)	D_1 (cm)	D_2 (cm)	D_3 (cm)	$D_{average}$ (cm)	Radius (cm)	$R^* d_{hkl}$ (nm-cm)
022	0.192	3.7	3.8	3.7	3.7	1.9	0.357

Eq. 8.23 can be rearranged to give, $L = R d_{hkl}/\lambda$. Given the fact that 100 keV electrons have a wavelength of 0.0037 nm, the value of the camera length is 96.4 cm.

The diffraction spots may then be indexed by reference to tables of data such as those contained in the JCPDS Powder Diffraction File hkl, corresponding to the measured d_{hkl} values. In doing so, a further step is required: the indexing must be such that the addition rule is satisfied. A given set or family of planes (hkl) will consist of a number of variants of identical d_{hkl} values. For example, in the cubic system, there are six variants of the "plane of the form" (100), all of which have identical d_{hkl} values.

Indexing involves selection of the appropriate variants such that the addition rule (Eq. 8.5) is satisfied. Note that there are six spots with identical d_{hkl} values closest to the center spot, and these are all reflections from planes of the form {022}. The particular indices have been chosen such that the indices of the remaining spots are determined correctly, and the zone axis, [111], is found by cross-multiplication of any pair. The *choice* only needs to be made once, and once made, all the other spots in the pattern can be indexed consistently by repeated application of the addition rule, all leading to a self-consistently indexed diffraction pattern, depending on the number of equivalent variants of the zone axis.

When diffraction takes place through use of a low-index zone axis, the symmetry of the diffraction pattern allows indexing to be done by inspection. In this case, the distance R from the transmitted beam to a specific hkl point is proportional to the reciprocal of the lattice spacing of the corresponding $\{hkl\}$ planes in real space. Hence, the ratio of squares of two specific reflections $(R_1/R_2)^2$ equals the ratio of the squares of the interplanar spacing. Also, the angle between the points $h_1k_1l_1$ and $h_2k_2l_2$ must equal the angle between the planes $(hkl)_1$ and $(hkl)_2$ in real space. Using this technique, let us index the single-crystal Ni diffraction pattern shown in Fig. 8.14. Nickel has an FCC structure. Inspection of the pattern reveals the fourfold symmetry, and hence the **ZA** is of the type $\langle 001 \rangle$. There are two sets of reflections with the same distance R from the transmitted beam. The ratio of $(R_1/R_2)^2$ equals the ratio of $(b/a)^2 = (1.414)^2 = 2$.

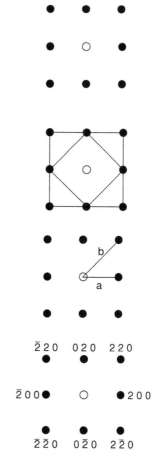

FIGURE 8.14. Diffraction pattern obtained from a single-crystal nickel sample using a [0 0 1] **ZA**.

This value of two is equal to the ratio of the squares of the interplanar spacing of two specific $\{hkl\}$ planes:

$$\left(\frac{R_b}{R_a}\right)^2 = 2 = \left(\frac{1\left/\sqrt{h_b^2 + k_b^2 + l_b^2}\right.}{1\left/\sqrt{h_a^2 + k_a^2 + l_a^2}\right.}\frac{a_o}{a_o}\right)^2 = \frac{\left(h_b^2 + k_b^2 + l_b^2\right)}{\left(h_a^2 + k_a^2 + l_a^2\right)}. \tag{8.24}$$

From the structure factor calculations in Section 7.7, the lowest order of allowed FCC reflections that satisfy Eq. 8.24 would be the $\{200\}$ and $\{220\}$ planes. Label the first $\{200\}$ and $\{220\}$ type reflections as (200) and (220), respectively. With the addition rule (Eq. 8.5), all the points can be indexed accordingly. To determine the **ZA**, one would take the cross-product of the two nonlinear vectors (corresponding to diffraction spots), yielding a value of [001]. The final step is to check for consistency; all diffraction spots must have the same **ZA**. Hence, using the Zone Law (Eq. 8.15), all points residing in that zone must satisfy the relationship $\mathbf{ZA} \cdot \mathbf{d}_{hkl}^* = 0$. If we use Eq. 8.9b, the angle between (200) and (220) equals 45°— the same value as the angle between the 200 and 220 reciprocal lattice points.

8.7.2   Polycrystalline Diffraction Patterns

For fine-grain polycrystalline materials, each grain has a **ZA** associated with it. Because λ is so small, many grains satisfy Bragg conditions. These have different orientations relative to the surface normal and the **ZA**. This difference in orientation corresponds to rotation about the origin in $^1/_R$ space and gives rise to ring patterns for polycrystalline materials. The simplest case is for BCC and FCC elemental metals. Since both are based on the cubic system, the camera constant equation can be rewritten as

$$L\lambda = R\frac{a_o}{\sqrt{h^2 + k^2 + l^2}}, \qquad R = \frac{L\lambda\sqrt{h^2 + k^2 + l^2}}{a_o}. \tag{8.25}$$

Each ring corresponds to a specific reflection, and the radii are easily measured from the plate. In a manner analogous to that done previously for X-ray spectra, we normalize the square of the radii relative to the square of the initial reflection. The relationship below shows how the normalized values of R squared correspond to the normalized sum of the squared h, k, and l values:

$$\left(\frac{R_i}{R_o}\right)^2 = \left(\frac{\dfrac{L\lambda\sqrt{h_i^2 + k_i^2 + l_i^2}}{a_o}}{\dfrac{L\lambda\sqrt{h_o^2 + k_o^2 + l_o^2}}{a_o}}\right)^2 = \frac{\left(h_i^2 + k_i^2 + l_i^2\right)}{\left(h_o^2 + k_o^2 + l_o^2\right)}. \tag{8.26}$$

For example, index the polycrystalline diffraction pattern shown in Fig. 8.15. Assume that the specimen is an elemental metal (with either an BCC or FCC structure). Given

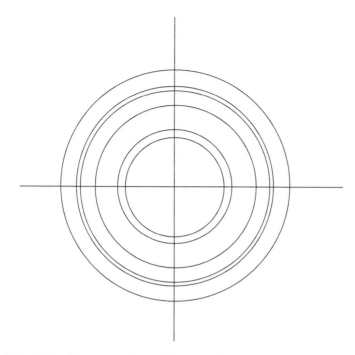

FIGURE 8.15. Diffraction pattern obtained from an unknown elemental BCC or FCC metal.

a camera constant of 0.629 cm-nm, Table 8.1 displays the approach used to identify the structure and to calculate the lattice. Initially for each ring, two perpendicular diameters are measured and the average is taken. This is done to minimize any stigmatism effects. From this value, the value of R— the distance from the transmitted spot —is obtained. The squared values of R are normalized relative to the initial value. Comparing these values to the normalized values in Table 7.1, one can identify the unknown elemental metal's structure as FCC. After indexing the pattern, the lattice parameter is calculated as 0.573 nm. Cullity's _Elements of X-ray Diffraction_ provides additional descriptions for the indexing of other crystalline structures.

TABLE 8.1. Approach used to index the polycrystalline ED example shown in Fig. 8.15.

	diameter 1	diameter 2	average	(cm)	R^2	R_i^2/R_o^2	$h^2 + k^2 + l^2$	(hkl)	(nm)
1	3.80	3.80	3.80	1.90	3.61	1.00	3	(1 1 1)	0.5734
2	4.40	4.35	4.38	2.19	4.79	1.33	4	(2 0 0)	0.5751
3	6.25	6.20	6.23	3.11	9.69	2.68	8	(2 2 0)	0.5716
4	7.28	7.30	7.29	3.65	13.29	3.68	11	(3 1 1)	0.5723
5	7.50	7.70	7.60	3.80	14.44	4.00	12	(2 2 2)	0.5734
6	8.80	8.80	8.80	4.40	19.36	5.36	16	(4 0 0)	0.5718
								avg =	**0.5729**

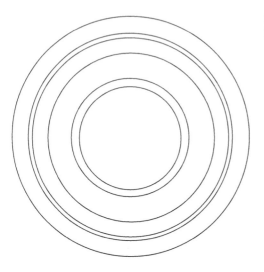

FIGURE 8.16. Diffraction pattern for Problem 8.2.

Problems

8.1. Calculate the wavelengths of 80 eV, 80 keV, and 1 MeV electrons.

8.2. Show that the electron diffraction pattern in Fig. 8.16 is from an FCC metal.

8.3. Give two examples of how TEM is used in the characterization of bulk materials. For your examples, what are the advantages and disadvantages of using TEM?

8.4. Show that the **ZA** has an orientation [0 0 1] for the diffraction pattern shown in Fig. 8.17. Use the rule of addition to index the diffraction pattern. By inspection, determine if the pattern is from an FCC or a BCC metal.

8.5. Describe how dark field images are generated and why their use is important in materials characterization. State any risk to image quality when using dark field imaging.

8.6. For the following statement, argue with supporting evidence whether the statement is true or false: "*One disadvantage of using the reciprocal lattice to determine which diffracted beams are possible is that the directions s and s_0 are not measurable; this is not the case for the scalar form of Bragg's Law.*"

8.7. Give two examples of how TEM is used in the characterization of bulk materials. For your examples, what are the advantages and disadvantages of using TEM?

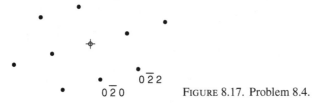

FIGURE 8.17. Problem 8.4.

8.8. Calculate the camera length for the diffraction pattern shown in Fig. 8.13. Note 100 keV electrons are used to obtain the pattern.

8.9. Calculate the camera constant for the diffraction pattern shown in Fig. 8.14.

References

1. E. M. Slayter and H. S. Slayter, *Light and Electron Microscopy* (Cambridge University Press, New York, 1992 and 1994).
2. G. Thomas and M. J. Goringe, *Transmission Electron Microscopy of Materials* (John Wiley and Sons, New York, 1979).
3. M. H. Loretto, *Electron Beam Analysis of Materials* (University Press, Cambridge, 1984).
4. T. Hahn, Ed., *International Tables For Crystallography*, 5th ed. (Kluwer Academic Publishers, Boston, 2002).
5. C. Hammond, *The Basics of Crystallography and Diffraction* (Oxford University Press, 2001).

9
Photon Absorption in Solids and EXAFS

9.1 Introduction

In surface and thin film analysis, the most important photon absorption process is the photoelectric effect. In this process, an incident photon of energy $\hbar\omega$ transfers all its energy to a bound electron in an atom. The energy E_e of the outgoing electron is

$$E_e = \hbar\omega - E_B,$$

where E_B is the binding energy of the electron in the atom. The binding energies of electrons are well known and distinct for each element, so the measurement of E_B for the atomic constituents in a solid represents a technique for materials analysis. In this chapter, we consider the binding energy of electrons in atoms and the cross section for photon absorption by a bound electron.

The photoelectric effect was instrumental in the early development of the quantum theory of matter. Today, it has been developed to a high degree of sophistication to yield not only an elemental analysis of materials but also a detailed description of the energies and momenta of electrons in solids.

The proper treatment of the photoelectric process requires a knowledge of the wave functions of the electrons within an atom. These wave functions result from solving the basic equation of quantum mechanics, the Schrödinger equation, which describes the properties of a quantum system through a wave equation. In this chapter, we review the Schrödinger equation and its solutions in order to arrive at a quantitative estimate of the cross section for the photo effect.

9.2 The Schrödinger Equation

The wave–particle duality of matter is expressed mathematically by the Schrödinger equation:

$$-\left[\frac{h^2}{2m}\nabla^2 + V(\mathbf{r})\right]\psi(\mathbf{r}, t) = i\hbar\frac{\partial\psi(\mathbf{r}, t)}{\partial t}, \tag{9.1}$$

where $\psi(\mathbf{r}, t)$ is the *wave function* that describes the motion of the particle under the influence of the potential $V(\mathbf{r})$. As in classical mechanics, a physics problem is *solved* when, under the influence of a given potential, the coordinates of the particle are expressed as a function of time. In quantum mechanics, a problem is solved when ψ is known as a function of $\mathbf{r}$ and t. There are very few interesting physical problems that can be solved exactly. Almost all processes that are necessary for materials analysis are approximate solutions to the Schrödinger equation using *perturbation theory*.

In Cartesian coordinates, the Schrödinger equation is written explicitly as

$$-\frac{\hbar^2}{2m}\left(\frac{\partial^2}{\partial x^2} + \frac{\partial^2}{\partial y^2} + \frac{\partial^2}{\partial z^2}\right)\psi + V(x, y, z)\psi = i\hbar\frac{\partial\psi}{\partial t}, \tag{9.2}$$

where ψ may be a function of x, y, z, and t.

For many examples it is sufficient to consider the one-dimensional Schrödinger equation

$$-\frac{\hbar^2}{2m}\frac{\partial^2\psi(x, t)}{\partial x^2} + V(x)\psi(x, t) = i\hbar\frac{\partial\psi(x, t)}{\partial t}. \tag{9.3}$$

The solution $\psi(x, t)$ can be written as the product of two functions:

$$\psi(x, t) = u(x)T(t). \tag{9.4}$$

This can be substituted into Eq. 9.3 to yield

$$\frac{1}{u}\left(-\frac{\hbar^2}{2m}\frac{\partial^2 u}{\partial x^2} + V(x)u\right) = \frac{i\hbar}{T}\frac{dT}{dt}. \tag{9.5}$$

Using the mathematical device of separation of variables, we note that the left-hand side depends only on x and the right-hand side only on t; therefore, both sides are proportional to a separation constant E. Then

$$T(t) = Ce^{-iEt/\hbar}, \tag{9.6}$$

where C is an arbitrary constant and the equation for $u(x)$ is

$$\left[-\frac{\hbar^2}{2m}\frac{\partial^2}{\partial x^2} + V(x)\right]u(x) = Eu(x). \tag{9.7}$$

The total solution is then

$$\psi(x, t) = Au(x)e^{-iEt/\hbar}, \tag{9.8}$$

where A is a normalization constant. Eq. 9.7 is known as the time-independent Schrödinger equation; the separation is valid provided that the potential V is not a function of time.

The philosophy underlying the Schrödinger equation is discussed in other books. Note, however, that $-(\hbar^2/2m)\cdot(\partial^2/\partial x^2)$ is associated with the kinetic energy, V with the potential energy, and Eq. 9.7 is often written in shorthand as

$$H\psi = E\psi, \tag{9.9}$$

where H (= kinetic energy + potential energy) is associated with the classical Hamiltonian. The meaning of $\psi(\mathbf{r}, t)$ is given through $|\psi(\mathbf{r}, t)|^2$, the probability of finding the particle at position $\mathbf{r}$ at time t. Eq. 9.7 is an example of an *eigenvalue equation*, where u is said to be the eigenfunctions of the operator $H = -(\hbar^2/2m) \cdot \nabla^2 + V$ and E is the eigenvalue. E is the energy of the system. The solution of Eq. (9.7) for many real potentials involves only discrete values of E, thus verifying the quantization assumptions in the Bohr theory given in Chapter 1.

In central force problems, the potential $V(x, y, z) = V(r)$, and we convert the Schrödinger equation to spherical coordinates (r, θ, ϕ):

$$-\frac{\hbar^2}{2m}\frac{1}{r^2}\frac{\partial}{\partial r}\left(r^2\frac{\partial u}{\partial r}\right) - \frac{\hbar^2}{2mr^2}\left[\frac{1}{\sin\theta}\frac{\partial}{\partial\theta}\left(\sin\theta\frac{\partial u}{\partial\theta}\right) + \frac{1}{\sin^2\theta}\frac{\partial^2 u}{\partial\phi^2}\right] + Vu = Eu. \tag{9.10}$$

Separation of variables is again possible,

$$u(r, \theta, \phi) = R(r)f(\theta)g(\phi), \tag{9.11}$$

and the radial part becomes

$$-\frac{\hbar^2}{2m}\frac{1}{r^2}\frac{d}{dr}\left(r^2\frac{dR}{dr}\right) + \left[\frac{l(l+1)\hbar^2}{2mr^2} + V(r)\right]R = ER, \tag{9.12}$$

where l is the *orbital* quantum number.

9.3   Wave Functions

There are two kinds of wave functions that we will need in order to illustrate the main points of this chapter: (1) a free particle wave function ($V = 0$) and (2) hydrogenic wave functions ($V = Z_1 Z_2 e^2/r$).

9.3.1   *Plane Waves* ($V = 0$)

Analysis techniques require an incident beam and outgoing radiation. The incident particle of energy E directed along the x-direction, say, is not under the influence of a potential. Then the solution to the appropriate Schrödinger equation,

$$-\frac{\hbar^2}{2m}\frac{d^2u}{dx^2} = Eu,$$

is

$$u(x) = Ae^{ikx}, \tag{9.13}$$

where

$$\frac{\hbar^2 k^2}{2m} = E. \tag{9.14}$$

This describes a particle moving in the positive x-direction with momentum $|\mathbf{p}| = \hbar k$. The total wave function $\psi(x, t)$ is given by

$$\psi(x, t) = Ae^{i(kx - \omega t)}, \tag{9.15}$$

where $E = \hbar\omega$. In both Eqs. (9.13) and (9.15), A is an arbitrary constant to be determined by the beam parameters.

9.3.2 Hydrogenic Wave Function [$V(r) = Z\,e^2/r$]

The solution to the Schrödinger equation in radial coordinates may be written as

$$u(r, \theta, \phi) = R(r)Y(\theta, \phi), \tag{9.16}$$

where $Y(\theta, \phi)$ is known as a *spherical harmonic*. The solution to the angular part of the differential equation results in quantum numbers, i.e., $Y(\theta, \phi)$ is not always a solution; it only satisfies physical conditions for certain integer values of the parameters l and m.

To see how this arises, consider the ϕ dependence of the Schrödinger equation:

$$\frac{d^2g}{d\phi^2} = -m^2g(\phi), \tag{9.17}$$

where m is a constant that arises in the separation of variables. Then

$$g(\phi) = Ae^{im\phi}, \tag{9.18}$$

but ϕ is the azimuthal angle, and the function must have the same value at $\phi = 0$ or $\phi = 2\pi$. This is true if

$$m = 0 \quad \text{or an integer.} \tag{9.19}$$

The separation of the three variables leads to three quantum numbers n, l, and m, and the solutions to the hydrogen atom problem are

$$u_{nlm}(r, \theta, \phi) = R_{nl}(r)f_{lm}(\theta)e^{im\phi}. \tag{9.20}$$

R_{nl} and f_{lm}, are functions that can be written explicitly for a particular value of n, l, and m. The first few wave functions are

$$u_{100} = \left[\frac{1}{\pi a^3}\right]^{1/2} e^{-\rho},$$

$$u_{200} = \frac{1}{8}\left[\frac{2}{\pi a^3}\right]^{1/2}(2 - \rho)e^{-\rho/2}, \tag{9.21}$$

where $\rho = Zr/a_0$ and $a = a_0/Z$. Table 9.1 lists the lowest-energy wave functions for the hydrogen atom, and Fig. 9.1 gives the radial distribution function in terms of $r^2 R_n^2$. This quantity gives the probability density of finding a particle within a shell of radius r. Figure 9.1 also shows the energy levels for the hydrogen atom; the same values are found in the Bohr treatment.

The above wave functions are properly normalized, i.e., the probability of finding a particle somewhere in space is unity. That is,

$$\int u^*u \, dr = 1, \tag{9.22}$$

where u^* is the complex conjugate of u.

TABLE 9.1. Hydrogenlike wave functions $u_{\min}(r, \theta\Phi)\rho = r/\sigma$ and $a = \alpha_0/Z$.

$n = 1, l = 0, m = 0$	$U_{100} = \left[\frac{1}{\pi a^3}\right]^{1/2} e^{-\rho}$
$n = 2, l = 0, m = 0$	$U_{200} = \frac{1}{8}\left[\frac{2}{\pi a^3}\right]^{1/2}(2 - \rho)e^{-\rho/2}$
$n = 2, l = 1, m = 0$	$U_{210} = \frac{1}{8}\left[\frac{2}{\pi a^3}\right]^{1/2}\rho e^{-\rho/2}\cos\theta$
$n = 2, l = 1, m = \pm 1$	$U_{211} = \frac{1}{8}\left[\frac{1}{\pi a^3}\right]^{1/2}\rho e^{-\rho/2}\sin\theta\, e^{i\phi}$
	$U_{21-1} = \frac{1}{8}\left[\frac{1}{\pi a^3}\right]^{1/2}\rho e^{-\rho/2}\sin\theta\, e^{-i\phi}$
$n = 2, l = 0, m = 0$	$U_{300} = \frac{1}{243}\left[\frac{2}{\pi a^3}\right]^{1/2}(27 - 18\rho + 2\rho^2)e^{-\rho/3}$
$n = 3, l = 1, m = 0$	$U_{310} = \frac{1}{81}\left[\frac{2}{\pi a^3}\right]^{1/2}\rho(6 - \rho)e^{-\rho/3}\cos\theta$
$n = 3, l = 1, m = \pm 1$	$U_{311} = \frac{1}{81}\left[\frac{1}{\pi a^3}\right]^{1/2}\rho(6 - \rho)e^{-\rho/3}\sin\theta e^{i\phi}$
	$U_{31-1} = \frac{1}{81}\left[\frac{1}{\pi a^3}\right]^{1/2}\rho(6 - \rho)e^{-\rho/3}\sin\theta e^{-i\phi}$
$n = 3, l = 2, m = 0$	$U_{320} = \frac{1}{486}\left[\frac{6}{\pi a^3}\right]^{1/2}\rho^2 e^{-\rho/3}(3\cos^2\theta - 1)$
$n = 3, l = 2, m = \pm 1$	$U_{321} = \frac{1}{81}\left[\frac{1}{\pi a^3}\right]^{1/2}\rho^2 e^{-\rho/3}\sin\theta\cos\theta e^{i\phi}$
	$U_{32-1} = \frac{1}{81}\left[\frac{1}{\pi a^3}\right]^{1/2}\rho^2 e^{-\rho/3}\sin\theta\cos\theta e^{-i\phi}$
$n = 3, l = 2, m = \pm 2$	$U_{322} = \frac{1}{162}\left[\frac{1}{\pi a^3}\right]^{1/2}\rho^2 e^{-\rho/3}\sin^2\theta e^{i2\phi}$
	$U_{32-2} = \frac{1}{162}\left[\frac{1}{\pi a^3}\right]^{1/2}\rho^2 e^{-\rho/3}\sin^2\theta e^{-i2\phi}$

Positions that satisfy the selection rule $\Delta l = \pm 1$.

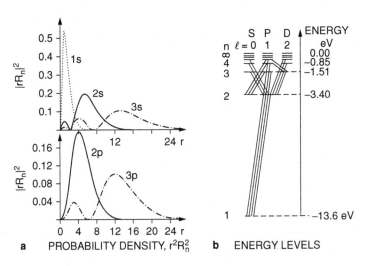

FIGURE 9.1. (a) Probability density for the hydrogenic wave function for different values of the quantum numbers n and l. (b) Energy levels in the hydrogen atom; the vertical lines represent transitions that satisfy the selection rule $\Delta l = \pm 1$.

9.4   Quantum Numbers, Electron Configuration, and Notation

Quantum mechanics requires the assignment of four quantum numbers to specify the state of the electron in an atom:

Principal quantum number	$n = 1, 2, \cdots$
Orbital quantum number	$l = 0, 1, 2, \ldots n - 1$
Magnetic quantum number	$m = 0, \pm 1, \pm 2, \ldots, \pm 1$
Spin quantum number	$m_s = +\,{}^1\!/_2, -\,{}^1\!/_2$

The internal angular momentum of the electron, the spin, leads to the fourth quantum number, m_s, which has values $\pm 1/2$. The existence of spin is responsible for the fine structure seen in high-resolution measurements of spectral lines. From the Pauli exclusion principle, only one electron can have a given set of quantum numbers; that is, no two electrons in an atom can have the same set of values of the quantum numbers $n, l, m,$ and m_s. Another set of quantum numbers can be assigned when the intersection between the orbital and spin is taken into account (i.e., the spin–orbit interaction).

An electron has both orbital angular momentum (the quantum number l) and spin angular momentum s. The resultant spin plus orbital angular momentum, $j = l + s$, has the value

$$A_j = \sqrt{j(j+1)}\hbar \ \ \text{with} \ j = |(1 \pm 1/2)|,$$

and the projection of the component around the polar axis has a quantized value of

$$(A_j)_z = m_j \hbar,$$

where m_j takes on integrally spaced values of $j, j - 1, \ldots, -j$. Thus for $j = 3/2, m_j = 3/2, 1/2, -1/2, -3/2$. In spectroscopic notation, the total angular momentum quantum number of an atomic state is written as a subscript, so a state with principal quantum number $2, l = 1,$ and $j = {}^1\!/_2$ is denoted as $2p_{1/2}$.

When spin–orbit splitting is considered, an appropriate set of quantum numbers is as follows:

Principal quantum number	$n = 1, 2, \cdots$
Orbital quantum number	$l = 0, 1, 2, \ldots, n - 1$
Angular momentum quantum number	$j = |1 \pm 1/2|$
Z-component of j quantum number	$m_j = j, j - 1, \ldots, -j$

Here the quantum numbers j and m_j are always half-integral.

The periodic table can be built by assigning electrons to quantum states characterized by the four quantum numbers $n, l, m,$ and m_s with no two electrons in any one atom having the same four quantum numbers. The Z electrons in the atom occupy the lowest energy states, with the energy primarily determined by the principal quantum number n and to a lesser degree by the orbital quantum number l with no appreciable difference due to different values of the spin substates m_s (spin–orbit splitting ignored). The filling of electron levels within an atom is often designated in terms of the principal quantum numbers, n, and the historical notation for angular momentum,

TABLE 9.2. Atomic versus ionic energy levels

		Atomic levels	Number of	Singly ionized atom
n	l	Electron shell	electrons	X-ray symbol
1	0	1s	2	K
2	0	2s	2	L_1
	1	2p	6	L_2
				L_3
3	0	3s	2	M_1
	1	3p	6	M_2
				M_3
	2	3d	10	M_4
				M_5
4	0	4s	2	N_1
	1	4p	6	N_2
				N_3
	2	4d	10	N_4
				N_5
	3	4f	14	N_6
				N_7
5	0	5s	2	O_1

where $l = 0$ corresponds to the letter s (for sharp), $l = 1$ to p (for principal), $l = 2$ to d (for diffuse), and $l = 3$ to f (fundamental). Thus He has an electronic configuration given by $1s^2$ (2 electrons in the $n = 1, l = 0$ shell) and neon has a configuration $1s^2, 2s^2, 2p^6$ (2 electrons in the $n = 1, l = 0$ shell; 2 electrons in the $n = 2, l = 0$ subshell and 6 electrons in the $n = 2, l = 1$ subshell). Table 9.2 lists the atomic levels, the electron shells, and X-ray symbols. When spin–orbit interactions are included, we find that the $p, d, \cdots$ shells can be further split (in energy) to give configuration of the sort 1s, 2s, $2p_{1/2}$, $2p_{3/2}$, where the subscript denotes the angular momentum $l \pm 1/2$, which is a result of summing the orbital angular momentum and spin angular momentum. The $2p_{3/2} - 2p_{1/2}$ splitting is approximately 1.5 eV for chlorine and is easily resolved in a standard XPS spectrometer. The splitting increases with increasing Z. Appendix 4 gives electron configurations of atoms along with ionization potentials (the energy to remove one electron from neutral atom), and Appendix 6 lists binding energies E_B.

9.5   Transition Probability

Analysis techniques involve one or more atomic transitions. Auger analysis involves the creation of a core hole, one atomic transition, and a subsequent Auger decay, the second atomic transition. Similarly, X-ray fluorescence (X-ray in, characteristic X-ray out) and the electron-microprobe (electron in, X-ray out) are examples of two transition processes. X-ray photoelectric spectroscopy is an example of a process that involves one atomic transition, the creation of an inner electron hole and an energetic photoelectron.

The most useful formula to calculate the probability of a transition comes from time-dependent perturbation theory. In first-order perturbation theory, the Hamiltonian acting on the system is written as

$$H = H_0 + H'. \tag{9.23}$$

H_0 is a Hamiltonian for which the Schrödinger equation can be solved, and H' contains an additional potential, i.e., an applied electric field.

There is a set of eigenfunctions that are the solutions to H_0 such that $H_0 u_n = E_n u_n$. As discussed in Section 9.10, the transition probability per unit time for a transition to the state k from the initial state m is given by

$$W = \frac{2\pi}{\hbar} \rho(E) |\langle \psi_k | H' | \psi_m \rangle|^2. \tag{9.24}$$

In this equation, $\rho(E)$ is the density of final states per unit energy,

$$\langle \psi_k | H' | \psi_m \rangle = \int \psi_k^* H' \psi \, d\tau = |H'_{km}|, \tag{9.25}$$

where ψ^* is the complex conjugate of ψ, $d\tau$ is a three-dimensional volume element (i.e., $r^2 dr \sin\theta \, d\theta \, d\phi$), and the wave function ψ_m is

$$\psi_m = e^{it\omega_m} u_m. \tag{9.26}$$

Eq. 9.24 is famous in quantum mechanics and is known as Fermi's Golden Rule. Its great advantage is that one need not know the wave functions for the true potential, $H_0 + H'$; rather, only the solutions to H_0 are required. Note that W has dimensions of (time)$^{-1}$:

$$W = \frac{1}{\text{energy} \cdot \text{time}} \cdot \frac{1}{\text{energy}} \cdot (\text{energy})^2 = \frac{1}{\text{time}}.$$

The derivation of Eq. 9.24 is given in Section 9.10.

9.6 Photoelectric Effect—Square-Well Approximation

The goal in this chapter is to make a quantitative calculation of the cross section for the photoeffect: i.e., an electron with binding energy E_B is irradiated with light of energy $\hbar\omega$ and is ejected with energy $\hbar\omega - E_B$. In this section, we calculate the relatively simple problem of the transition probability and then the cross section for an electron bound in a one-dimensional square well. In the next section, we consider the more realistic case of the photoeffect in a hydrogenic atom.

A flux of incident photons is represented by an electromagnetic field, and the perturbation is

$$H'(x, t) = H'(x)e^{i\omega t} = e\mathscr{E}xe^{-i\omega t}, \tag{9.27}$$

where the electric field $\mathscr{E}$ that acts on the particle is uniform in space (wavelength greater than atomic dimensions) but harmonic in time. This perturbation represents the potential energy of the electron in the field of a photon of frequency ω.

The transition rate between the initial and final state is given in Eq. 9.24 as

$$W = \frac{2\pi}{\hbar} \rho(E) \left| H'_{f_1} \right|^2,$$

where the initial state is the wave function ψ_i of a particle bound in the well and the final state is an outgoing one-dimensional plane wave, e^{-ikx}.

To calculate the density of final states $\rho(E)$, we consider the system to have dimension of length L and require periodic boundary conditions so that $\psi(x_0) = \psi(x_0 + L)$. The normalized states are

$$\psi_E = \frac{1}{\sqrt{L}} e^{ikx}, \tag{9.28}$$

where $kL = (2mE)^{1/2} L/\hbar = 2\pi N$ (i.e., periodic boundary conditions).

The density of states is the number of states with energy between E and $E + \Delta E$. Then

$$\rho(E)\Delta E = \Delta N = \frac{L}{2\pi} \Delta k,$$

$$\rho(E) = \frac{L}{2\pi} \cdot \frac{\Delta k}{\Delta E}. \tag{9.29}$$

For a free particle, $E = \hbar^2 k^2/2m$ and $\Delta E = (\hbar^2 k/m) \cdot \Delta k$, so

$$\rho(E) = \frac{L}{2\pi\hbar} \left(\frac{2m}{E} \right)^{1/2}, \tag{9.30}$$

where we have included a factor of 2 for positive and negative values of N. The matrix element in the transition probability is

$$H'_{f_1} = \int \frac{1}{\sqrt{L}} e^{ikx} e \mathscr{E} x \psi_i(x) \, dx, \tag{9.31}$$

where $\hbar k$ is the momentum of the final state.

In order to calculate the matrix element for a specific example, we consider the electron to be weakly bound in a narrow potential well. For the one-dimensional well, the wave function u_0 outside the well must satisfy

$$\frac{-h^2}{2m} \frac{d^2 u_0}{dx^2} = -E_B u_0,$$

which leads to exponentially decaying solutions of the form

$$u_0 = C e^{\pm k_i x},$$

where k_i is the momentum associated with the bound state, $k_i = (2mE_B)^{1/2}/\hbar$ and $C = \sqrt{k_i}$. For ease of calculation, we extrapolate the exterior wave function to the origin, as indicated in Fig. 9.2. Now, the normalized initial wave function ψ_i is

$$\psi_i(x) \cong \left[\sqrt{k_i} \right] e^{-k_i |x|}. \tag{9.32}$$

FIGURE 9.2. Schematic of the photoeffect from an electron bound in
a square well. The full curve shows
the approximate wave function used
in the calculation, while the dashed
line shows the true wave function in
the region between $x = \pm a$.

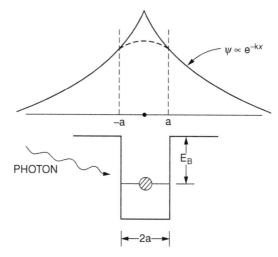

With this simplification, we have

$$\langle \psi_f | x | \psi_i \rangle = \frac{1}{\sqrt{L}} \int_{-\infty}^{\infty} e^{ikx} x \psi_i(x) \, dx,$$

$$= \frac{1}{\sqrt{L}} k_i^{1/2} \left[\frac{4ikk_i}{(k^2 + k_i^2)^2} \right]. \tag{9.33}$$

The transition rate is then

$$W = \frac{2\pi}{\hbar} \rho(E) |H'_{fi}|^2,$$

$$= \frac{4e^2 \mathcal{E}^2 \hbar}{m} \left(\frac{E_B}{E} \right)^{1/2} \frac{E E_B}{(E + E_B)^4},$$

where $\mathcal{E}$ is the external electric field. We consider the case that $\hbar\omega \gg E_B$ so that $E \cong \hbar\omega$. Then

$$W = \frac{4e^2 \mathcal{E}^2 \hbar}{m} \frac{E_B^{3/2}}{E^{7/2}}. \tag{9.34}$$

For analytical purposes, we are interested in the probability of the event per incident photon, i.e., the cross section for the process. This quantity σ is related to W by determining the flux F of photons in the oscillating electric field.

From classical electromagnetic theory, we have that the power density in the field is $c\mathcal{E}^2/2$. The power density is the energy/area-sec. Thus the flux F, the number of photons/area/s, is given by $c\mathcal{E}^2/2\hbar\omega$. Dimensionally, the cross section is

$$\sigma = \frac{W}{F} = \frac{\text{Transition probability/time}}{\text{No. of photons/area/time}},$$

or

$$\sigma = \frac{8e^2\hbar}{mc}\frac{E_B^{3/2}}{E^{5/2}}. \tag{9.35}$$

The cross section decreases with increasing photon energy ($E = \hbar\omega$) as $E^{5/2}$.

9.7 Photoelectric Transition Probability for a Hydrogenic Atom

In this section, we describe a calculation of the cross section for the photoelectric effect using hydrogenic wave functions in three-dimensional space. The formula for the transition probability is given by Eq. 9.24. The relevant wave functions for the initial and final state are given by

$$\psi_i = \frac{1}{\sqrt{\pi a^3}}e^{-\rho}$$

and

$$\psi_f = \frac{1}{\sqrt{V}}e^{ik \cdot r}, \tag{9.36}$$

where the initial state, ψ_i, describes a ground state hydrogenic wave function in an atom of atomic number Z and the final state is the usual outgoing plane wave of final energy $E_f = \hbar^2 k^2/2m$ normalized to a volume V. The binding energy of the electron, E_B, is expressed as $Z^2 e^2/2a_0$ (see Eq. 1.17), and in this calculation we assume the energy of the incoming photon, $\hbar\omega \gg E_B$. Here, the three-dimensional density of states, $\rho(E) = (V/2\pi^2)(2m/\hbar^2)^{3/2}E^{1/2}$, is used.

The transition probability can be calculated explicitly if the perturbation potential used is

$$H' = -e\mathscr{E}x e^{i\omega t}.$$

In that case, the final result for the photoeffect cross section σ_{ph} yields

$$\sigma_{ph} = \frac{288\pi}{3}\frac{e^2\hbar}{mc}\frac{E_B^{5/2}}{E^{7/2}},$$

a result similar to the square-well one-dimensional calculation carried out explicitly in the previous section.

Using a more sophisticated description of the perturbation potential, but precisely the same wave functions and assumption of $\hbar\omega \gg E_B$, Schiff (1968) shows that

$$\sigma_{ph} = \frac{128\pi}{3}\frac{e^2\hbar}{mc}\frac{E_B^{5/2}}{E^{7/2}}. \tag{9.37}$$

The value of $e^2\hbar/mc = 5.56 \times 10^{-4}$ eV-nm^2, so for convenience we write

$$\sigma_{ph} = \frac{7.52 \times 10^{-2}\text{ nm}^2}{\hbar\omega} \times \left(\frac{E_B}{\hbar\omega}\right)^{5/2}, \tag{9.37'}$$

by setting the incoming photon energy $\hbar\omega$ in eV equal to the energy E of the outgoing electron, since $E_B \ll \hbar\omega$ (i.e., $E = \hbar\omega - E_B \cong \hbar\omega$).

As an example, the photoelectric cross section for Fe K_α radiation ($\hbar\omega = 6.4 \times 10^3$ eV, Appendix 7) incident on K-shell electrons in Al ($E_B{}^K = 1.56 \times 10^3$ eV, Appendix 6) has a value

$$\sigma_{ph} = \frac{7.45}{6.4 \times 10^3} \times 0.01 \times \left(\frac{1.56}{8.4}\right)^{5/2} \text{nm}^2 = 3.4 \times 10^{-21} \text{ cm}^2.$$

For the total absorption, all electrons in all shells must be considered.

The cross section for impact ionization by electrons is given in Chapter 6 (Eq. 6.11) for $E > E_B$ as $\sigma_e = \pi(e^2)^2/E_B E$, where E is the energy of the incident electron. For the same conditions, $E = 6.4 \times 10^3$ eV and $E_B = 1.56 \times 10^3$ eV, the electron impact ionization cross section σ_e has a value

$$\sigma_e = \frac{\pi(1.44\,\text{eV-nm})^2}{6.4 \times 1.56 \times 10^6} = 6.5 \times 10^{-21} \text{ cm}^2,$$

which is a factor of two greater than that for the photoelectron cross section.

The electron impact cross section depends inversely on the energy of the incident particle while the photo-effect cross section is a strong function, $\sigma \propto (\hbar\omega)^{-7/2}$, of the incident photon energy for cases where $\hbar\omega \gg E_B$. Thus, in most cases, the values of σ_e are significantly greater than that of the photoelectric cross section. The primary advantage of using electrons as a method of creating inner-shell vacancies is not the increase in the cross section but rather that an electron beam can be obtained with orders-of-magnitude greater intensity than is possible with an X-ray source in a laboratory system. An electron beam can also be focused and scanned for the analysis of submicron regions.

9.8   X-ray Absorption

In the previous section, we have been concerned with photoelectric absorption. This is but one of three processes that lead to attenuation of a beam of high-energy photons penetrating a solid: photoelectron production, Compton scattering, and pair production. In the Compton effect, X-rays are scattered by the electrons of an absorbing material. The radiation consists of two components, one at the original wavelength λ and one at a longer wavelength (lower energy). The problem is generally treated as an elastic collision between a photon with momentum $p = h/\lambda$ and a stationary electron with rest energy mc^2. After scattering at an angle θ, the photon wavelength is shifted to larger values by an amount $\Delta\lambda = (h/mc)(1 - \cos\theta)$, where $h/mc = 0.00243$ nm is known as the Compton wavelength of the electron.

If the photon energy is greater than $2mc^2 = 1.02$ MeV, the photon can annihilate with the creation of an electron–positron pair. This process is called *pair production*. Each of the three processes—photoelectric, Compton scattering, and pair production—tend to dominate in a given region of photon energies, as shown in Fig. 9.3. For X-ray and low-energy gamma rays, photoelectric absorption makes the dominant contribution to the attenuation of the photons penetrating the material. It is this energy regime that is of primary concern for atomic processes in materials analysis.

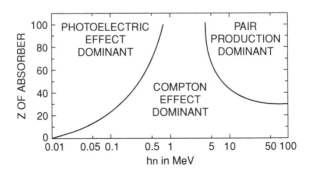

FIGURE 9.3. The relative importance of the three major types of photon interactions. The lines show the values of Z and $\hbar\omega$ for which the neighboring effects are equal.

The intensity I of X-rays transmitted through a thin foil of material for an incident intensity I_0 follows an exponential attenuation relation

$$I = I_0 e^{-\mu x} = I_0 \exp[-\mu/\rho]\rho x, \qquad (9.38)$$

where ρ is the density of the solid (g/cm^3), μ is the linear attenuation coefficient, and μ/ρ is the mass attenuation coefficient given in cm^2/g. Figure 9.4 shows the mass absorption coefficient in Ni as a function of X-ray wavelength. The strong energy dependence of the absorption coefficient follows from the energy dependence of the photoelectric cross section. At the K absorption edge, photons eject electrons from the K shell. At wavelengths longer than the K edge, absorption is dominated by the photoelectric process in the L shells; at shorter wavelengths where $\hbar\omega \gtrsim E_B(K)$, photoelectric absorption in the K shell dominates.

Both X-ray photoelectron spectroscopy (discussed in Chapter 9) and X-ray absorption depend on the photoelectric effect. The experimental arrangements are shown in the upper portion of Fig. 9.5 (XPS on the left side and X-ray absorption on the right side). In XPS, a bound electron such as the K-shell electron shown in Fig. 9.5 is promoted to a free state outside the sample. The kinetic energy of the photoelectron is well defined, and sharp photopeaks appear in the photoelectron spectrum. In X-ray absorption

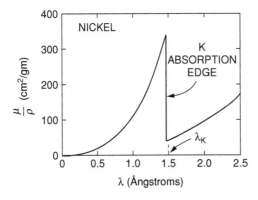

FIGURE 9.4. The mass absorption coefficient μ/ρ (cm^2/g) of Ni versus λ.

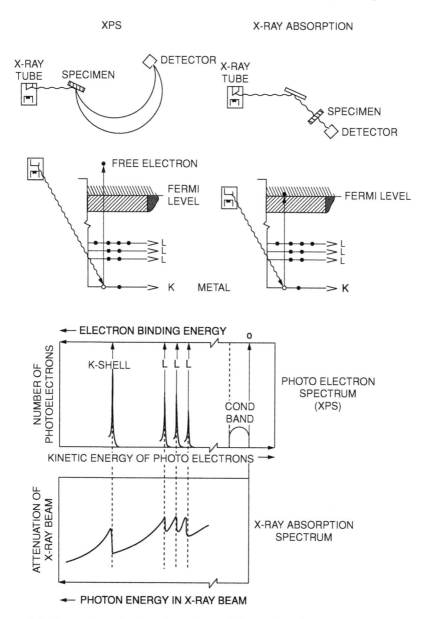

FIGURE 9.5. Comparison of X-ray absorption and X-ray photoelectron spectroscopy. [From Siegbathn et al., in *ESCA* (Almguist and Wiksells, Uppsala, Sweden, 1967).]

spectra, an edge occurs when a bound electron is promoted to the first unoccupied level allowed according to the selection rules. With metallic samples, this unoccupied level is at or just above the Fermi level. In X-ray absorption, the absorption is measured as a function of X-ray energy, whereas in XPS one irradiates with constant-energy photons and measures the kinetic energy of the electrons.

The mass absorption coefficient μ/ρ for electrons in a given shell or subshell can be calculated from the photoelectric cross section σ:

$$\frac{\mu}{\rho} = \frac{\sigma(\text{cm}^2/\text{electrons}) \times N(\text{atoms/cm}^3)^2 \cdot n_s(\text{electrons/shell})}{\rho(\text{g/cm}^3)}, \tag{9.39}$$

where ρ is the density, N the atomic concentration, and n_s the number of electrons in a shell. For example, for Mo K_α radiation ($\lambda = 0.0711$ nm and $\hbar\omega = 17.44$ keV) incident on Ni, which has a K-shell binding energy of 8.33 keV, the value of the photoelectric cross section per K electron is

$$\sigma_{\text{ph}} = \frac{7.45 \times 10^{-16} \text{ cm}^2}{17.44 \times 10^3} \left(\frac{8.33}{17.44}\right)^{5/2} = 6.7 \times 10^{-21} \text{ cm}^2.$$

The atomic density of Ni is 9.14×10^{22} atom/cm^3 and the density is 8.91 g/cm^3. The mass absorption coefficient μ/ρ for K-shell absorption ($n_s = 2$ for the K-shell) is

$$\frac{\mu}{\rho} = \frac{6.7 \times 10^{-21} \times 9.14 \times 10^{22} \cdot 2}{8.91} = 138 \text{ cm}^2/\text{g}.$$

In this calculation, the contribution of the L-shell electrons was neglected. For photon energies greater than the K-shell binding energy, the photoelectric cross section for the L-shell is at least an order of magnitude smaller than that of the K shell; of course, this is the major factor in the sharp increasing absorption when one crosses the K absorption edge. In the present case of MoK_α radiation on Ni, if we assume an average binding energy of 0.9 keV for the L_1, L_2, and L_3 shells, the photoelectric cross section per electron is a factor of 3.8×10^{-3} smaller than that of the K-shell electrons due to the $(E_B/\hbar\omega)^{5/2}$ term.

The calculated value, 138 cm^2/g, is greater than the measured value of 47.24 (Appendix 7). The major difficulty in the mass absorption calculation above was that the energy E of Mo K_α radiation is only twice that of the K-shell binding energy E_B, and the derivation of Eq. 9.37 was based on $\hbar\omega \gg E_B$. For Cu K_α radiation with $E = 8.04$ keV, the photon energy is about 10 times that of the L-shell binding energy, and the calculated photoelectric cross section ($\sigma = 3.1 \times 10^{-21}$ cm^2) for the L shell gives a value of $\mu/\rho = 32$ cm^2/g, a value close to the tabulated value of 48.8 cm^2/g.

Measured values of the mass absorption coefficient for different radiation are tabulated in Appendix 8 and displayed in Fig. 9.6a for $Z = 2 - 40$. For a given element, the absorption coefficient can vary over two orders of magnitude depending on the wavelength of the incident radiation. The strong photon energy dependence ($\hbar\omega^{-7/2}$) of the absorption coefficient is illustrated in Fig. 9.6b.

The tenfold change in the absorption coefficient on either side of the K edge, as shown in Fig. 9.4, represents a major change in transmitted intensity for thin foils because of the exponential nature of the transmission factor I/I_0. If the transmission factor of a particular sheet is 0.1 for a wavelength just longer than λ_K, then for a wavelength just shorter, the transmission is reduced by a factor of about $\exp(-10)$. This effect has been used to design filters for X-ray diffraction experiments that require nearly monochromatic radiation. As shown in Fig. 9.7a, the characteristic radiation from the K shell contains a strong K_α line and a weaker K_β line (the K_β/K_α emission ratio is discussed in Section 11.10). The K_β line intensity relative to that of the K_α line can be decreased by passing the beam through a filter made of material whose absorption

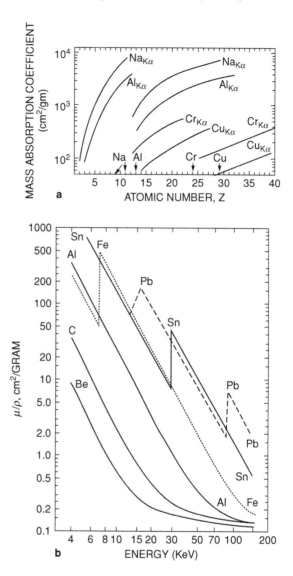

FIGURE 9.6. Mass absorption co-efficient (a) for elements from $Z = 2 - 40$ for K_α, radiation from a variety of sources, and (b) as a function of energy for different absorbers.

edge lies between the K_α and K_β wavelength of the target material. For metals with Z near 50, the filter will have an atomic number one less than that of the target. As shown in Fig. 9.7, a Ni filter has a strong effect on the ratio of the Cu K_α and K_β lines, where μ/p has a value of 48 for Cu K_α and 282 for K_β radiation.

9.9   Extended X-ray Absorption Fine Structure (EXAFS)

In the previous sections, the emphasis was upon the photoelectric cross section and absorption edges without consideration of the fine structure that is found at energies above the absorption edges. Figure 9.8 is a schematic representation of an X-ray absorption of

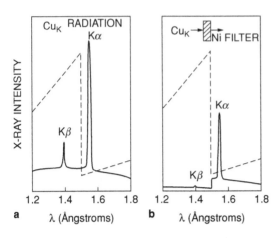

FIGURE 9.7. Comparison of the Cu radiation before and after passage through an Ni filter. The dashed line represents the mass absorption co-efficient of nickel. [Adapted from Cullity, 1978.]

μX versus the energy of the incident radiation plotted over an energy region extending about 1 keV above the K absorption edge. In this energy region, there are oscillations in absorption. The term *extended X-ray absorption fine structure* (EXAFS) refers to these oscillations, which may have a magnitude of about 10% of the absorption coefficient in the energy region above the edge. The oscillations arise from interference effects due to the scattering of the outgoing electron with nearby atoms. From analysis of the absorption spectrum for a given atom, one can assess the types and numbers of atoms surrounding the absorber. EXAFS is primarily sensitive to short-range order in that it probes out to about 0.6 nm in the immediate environment around each absorbing species. Synchrotron radiation is used in EXAFS measurements because it provides an intense beam of variable-energy, monoenergetic photons.

For an incoming photon of energy $\hbar\omega$, a photoelectron can be removed from a K shell of atom i and have a kinetic energy $\hbar\omega - E_B^K$. The outgoing electrons can be

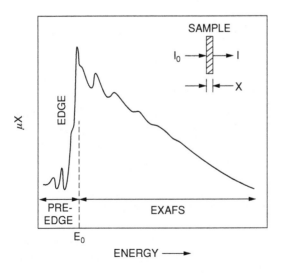

FIGURE 9.8. Schematic of the transmission experiment and the resulting X-ray absorption μx versus E for an atom in a solid.

FIGURE 9.9. Schematic of the EXAFS process illustrating an emitted e$^-$ scattering from a nearby atom at a distance R_j.

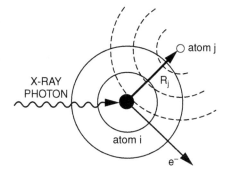

represented as a spherical wave (Fig. 9.9), which has a wave number $k = 2\pi/\lambda$ given by

$$k = \frac{p}{\hbar} = \frac{\sqrt{2m(\hbar\omega - E_B^K)}}{\hbar} \tag{9.40}$$

and a wave function y of the form

$$\psi = \frac{\psi_0 e^{i\mathbf{k}\cdot\mathbf{r}}}{r}. \tag{9.41}$$

Note that this is a different final state term than used in the calculation of σ_{ph}, since low-energy electrons are well represented by this spherical wave. When the outgoing wave from atom i arrives at atom j a distance R_j away, it can be scattered through 180° so that its wave function is

$$\psi_j = \frac{\psi_0 f e^{i\mathbf{k}\cdot\mathbf{R}_j + \phi_a}}{R_j};$$

where f is an atomic scattering factor and ϕ_a is a phase shift. When the scattered wave arrives back at atom i, it has a wave function ψ_{ij},

$$\psi_{ij} = \psi_0 \frac{f e^{i\mathbf{k}\cdot\mathbf{R}_j + \phi_s}}{R_j} \frac{e^{i\mathbf{k}\cdot\mathbf{R}_j + \phi_j}}{R_j};$$

that is, the outgoing photoelectron wave from atom i is backscattered with amplitude f from the neighboring atom, thereby producing an incoming electron wave. It is the interference between the outgoing and incoming waves that gives rise to the sinusoidal variation in the absorption coefficient.

The net amplitude of the wave at atom will be $\psi_0 + \psi_{ij}$,

$$\psi_0 + \psi_{ij} = \psi_0 \left(1 + \frac{f}{R_j^2} e^{i 2\mathbf{k}\cdot\mathbf{R}_j + \phi_i + \phi_a} \right),$$

and the intensity $I = \psi\psi^*$ will have the form

$$I = \psi_0 \psi_0^* \left(1 + \frac{2f \sin(2\mathbf{k}\cdot\mathbf{R}_j + \phi_{ij})}{R_j^2} + \text{higher terms} \right),$$

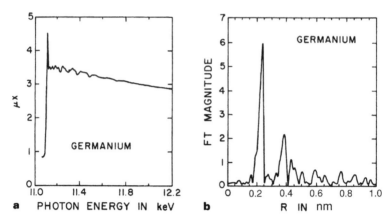

FIGURE 9.10. (a) X-ray absorption spectrum of crystalline Ge at a temperature of 100 K. The sharp rise near 11 keV is the K edge, and the modulation in μx above the edge is the EXAFS. (b) Fourier transform of (a) showing the nearest neighbor and second nearest neighbor distances.

where ϕ_{ij} represents the phase shifts. There are additional terms to account for the fact that the atoms have thermal vibrations and that the electrons which suffer inelastic losses in their path between atoms will not have the proper wave vector to contribute to the interference process. This latter factor is usually accounted for by use of an exponential damping term, $e^{(-2R_j/\lambda)}$, where λ is the electron mean free path. The damping term is responsible for the short-range order description, while the sinusoidal oscillation is a function of the interatomic distances $(2kR_j)$ and the phase shift (ϕ_{ij}). The important part of this equation is the term proportional to $\sin(2kR_j + \phi_{ij})$. By measuring $I(k)$ and taking a Fourier transform of the data with respect to k, one can extract R_j. The data and the transform are illustrated in Fig. 9.10.

The ability of EXAFS to determine the local structure around a specific atom has been used in the study of catalysts, multicomponent alloys, disordered and amorphous solids, and dilute impurities and atoms on a surface. In surface EXAFS (SEXAFS), the technique has been used to determine the location and bond length of absorbed atoms on clean single-crystal surfaces. EXAFS is an important tool in structural studies; the requirement for strong radiation sources leads to these types of experiments being carried out at synchrotron radiation facilities.

9.10 Time-Dependent Perturbation Theory

9.10.1 Fermi's Golden Rule

In this section, we give a brief treatment of perturbation theory, which leads to the basic formula for the transition probability of a quantum system. It is the formula that is the starting point for many of the derivations of cross sections given in this book.

Consider a system with a Hamiltonian H given by

$$H = H_0 + H',$$

$\qquad(9.42)$

where H_0 is a time-independent operator with an eigenvalue ψ_0. H_0 could be, for example, the Hamiltonian that describes a hydrogenic atom, while H' may be a time-dependent perturbation, i.e., an oscillating electric field. The wave function ψ_0 satisfies the Schrödinger equation such that

$$i\hbar\frac{\partial \psi_0}{\partial t} = H_0\psi_0. \tag{9.43}$$

Since H_0 is time independent, we can write

$$\psi_0 = u(x, y, z)e^{-iE_0t/\hbar} \tag{9.44}$$

or

$$\psi_0 = \sum_n a_n^0 u_n^0 e^{-iE_n^0 t/\hbar}, \tag{9.45}$$

where

$$H_0 u_n^0 = E_n^0 u_n^0, \tag{9.46}$$

with the u_n^0 being an orthonormal set of eigenvectors and the constants a_n^0 independent of time.

For the perturbed Hamiltonian, we write

$$H\psi = i\hbar\frac{\partial \psi}{\partial t} = (H_0 + H')\psi \tag{9.47}$$

and

$$\psi = \sum_n a_n(t)u_n^0 e^{-E_n^0 t/\hbar}, \tag{9.48}$$

where the coefficients a_n are now a function of time.

Substituting of (9.48) into (9.47), multiplying by the complex conjugate u_n^{0*}, and using the orthonormality relation yields

$$\frac{da_s}{dt} = \dot{a}_s = -\frac{i}{\hbar}\sum_n a_n(t)H_{sn}' e^{i(E_s^0 - E_n^0)t/\hbar}, \tag{9.49}$$

where

$$H'_{sn} = \int u_s^{0*} H' u_n^0 d\tau,$$

the integral being over all space. We approximate the solution by noting that if the perturbation is small, the time variation of $a_n(t)$ is slow; then $a_n(t) \cong a_n(0)$ and

$$a_s(t) - a_s(0) = -\frac{i}{\hbar}\sum_n a_n(0)\int_0^t H'_{sn}(t)e^{i\omega_{sn}t}dt, \tag{9.50}$$

where $\hbar\omega_{sn} = E_s^0 - E_n^0$.

A special case is if the system is in state n at $t = 0$; then $a_n(0) = 1$ and all other a's are zero. For the case $s \neq n$, Eq. 9.50 then gives

$$a_s(t) = \frac{i}{\hbar}\int_0^t H'_{sn}(t)e^{i\omega_{sn}t}dt, \tag{9.51}$$

The perturbation $H'(t)$ can induce transitions from the state n to any state s, and the probability of finding the system in the state s at time t is $|a_s(t)|^2$.

If H' is independent of time, then

$$a_s = -H'_{sn} \frac{(e^{i\omega_{sn}t} - 1)}{\hbar\omega_{sn}}, \tag{9.52}$$

and

$$|a_s(t)|^2 = 4\frac{H'^2_{sn}\sin^2(\omega_{sn}t/2)}{\hbar^2\omega^2_{sn}}, \tag{9.53}$$

and is valid if $|a_s(t)| < 1$.

In many applications, the result of a perturbation is a particle in the continuum, i.e., a free particle. Then an explicit final state is not appropriate, but the density of final states is relevant. We call $\rho(E)$ the density of final states (number of energy levels/per unit energy interval) and assume that H'_{sn} is the same for all final states.

The transition probability $P(t)$ is then given by

$$P(t) = \sum_s |a_s(t)|^2 = 4|H'_{sn}|^2 \sum_S \frac{\sin^2(\omega_{sn}t/2)}{\hbar^2\omega^2_{sn}}. \tag{9.54}$$

For a continuum, we transform the sum to an integral and note that the number of states in energy interval dE_s is $\rho(E_s)dE_s$; then

$$P(t) = 4|H'_{sn}|^2 \int_{-\infty}^{\infty} \rho(E_s)\frac{\sin^2(\omega_{sn}t/2)}{\hbar^2\omega^2_{sn}} d\hbar\omega_{sn}. \tag{9.55}$$

The major contribution of the integral comes from $\omega_{sn} = 0$ (it is a bit like a delta function), and noting that $\int \sin^2\alpha x/x^2\,dx = \pi\alpha$, we have

$$P(t) = 4|H'_{sn}|^2 \rho(E_n)\frac{\pi t}{2\hbar}, \tag{9.56}$$

so the rate of the transition, W, is given by

$$W = \frac{2\pi}{\hbar}\rho(E_n)|H'_{sn}|^2. \tag{9.57}$$

This is a famous and important formula denoted as the *Golden Rule* of Fermi.

9.10.2 Transition Probability in an Oscillating Electric Field

Another important application of perturbation theory is when the perturbing field has a time dependence of the form $e^{i\omega t}$. Consider for example an oscillating electric field directed along the x-axis such that

$$\mathscr{E} = \mathscr{E}\cos\omega t = \frac{\mathscr{E}_0}{2}(e^{i\omega t} + e^{-i\omega t}) \tag{9.58}$$

and

$$H'(t) = \frac{e}{2}\mathscr{E}_0 x(e^{i\omega t} + e^{-i\omega t}). \tag{9.59}$$

Inserting $H'(t)$ into (9.51) gives

$$a_s(t) = \frac{i}{\hbar} e^{\frac{\mathscr{E}_0 x_{sn}}{2}} \int_0^t (e^{i(\omega_{sn}+\omega)t} + e^{-i(\omega_{sn}-\omega)t})\, dt. \tag{9.60}$$

If $\omega_{sn} \neq \omega$, the integral averages to zero. If $\omega_{sn} \cong \omega$, we have

$$|a_s(t)|^2 = \frac{e^2 \mathscr{E}_0^2}{\hbar^2} |x_{sn}|^2 \frac{\sin^2[(t/2)(\omega - \omega_{sn})]}{(\omega - \omega_{sn})^2}, \tag{9.61}$$

where $|x_{sn}| = \int u_s^* x u_n \, d\tau$. We recognize that $\mathscr{E}_0^2/2$ is the energy density in the field, which can be represented by $\rho(\omega)$ to obtain a transition probability:

$$T_{ns} = |a_s(t)|^2 = 2e^2 |x_{sn}|^2 \int_0^\infty \frac{\sin^2[(t/2)(\omega - \omega_{sn})]}{\hbar^2 (\omega - \omega_{sn})} \rho(\omega)\, d\omega. \tag{9.62}$$

Assume that the distribution $\rho(\omega)$ varies much more slowly than the sharply peaked function by which it is multiplied. Then $\rho(\omega)$ is nearly constant over the small range of values of ω for which the integrand is nonzero, so we can replace $\rho(\omega)$ by its value at $\omega = \omega_{sn}$, and remove it from the integral, with no loss of accuracy. With the further substitution $z = \frac{1}{2}(\omega - \omega_{sn})t'$, the expression becomes

$$|a_{sn}(t')|^2 = \frac{e^2 \rho(\omega_{sn}) |x_{sn}|^2 t'}{\hbar^2} \int_{-\infty}^\infty \frac{\sin^2 z}{z^2}\, dz,$$

or, since

$$\int_{-\infty}^\infty \frac{\sin^2 z}{z^2}\, dz = \pi,$$

the transition probability is

$$T_{sn} = |a_{sn}(t')|^2 = \frac{\pi e^2 \rho(\omega_{sn}) |x_{sn}|^2 t'}{\hbar^2}, \tag{9.63}$$

where $x_{sn} = \int u_s^* x u_n \, d\tau$. Eq. 9.63 holds for radiation that is polarized along the x-axis. In the general case, when radiation is incident upon the atom from all directions and with random polarization, T_{sn} must include equal contributions from x_{sn}, y_{sn}, and z_{sn}, so

$$T_{sn} = \frac{\pi e^2 \rho(\omega_{sn}) t'}{3\hbar^2} |\langle \psi_s | \mathbf{r} | \psi_n \rangle|^2, \tag{9.64}$$

where the factor of 3 has been introduced into the denominator because each polarization direction is assumed to contribute one-third of the intensity and $|x_n|^2 + |y_{sn}|^2 + |z_{sn}|^2 = |\langle \psi_s | \mathbf{r} | \psi_n \rangle|^2 = |r_{sn}|^2 \cdot \langle \psi_s | \mathbf{r} | \psi_n \rangle$ is known as the dipole matrix element and Eq. 9.64 is the dipole approximation to the transition probability.

9.10.3   Spontaneous Transitions

Spontaneous transitions are those that occur in the absence of an external field. An example is the transition from an excited state to a ground state. To calculate such phenomena, we rely on a treatment developed in 1917 by Einstein, which enables one to calculate the rate of spontaneous transitions from knowledge of the rate of induced transitions calculated in Section 9.10.2.

Consider a collection of atoms in thermal equilibrium; each atom must be emitting and absorbing radiation at the same rate. Let P_{ns} be the probability that a given atom will go from the nth state to the sth state in a short time dt. This *probability*, P_{ns}, must be proportional to the *probability* P_n that the atom is in the nth state to begin with, multiplied by the transition probability to go from state n to state s, T_{ns} (given by Eq. 9.64):

$$P_{ns} = T_{ns} P_n.$$

In view of Eq. 9.64, we may write this expression as

$$P_{ns} = A_{ns} \, \rho(\omega_{ns}) P_n \, dt, \tag{9.65}$$

where $\rho(\omega_{ns})$ is defined as before, dt takes the place of t' as the time interval, and A_{ns} is equal to all the other factors in Eq. 9.64.

The probability of a downward transition from state s to state n may be written as

$$P_{sn} = A_{sn} \rho(\omega_{sn}) P_s \, dt,$$

but because of the symmetry of the equations leading to Eq. 9.64, we know that $A_{sn} = A_{ns}$ and $\omega_{sn} = \omega_{ns}$, so

$$P_{sn} = A_{sn} \rho(\omega_{ns}) P_s \, dt. \tag{9.66}$$

Notice that P_{ns} does not equal P_{sn}, because $P_n > P_s$; the nth state, being lower in energy, is more heavily populated, according to the Boltzmann factor.

The system is in equilibrium, so the total number of transitions from n to s must equal the total number of transitions from s to n. Since the induced transition probabilities are unequal ($P_{ns} \neq P_{sn}$), there must be additional transitions from state s to state n that are spontaneous. The spontaneous transition probability by definition does not depend on the energy density of the externally applied field; this probability may be written $B_{sn} P_s \, dt$, where B_{sn} is the spontaneous transition probability. The coefficients A and B are referred to as the *Einstein coefficients*.

The total transition probability from s to n is therefore equal to $P_{sn} + B_{sn} P_s \, dt$, and this total should equal the transition probability P_{ns}. Thus, from Eqs. 9.65 and 9.66, we have

$$\rho(\omega_{ns}) P_n A_{ns} = \rho(\omega_{ns}) P_n A_{ns} + P_s B_{sn}$$

or

$$B_{sn} = \rho(\omega_{ns}) A_{ns} \left[\frac{P_n}{P_s} - 1 \right].$$

But the population of a state of energy E is proportional to the Boltzmann factor $e^{-E/kT}$, so the ratio P_n/P_s may be written

$$\frac{P_n}{P_s} = e^{(E_s - E_n)/kT}$$

$$= e^{\hbar \omega_m/kT}.$$

Thus

$$B_{sn} = \rho(\omega_{ns}) A_{ns} \left(e^{\hbar \omega_m/kT} - 1 \right). \tag{9.67}$$

The factor $\rho(\omega_{ns})$ in Eq. 9.67 is misleading because B_{sn} by definition does not depend on the energy density of the radiation field. We can eliminate this factor by using an expression for the energy density inside the cavity. Planck's law for thermal radiation gives

$$\rho(\omega) = \frac{\hbar \omega^3}{\pi^2 c^2 \left(e^{\hbar \omega_m/kT} - 1 \right)}. \tag{9.68}$$

Substitution of this expression into Eq. 9.67 yields an expression for B_{sn} involving only A_{ns} and known constants:

$$B_{sn} = \frac{\hbar \omega_{ns}^3}{\pi^2 c^3} A_{ns}. \tag{9.69}$$

The transition rate for a spontaneous transition from a filled to an empty state is

$$W = \frac{4}{3} \frac{e^2}{\hbar} \left[\frac{\omega_{ns}}{c} \right]^3 |\langle \psi_s | \mathbf{r} | \psi_n \rangle|^2, \tag{9.70}$$

where $e^2 = 1.44$ eV-nm. This can be expressed as

$$W = 0.38 \times 10^{14} (\hbar \omega)^3 \cdot |\langle \psi_s | \mathbf{r} | \psi_n \rangle|^2,$$

where $\hbar \omega$ is in units of keV and the matrix element has dimensions of $(0.01 \text{ nm})^2$. A typical value of W is 10^{15}/s for elements in the middle of the periodic table.

In summary,

$$W = \frac{2\pi}{\hbar} |H'_{sn}|^2 \rho(E_n); \text{ static perturbation [Eqs. 9.24 and 9.57]}$$

$$W = \frac{\pi e^2 \rho(\omega_{sn})}{3\hbar^2} |\mathbf{r}_{sn}|^2; \text{ time dependent, } H' = \mathscr{E}_0 \cos \omega t \text{ [Eq. 9.64]}$$

$$W = \frac{4}{3} \frac{e^2}{\hbar} \left[\frac{\omega_{sn}}{c} \right]^3 |\mathbf{r}_{sn}|^2; \text{ spontaneous transitions [Eq. 9.70]}$$

Problems

9.1. For Cu K_a radiation (E = 8.04 keV) incident on Al:

(a) Calculate the photoelectric cross section σ_{ph} for the K shell of Al and compare its value with the electron impact ionization cross section σ_e for 8.04 keV electrons.

(b) Calculate the mass absorption coefficient for Al based on only K-shell absorption and compare with the values tabulated in Appendix 8. The Al L shells have 8 electrons compared to 2 in the Al K shell.

(c) Estimate the contribution of L-shell electrons to the mass absorption coefficient for 8.04 keV electrons.

9.2. Consider 5.41 keV radiation [photon (Cr K_α) and electron] incident on Si. Compare the values in microns of the linear absorption coefficient, the electron range, and the electron mean free path.

9.3. You use an Ni filter to attenuate the Cu K_β radiation from a Cu X-ray source. If the Ni filter thickness is sufficient to a attenuate the K_B radiation by a factor of 1000, how much will the K_α radiation be attenuated. How thick is the Ni filter? Estimate to zero order how much the Cu L_α radiation is attenuated.

9.4. Be is used as a "window" material that allows X-rays to enter the X-ray detector with minimum attenuation. The absorption of X-rays from elements with $Z < 10$ is one of the limitations of microprobe analysis. Estimate the absorption of carbon K_α X-rays by a 7-μm thick Be window.

9.5. Consider a photon of energy $\hbar\omega$ inducing a photoeffect process with an electron of binding energy E_B bound in an atom of mass M_n. Assume that the electron is emitted in the forward direction, i.e., along the same direction as the incident photon. In the approximation that $\hbar\omega \gg E_B$, show that the energy of the recoiling atom, E, is given by

$$E_r = \frac{(\hbar\omega)^2}{2M_n c^2} + \frac{m_e}{M_n}\hbar\omega = \frac{\hbar\omega}{M_n c^2}\sqrt{2M_e c^2 \hbar\omega}$$

where m_e is the electron mass. Evaluate this expression for $\hbar\omega = 1$ keV, 100 keV, and $M_n = 28$ (Silicon) and compare the result to 14 eV, the binding energy of Si in the Si lattice. Such considerations are important in determining the *destructiveness* of a given analysis technique.

9.6. Using conservation of momentum and energy, show that the photoeffect cannot occur with a free electron. Consider the nonrelativistic case, $\hbar\omega < m_e c^2$. The photoeffect occurs with photon irradiation of solids, since all electrons in solids are bound to some degree.

References

1. E. E. Anderson, *Modern Physics and Quantum Mechanics* (W. B. Saunders, Philadelphia, 1971).
2. B. D. Cullity, *Elements of X-ray Diffraction*, 2nd ed. (Addison-Wesley, Reading, MA, 1978).
3. P. A. Lee, P. H. Citrin, P. Eisenberger, and P. M. Kincaid, "Extended X-ray Absorption Fine Structure," *Rev. Mod. Phys.* 53, 769 (1981).
4. R. Saxon, *Elementary Quantum Electrodynamics* (Holden-Day, San Francisco, 1968).
5. L. I. Schiff, *Quantum Mechanics*, 3rd ed. (McGraw-Hill Book Co., New York, 1968).
6. R. L. Sproull and W. A. Phillips, *Modern Physics*, 3rd ed. (John Wiley and Sons, New York, 1980).
7. P. A. Tipler, *Modern Physics* (Worth Publishers, New York, 1978).

10
X-ray Photoelectron Spectroscopy

10.1   Introduction

In this chapter, electronic structure is the dominant theme. Photons with energies of up to 10 keV interact with the atomic electrons primarily via the photon absorption process (Chapter 9). The photoelectric process is a direct signature of the photon interaction with the atom and is the basis of one of the major analytical tools—photoelectron spectroscopy. This is referred to as UPS when ultraviolet light is incident on the sample and XPS when X-rays are used. Another acronym is ESCA (for Electron Spectroscopy for Chemical Analysis); in this case, the main concern is the chemical bonding.

The energy spectrum of electromagnetic radiation along with the common nomenclature is shown in Fig. 10.1. In material analysis, the photon energy range of interest corresponds to the ultraviolet (UV) and X-ray region. In practice, it extends from 10 eV, close to the binding energy (13.6 eV) of the electron in the hydrogen atom, to energies of around 0.1 MeV. At these energies, photons can penetrate within the solid and interact with the inner-shell electrons. Lower-energy photons are used to establish the visible spectra associated with the outermost, less tightly bound electrons. These outermost electrons are involved in chemical bonding and are not associated with specific atoms and hence are not useful for elemental identification. Photon-induced spectroscopies have undergone a major advance due to the advent of electron synchrotrons, which produce an intense source of monochromatic photons over a broad range of energies for materials science. Most laboratory instruments produce X-rays in the 1–10 keV region, which is the main region discussed in this chapter.

10.2   Experimental Considerations

The basic processes of interest in photoelectron spectroscopy are the absorption of a quantum of energy $\hbar\omega$ and the ejection of an electron, the photoelectron, whose kinetic energy, referenced to an appropriate zero of energy, is related to the binding energy of an electron in the target atom. In this process, an incident photon transfers its entire energy to the bound electron, and element identification is provided by the measurement of the energy of the electrons that escape from the sample without energy loss. As indicated in Fig. 10.2, photoelectron spectroscopy requires both a source of

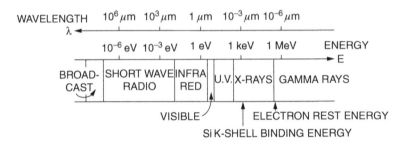

FIGURE 10.1. Electromagnetic spectrum indicating the region used for photoelectron spectroscopy. Ultraviolet photoelectron spectroscopy corresponds to incident photons in the UV region. X-ray electron spectroscopy corresponds to incident photons in the X-ray region.

monochromatic radiation and an electron spectrometer. As is common to all the electron spectroscopies where the escape depth is 1–2 nm, careful sample preparation and clean vacuum systems are required.

10.2.1 Radiation Sources

A convenient source of characteristic X-rays is provided by electron bombardment of Mg or Al targets. The relative intensity of the bremsstrahlung or X-ray continuum to characteristic X-rays is less important in the production of these soft X-rays (~ 1 keV) than for hard X-rays, i.e., from Cu bombardment. For Mg and Al, about one-half of the X-rays produced by electron bombardment are the K_α X-rays. The contribution from the continuous spectrum is hardly noticeable, since the bremsstrahlung spectrum is distributed over several keV while the K X-rays are concentrated in a peak of ~ 1 eV FWHM. In addition to the two K_α lines ($K_{\alpha 2}$ corresponds to a $2p_{1/2} \rightarrow$ 1s transition and $K_{\alpha 1}$ corresponds to a $2p_{3/2} \rightarrow$ 1s transition, indicated in Fig. 10.3a), there are lower intensities of higher-energy characteristic lines that correspond to two electron

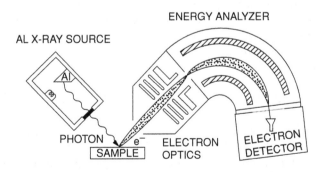

FIGURE 10.2. Schematic of the basic apparatus used in X-ray photoelectron spectroscopy. X-rays are produced at the Al anode by bombardment of electrons created at the filament. The X-rays impinge on a sample, producing photoelectrons that are detected after analysis in the electron energy analyzer.

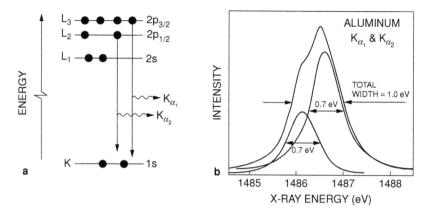

FIGURE 10.3. The two components of the K_α spectrum ($K_{\alpha 1} + K_{\alpha 2}$) which comprise the K_α spectrum of Al. [Spectrum from Sieqbanhn et al., 1967]

excitations (a 1s ionization plus a 2p ionization) in the Al target. For most applications, however, the spectrum is sufficiently clean for analysis purposes. If higher-energy resolution is required in the photon source, a monochromator (Fig. 10.4) must be used, with a corresponding decrease in efficiency. X-ray monochromators usually make use of crystal diffraction to energy select the beam.

As shown in Fig. 10.3, the Al $K_{\alpha 1,2}$ lines consist of two components separated by the 0.4 eV spin–orbit splitting of the 2p state. The $K_{\alpha 1}$ line that arises from the four electrons in the $2p_{3/2} \rightarrow 1s$ transition has about twice the intensity as that from the two electrons in the $2p_{1/2}$ state. For Mg K_α X-rays, a somewhat better resolution (≈ 0.8 eV) can be obtained. The K_α lines from Cr (~ 5 keV) and Cu (~ 8 keV) have energy widths ≥ 2.0 eV, Mo (~ 17 keV) has an energy width of about 6 eV, and all are not suitable for high-resolution studies without further energy selection.

Ultraviolet photoemission spectroscopy (UPS) generally uses resonance light source such as He discharge lamp with energies in the 16–41 eV range. The energies are sufficient to allow analysis of the valence band density of states of most solids. The

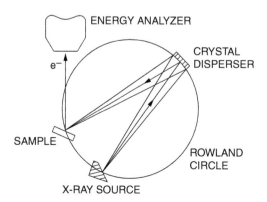

FIGURE 10.4. Schematic of an X-ray monochromation system.

intensity of the light sources is high and the energy widths are sharp. The energy resolution in these experiments is generally limited by the electron analyzer, in contrast to the case with X-ray sources. UPS studies are primarily directed at examining electron configurations in the valence shells or bonding orbits of a solid rather than determining elemental composition.

The use of synchrotron radiation from electron storage rings provides a continuous spectrum with intensities far in excess of the characteristic X-ray lines or resonance light sources. The use of polarized, tunable radiation from the synchrotron is a distinct advantage in experimental investigations. However, the limited access to synchrotron facilities restricts the applicability of synchrotron radiation in routine sample analysis.

10.2.2 Electron Spectrometers

The energy of photoelectrons is determined by their deflection in electrostatic or magnetic fields. Magnetic deflection analyzers such as those used in β-ray spectroscopy or early XPS measurements (Siegbahn et al.,1967) are difficult to use in routine analysis, and electrostatic analyzers are the instruments found in most laboratory systems. There are two general operating modes for analyzers: deflection and reflection (mirror). In deflectors, electrons travel along equipotential lines, and in mirror-type analyzers, the electrons travel across equipotentials. In the deflection type, shown in Fig. 10.2, a potential is applied across two concentric sectors, and the electrons pass through the analyzer without a change in energy. In the mirror-type analyzer, the electrons travel across potential lines and are reflected away from the reflecting electrode into the analyzer exit.

A common type of mirror analyzer is the cylindrical mirror analyzer (CMA) with angular entrance and exit slits so that the entire spectrometer has cylindrical symmetry (Fig. 10.5). The deflection is caused by the potential difference (set by the analyzer control) between the inner and outer cylinder. The CMA shown in Fig. 10.5 is a double pass with essentially two CMAs in series. The spherical retarding grids are used to scan the spectrum while the CMA is operated at constant pass energy in order to maintain a constant energy resolution.

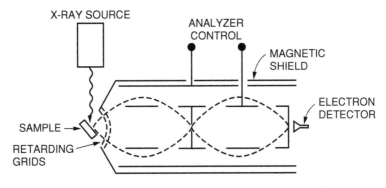

FIGURE 10.5. Schematic of a double pass cylindrical mirror analyzer (CMA) used in photoelectron spectroscopy.

The detection system is based upon the gain provided by electron multipliers, commonly a channel electron multiplier or channeltron. These channeltrons have a conelike opening and a continuous tube of high-resistivity, semiconducting glass with a high secondary emission coefficient. A high electric field is applied along the tube, and incident electrons create a shower of secondary electrons that in turn hit the tube walls and create further secondaries. A gain of 10^8 can be achieved with the output taken through an amplifier-rate meter system.

10.3 Kinetic Energy of Photoelectrons

In photoelectron spectroscopy of solids, one analyzes the kinetic energy of electrons ejected when a solid is irradiated with monoenergetic photons of energy $\hbar\omega$. The relevant energy conservation equation is

$$\hbar\omega + E_{\text{tot}}^{\text{i}} = E_{\text{kin}} + E_{\text{tot}}^{\text{f}}(k), \tag{10.1}$$

where $E_{\text{tot}}^{\text{i}}$ is the total energy of the initial state, E_{kin} is the kinetic energy of the photoelectron, and $E_{\text{tot}}^{\text{f}}(k)$ is the total final energy of the system after ejection of the photoelectron from the kth level. Contributions from a recoil energy E_r can be neglected (see problem 10.5). Only for the lightest atoms (H, He, and Li) is E_r significant when compared to the instrumental linewidths in XPS spectra. The binding energy of the photoelectron is defined as the energy required to remove it to infinity with a zero kinetic energy. In XPS measurements $E_B^{\text{V}}(k)$, the binding energy of an electron in the kth level referred to the local vacuum level is defined as

$$E_B^{\text{V}}(k) = E_{\text{tot}}^{\text{f}} - E_{\text{tot}}^{\text{i}}. \tag{10.2}$$

Substituting Eq. 10.1 into Eq. 10.2 results in the photoelectric equation

$$\hbar\omega = E_{\text{kin}} + E_B^{\text{V}}(k). \tag{10.3}$$

Binding energies are expressed relative to a reference level. In gas phase photoemission, binding energies are measured from the vacuum level. In the study of solids, the Fermi level is used as a reference.

In the case of a solid specimen, an electrical contact is made to the spectrometer. For metallic samples, the resulting energy levels are shown in Fig. 10.6. Because the sample and spectrometer are in thermodynamic equilibrium, their electrochemical potentials or Fermi levels are equal. In passing from the sample surface into the spectrometer, the photoelectron will feel a potential equal to the difference between the spectrometer work function ϕ_{spec} and the sample work function ϕ_s. Thus, the electron kinetic energy, E_{kin}^1, at the sample surface is measured as E_{kin} inside the spectrometer analyzer:

$$E_{\text{kin}} = E_{\text{kin}}^1 + (\phi_s - \phi_{\text{spec}}). \tag{10.4}$$

From Fig. 10.6, it can be seen that the binding energy in a metallic specimen may be determined relative to the common Fermi level as follows:

$$\hbar\omega = E_B^{\text{F}}(k) + E_{\text{kin}} + \phi_{\text{spec}}, \tag{10.5}$$

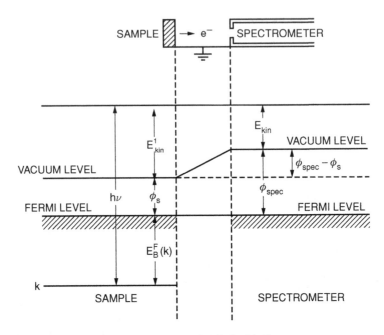

FIGURE 10.6. Schematic of the relevant energy levels for binding energy measurements. Note that the conducting specimen and spectrometer housing are in electrical contact and thus have common Fermi levels. The incoming photons, energy $h\nu$, create an electron of kinetic energy E relative to the vacuum level of the sample. The electron is detected by spectrometer with work function Φ_{spec} so that the measured energy $E_{kln} = E_{kln}^{1} - (\phi_{spec} - \phi_{s})$.

where $E_B^F(k)$ is the binding energy referred to the Fermi level. Notice that the sample work function ϕ_s is not involved but that of the spectrometer is.

When analyzing insulating samples, more care is required because of sample charging and the uncertainty in the location of the Fermi level within the band gap. One approach is to deposit a thin film of Au (or other metal) on the surface of the sample and use one of the known Au core levels to define the energy scale. Alternatively, the energies can be referenced to a well-defined feature of the electronic structure such as the valence band edge, which can be located in the XPS spectra.

In the following, we will use the symbol E_B to indicate the binding energy without specifying the energy reference level. Although the Fermi level is most commonly used for metals and metallic substances such as silicides, a well-defined reference level is not found in semiconductors and insulators. This ambiguity along with sample charging indicates that care must be taken in evaluating spectra.

10.4 Photoelectron Energy Spectrum

The major features of the energy spectrum of X-ray-excited photoelectrons are illustrated in Fig. 10.7 for Mg $K_\alpha(E = 1.25$ keV) irradiation of Ni. The spectrum exhibits the typical appearance of sharp peaks and extended tails through the allowed energy

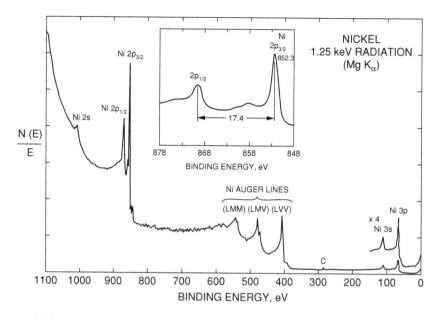

FIGURE 10.7. The energy spectrum of electrons from the 1.25 keV photon irradiation—Mg K_α – of nickel. The energy axis is in terms of binding energy, i.e., $h\nu - E_{kin}$. The vertical axis is denoted as $N(E)/E$ to denote that the admittance of the spectrometer decreases as $1/E$. [From *Phi Handbook*, Eden Prairie, MN.]

range. The peaks correspond to the energies of characteristic electrons that escape from the solid without undergoing energy loss. The higher-energy tails correspond to electrons that have undergone inelastic scattering and energy loss on their outward path, thus emerging with lower kinetic energy (apparently higher binding energy).

The energy of the Mg K_α line is not sufficient to eject K-shell electrons from Ni but can create vacancies in the L and M shells. The 2s and 2p as well as the 3s and 3p lines are clearly seen. The most prominent lines are the $2p_{1/2}$ and $2p_{3/2}$. Photoemission from p, d, and f electronic states, with nonzero orbital angular momentum produces a spin–orbit doublet such as the $2p_{1/2} - 2p_{3/2}$ lines shown in the inset to Fig. 10.7. The two lines correspond to final states, with $j_+ = 1 + m_s = 3/2$ and $j_- = 1 - m_s = 1/2$. The intensity ratio of the lines is given by the ratio $(2j_- + 1)/(2j_+ + 1)$, which gives a ratio of line intensities of 1:2 for $p_{1/2}$ to $p_{3/2}$, 2:3 for $d_{3/2}$ to $d_{5/2}$, and 3:4 for $f_{5/2}$ to $f_{7/2}$.

After emission of a core electron such as the 2s or 2p electrons from the L shell, a hole is left in the core shell. The hole can be filled by an electron from the M shell or valence band (V), with another M or V electron carrying away the energy. This Auger process is the dominant deexcitation process for elements lighter than $Z \approx 35$ and is described in more detail in Chapter 12. The Auger lines LMM, LMV, and LVV are clearly visible in the XPS spectrum of Fig. 10.7. Since the Auger lines are element specific, they can also be used in element identification. As is the case for photoelectron lines, each Auger line is accompanied by a low-energy tail corresponding to electrons that have lost energy on the outward path. The energy of an Auger line is independent of the incident photon energy, while the energy of the photoelectron line varies linearly with the incident photon energy.

10.5   Binding Energy and Final-State Effects

X-ray photoelectron spectroscopy is a straightforward and useful technique for the identification of atomic species at the surface of a solid. Adjacent elements throughout the periodic chart can easily be distinguished. The binding energies of adjacent elements are shown in Fig. 10.8 for the $2s(L_1)$ lines of elements in the third period of the periodic table. Electron spectra for a variety of elements in the energy range of 600–20 eV are shown in Fig. 10.9. The spin–orbit splittings are apparent for each group of elements. An overview of the variation in binding energy with atomic numbers is shown in Fig. 10.10. The binding energies increase as the square of atomic number. For photon energies around 1 keV, only the outer M or N shells can be ionized for $Z > 30$. A compilation of binding energies is given in Appendix 6.

As indicated by Eq. 10.2, the binding energy as measured in XPS is the difference in the total energy between the initial and final state of the system from which one electron has been removed. This binding energy is not the same as the eigenvalue that one would calculate from an atom in its initial state with all orbitals occupied. In photoemission, the outer shells of the atom readjust when an inner electron is removed, because the Coulomb attraction of the positive nucleus is then less effectively screened. The difference between an initial-state calculation with occupied orbitals and the experiment can be viewed as the following sequence: After the bound electron absorbs an energy $\hbar\omega$ from the photon, it loses some of its kinetic energy in overcoming the Coulomb attraction of the nucleus — that is, it loses kinetic energy. The outer orbitals readjust, lowering the energy of the final state and giving this additional energy to the outgoing electron. As a result of the formation of a vacancy in an inner shell by photoelectron ejection, the readjustment in the orbital of the

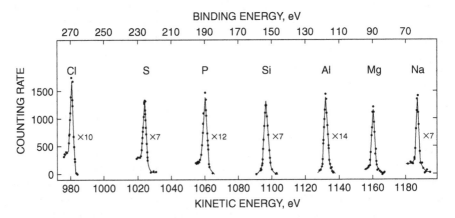

FIGURE 10.8. Electron lines from the L_1 subshells of the third period elements (Sodium to chlorine) excited with magnesium K_α radiation (1.25 keV). [From Siegbahn et al., 1967]

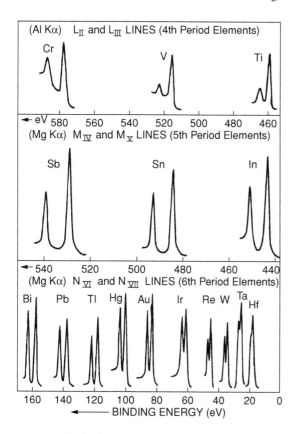

FIGURE 10.9. Electron spectra from a large variety of elements in the electron binding energy range of ~600 eV to ~20 eV. The incident photon radiations are Mg and Al K_α. [From Siegbahn et al., 1967.]

outer-shell electron may not necessarily proceed to the ground state of the hole-state atom. The outer electron may go into an excited state (electron shakeup) or into a continuum state (electron shakeoff), and hence less additional energy is given to the outgoing electron. These transitions that produce excited final states result in satellite structure at the high binding energy (lower kinetic energy) side of the photoemission line.

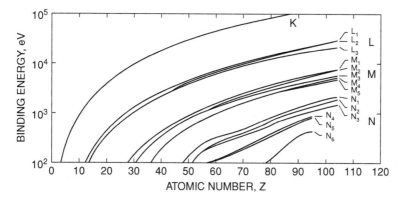

FIGURE 10.10. Binding energies of the elements.

FIGURE 10.11. The chemical shift in binding energy of the Si 2p line for elemental Si and SiO$_2$. The spectra are taken with Al K_α radiation. [From *Phi Handbook*, Eden, Prairie, MN.]

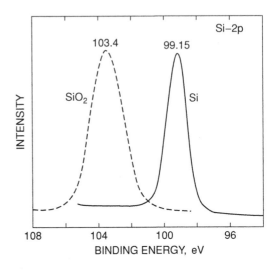

10.6   Binding Energy Shifts—Chemical Shifts

The exact binding energy for an electron in a given element depends on the chemical environment of that element. If we consider a core level, the energy of an electron in this core state is determined by the Coulomb interaction with the other electrons and the attractive potential of the nuclei. Any change in the chemical environment of the element will involve a spatial redistribution of the valence electron charges of this atom and the creation of a different potential as seen by a core electron. This redistribution affects the potential of the core electrons and results in a change in their binding energies.

The shift in the binding energies of core electrons as a function of the chemical environment is demonstrated in Fig. 10.11 for the Si 2p line. The measured binding energy of the Si 2p level shifts by more than 4 eV when the matrix is changed from Si to SiO$_2$. The existence of chemical shifts in XPS has led directly to analytical applications. The early work of the Uppsala group (Siegbahn et al., 1967) showed that core electron binding energies in molecular systems exhibit chemical shifts that are simply related to the covalency.

The concept of chemical shifts is based on the idea that the inner electrons feel an alteration in energy due to a change in the valence-shell contribution to the potential based on the outer-electron chemical binding. In the simplest picture, valence electrons are drawn either from or toward the nucleus, depending on the type of bond. The greater the electronegativity of the surrounding atoms, the more the displacement of electronic charge from the atom and the higher the observed binding energies of the core electrons. For example, Fig. 10.12 shows the binding energy shifts of the carbon atoms in ethyl trifluoroacetate, C$_4$F$_3$O$_2$H$_5$. Each carbon atom is in a different chemical environment and yields a slightly different XPS line. The binding energy shifts cover a change of about 8 eV.

An example of the shift in the Ni 2p XPS spectra as a result of the formation of Ni$_2$Si and NiSi is shown in Fig. 10.13. Here the shift in the Ni signal represents 1.1 eV

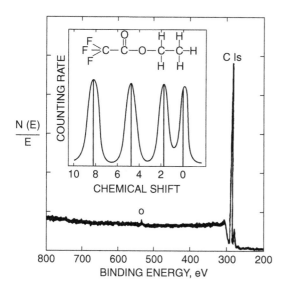

FIGURE 10.12. Carbon Is chemical shifts in ethyl trifluoroacetate. The four carbon lines correspond to the four carbon atoms within the molecule. [Adapted from Ghosh, 1983.]

for the transition from Ni to NiSi. The decrease in the peak intensity is due to the decrease in the amount of Ni atoms/cm^2 contained within the escape depth for the Ni 2p electrons as the compound becomes richer in Si. This is an example in which information on stoichiometry change comes primarily from intensity variations, with only minor chemical shift changes.

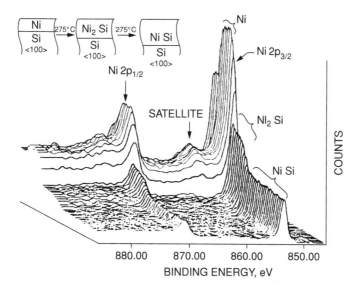

FIGURE 10.13. Three-dimensional plot of the Ni$_2$p XPS spectra, with the z-axis representing time during heat treatment. The spectra illustrate the planar growth of different forms of nickel silicide, as shown in the inset. Here chemical shifts are relatively small, but the change in intensity indicates a change in composition. [From P.J. Grunthaner, Ph.D. thesis, Caltech, 1980.]

10.7   Quantitative Analysis

Line intensities or the area of the photoelectron peaks are of interest for quantitative analysis. The intensity of a given line depends on a number of factors, including the photoelectric cross section σ, the electron escape depth λ, the spectrometer transmission, surface roughness or inhomogeneities, and the presence of satellite structure (which results in a decrease in the main peak intensities). The flux of X-rays is essentially unattenuated over the depths from which XPS signal peaks originate because the absorption lengths of X-rays are orders of magnitude larger than the escape depth of the electrons. The probability P_{pe} per incident photon for creating a photoelectron in a subshell k is

$$P_{pe} = \sigma^k Nt, \tag{10.6}$$

where Nt is the number of atoms/cm^2 in a layer of thickness t and σ^k is the cross section for ejecting a photoelectron from a given orbital k. Some of the basic concepts underlying the determination of the photoelectric cross section are given in Chapter 9. The calculations of Scofield for the photoelectric cross section at 1.5 keV for different subshells are presented in Fig. 10.14 in units of barns (one barn is 10^{-24} cm^2). These calculations give an overview of the large variation in cross sections that can be found in the analysis of a given material. Empirical studies indicate that these cross sections follow the strong Z dependence shown in Fig. 10.14, but the values may be in error by a factor of 2 or greater.

As discussed in Chapter 6, the number of electrons that can escape from a solid without undergoing an elastic collision decreases with depth as $\exp(-x/\lambda)$, where λ is the mean free path. The *universal* curve for escape depth versus energy is given in Fig. 6.4. The number of atoms per cm^2 that can produce a detectable photoelectron is then $N\lambda$, so the probability P_d per incident photon of creating a detectable photoelectron

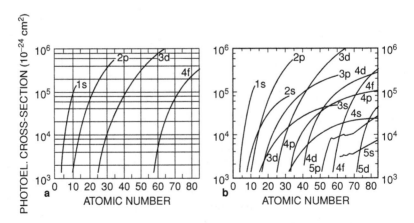

FIGURE 10.14. Calculations of the photoelectric cross section for different subshells throughout the periodic chart. The incident radiation is 1.5 keV; (a) the dominant shells most used in XPS; (b) the complete set of subshells. [From J.H. Scofield, *J. Electron Spectrosc.* 8, 129, 1976, with permission from Elsevier.]

from a subshell k is given by

$$P_d = \sigma^k N \lambda. \tag{10.7}$$

Not all the photoelectrons from a given subshell contribute to the photopeak, which corresponds to the ground-state configuration for a single vacancy in an inner shell. The influence of excited states (electron shakeup and shakeoff) is to decrease the intensity of the photopeak. The efficiency y in the production of a photopeak signal can vary over values of 0.7–0.8 for free atoms and, more importantly, can have a strong dependence on the chemical environment.

Finally, the instrumental efficiency T is a function for the kinetic energy E of the electron and usually varies as E^{-1}. For example, Fig. 10.7 presents the energy spectrum $N(E)$ of the photoelectrons as $N(E)/E$ to compensate for the transmission efficiency.

In chemical analysis, one is generally interested in the relative concentrations, n_A/n_B, of elements A and B in a sample, so that only the ratio of the areas of lines (the intensity ratio I_A/I_B) is required. The composition ratio is then

$$\frac{n_A}{n_B} = \frac{I_A}{I_B} \frac{\sigma_B \lambda_B y_B T_B}{\sigma_A \lambda_A y_A T_A}. \tag{10.8}$$

If the photopeaks have about the same energy so that $\lambda_A \cong \lambda_B$ and $T_A \cong T_B$ and the photopeak efficiencies are about equal, then the composition ratio can be approximated by

$$\frac{n_A}{n_B} = \frac{I_A}{I_B} \frac{\sigma_B}{\sigma_A}. \tag{10.9}$$

This approach assumes that the sample is flat and homogeneous, that the photoelectrons are emitted isotropically, and that the sample surface is clean, without a layer of surface contamination.

The sensitivity to the detection of trace elements depends upon the cross section of the element and the background of the signals from the other elements. Under favorable conditions, elemental analysis in bulk samples can reach sensitivities of 1 part in 1000. XPS measurements are extremely sensitive to the presence of surface layers. One often can detect as little as 0.01 monolayers of an element. The main use of XPS in materials analysis is determination of the chemical binding of atoms in the surface region of a solid.

Problems

10.1. (a) Compare the tabulated values of the binding energy, E_B (Appendix 6), for the 1s and 2s levels with the prediction of the Bohr theory (Eq. 1.15) for atoms with $Z = 10, 20, 30$, and 40. Does the Bohr theory serve as a useful approximation (to within 10%) of the binding energies?

   (b) Using the Pauli exclusion principle and quantum numbers with spin–orbit splitting, sketch an energy-level diagram for copper based on Bohr values of the binding energy. Indicate the number of electrons in each level and verify that the ratio of electrons in the d levels, $d_{3/2}$ to $d_{5/2}$, is $2/3$.

10.2. For Al K_α X-rays incident on the compound NiSi, calculate the ratio of yields using Eq. 10.8, assuming that the ratios of photopeak and instrument efficiencies are equal. Use Fig. 10.14 and 6.4 to estimate the cross sections and escape depths:

   (a) Ni 2s to Ni 2p;
   (b) Ni 2s to Ni 3s;
   (c) Ni 2p to Si 2p.

10.3. Consider incident radiation on Al at the same energy: 5.41 keV electrons and Cr K_α X-rays.

   (a) Calculate the cross section for creating Al K-shell vacancies for electrons (Eq. 6.11) and photons (Eq. 10.37).
   (b) Calculate the electron range (Eq. 6.25) and linear absorption coefficient (Eq. 10.35). Compare the calculated value of the mass absorption coefficient with the value listed in Appendix 8.
   (c) What is the energy of the K-shell photoelectrons? Calculate the value of the escape depth λ (Eq. 6.20) and compare it with the value estimated from Fig. 6.4.

10.4. For Al K_α radiation incident on Cu, calculate the ratio of 2s to 3s photoelectron yields based only on cross sections (Fig. 10.15) and escape depths.

10.5. Estimate the binding energy shift between the Li ($Z = 3$) atom and LiF in the following way (the Li atom has an electron configuration of $1s^2 2s^1$):

   (a) Calculate the probability of finding the 2s electron within the 1s orbit; i.e., calculate

$$\Delta q = \int_{r=0}^{r=a_0/Z} \psi_{2s}^2 r^2 \, dr \, \sin\theta d\theta d\phi,$$

   where ψ_{2s} is a hydrogenic wave function (Chapter 9) and a_0/Z is the Bohr radius of the 1s shell. (Note that $0 < \Delta q < 1$).
   (b) Estimate the binding energy of the 1s shell in the atom using the Bohr model with $Z_{\text{eff}}^{\text{Li}} = Z - \Delta q$.
   (c) In LiF, the outer electron is essentially on the F atom. Assume there is no contribution to the 1s screening from the 2s electron in LiF and estimate the binding energy shift using the Bohr model.

References

Atomic Physics and Quantum Mechanics

1. E. E. Anderson, *Modern Physics and Quantum Mechanics* (W. B. Saunders, Philadelphia, 1971).
2. J. D. McGervey, *Introduction to Modern Physics* (Academic Press, New York, 1971).
3. F. K. Richtmyer, E. H. Kennard, and J. N. Cooper, *Introduction to Modern Physics*, 6th ed. (McGraw-Hill Book Co., New York, 1969).
4. L. I. Schiff, *Quantum Mechanics*, 3rd ed. (McGraw-Hill Book Co., New York, 1968).

5. R. L. Sproull and W. A. Phillips, *Modern Physics*, 3rd ed. (John Wiley and Sons, New York, 1980).
6. P. A. Tipler, *Modern Physics* (Worth Publishers, New York, 1978).
7. R. T. Weidner and R. L. Sells, *Elementary Modern Physics*, 3rd ed. (Allyn and Bacon, Boston, MA, 1980).

X-Ray Photoelectron Spectroscopy

1. D. Briggs, Ed., *Handbook of X-ray and Ultraviolet Photoelectron Spectroscopy* (Heydon and Son, London, 1977, 1978).
2. D. Briggs and M. P. Seah, *Practical Surface Analysis by Auger and X-ray Photoelectron Spectroscopy* (John Wiley and Sons, New York, 1983).
3. M. Cardona and L. Ley, Eds., *Photoemission in Solids I and II, Topics in Applied Physics*, Vols. 26 and 27 (Springer-Verlag, New York, 1978 and 1979).
4. T. A. Carlson, *Photoelectron and Auger Spectroscopy* (Plenum Press, New York, 1975).
5. G. Ertl and J. Kuppers, *Low Energy Electrons and Surface Chemistry* (Verlag Chemie International, Weinheim, 1974).
6. P. K. Ghosh, *Introduction to Photoelectron Spectroscopy* (Wiley-Interscience Publishers, New York, 1983).
7. H. Ibach, Ed., *Electron Spectroscopy for Surface Analysis, Topics in Current Physics*, Vol. 4 (Springer-Verlag, New York, 1977).
8. K. D. Sevier, *Low Energy Electron Spectrometry* (Wiley-Interscience Publishers, New York, 1972).
9. K. Siegbahn, C. N. Nordling, A. Fahlman, R. Nordberg, K. Hamrin, J. Hedman, G. Johansson, T. Bergmark, S. E. Karlsson, I. Lindgren, and B. Lindberg, ESCA, *Atomic, Molecular, and Solid State Structure Studied by Means of Electron Spectroscopy* (Almqvist and Wiksells, Uppsala, Sweden, 1967).

11
Radiative Transitions and the Electron Microprobe

11.1   Introduction

In previous chapters, we have calculated the cross section for creating an inner-shell vacancy by irradiation with X-rays (photoelectric cross section, Chapter 9) and energetic electrons (impact ionization cross section, Chapter 6). After a vacancy is created, an electron can make a transition from an outer shell to fill the vacancy with the emission of a photon. This process is referred to as the *spontaneous emission of radiation*. In this chapter, we consider the energies of X-ray transitions and calculate radiative transition rates. We will use the formula for the radiative transition rate from an initial state i to a final state f (derived in Chapter 9):

$$W = \frac{4}{3} \frac{(\hbar\omega_{f_1})^3}{(\hbar c)^3} \frac{(e^2)}{\hbar} |\langle \psi_f | \mathbf{r} | \psi_i \rangle|^2 , \tag{11.1}$$

where $\hbar\omega_{f_1} = E_{\mathrm{B}}^i - E_{\mathrm{B}}^f$ is the energy of the emitted radiation, where E_{B} is the binding energy of the initial or final state. The transition rate increases strongly with photon energy or, for a given transition, increases rapidly with Z. Evaluation of the matrix elements indicates that $W = 0$ for some of the transitions, and hence selection rules for allowed transitions can be derived. These considerations are used in the description of the electron microprobe, where bombardment of a solid by a beam of electrons leads to the emission of X-rays with energies characteristic of the atomic species. The relative advantages of proton-induced X-ray excitation will be discussed in Section 11.9.

Consider an excited atom with a hole in the K or L shell. A straightforward means of deexcitation is the transition of an electron from an occupied state in a shell to the empty state (hole) with the emission of X-rays as shown by the L_3 to K transition (K_{α_1} X-ray) in Fig. 11.1.

X-ray emission is mainly due to dipole radiation, and electron transition selection rules are obeyed ($\Delta l = \pm 1$, $\Delta j = 0, \pm 1$); the energy is given by the difference in binding energies,

$$\hbar\omega(K_\alpha) = E_{\mathrm{B}}^{\mathrm{K}} - E_{\mathrm{B}}^{\mathrm{L}_3} = h\nu . \tag{11.2}$$

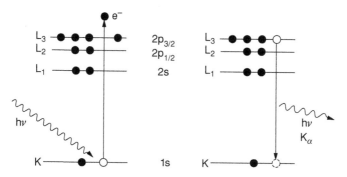

FIGURE 11.1. Schematic of photon interaction process with atoms: (a) photon absorption in which an electron is emitted with energy, E, given by $E = \hbar\omega - E_B$, where E_B is the electron binding energy of the shell; (b) photon (X-ray) emission where the L-shell electron makes a transition to fill the vacancy in the K shell.

11.2 Nomenclature in X-Ray Spectroscopy

The arrows in Fig. 11.2 show the allowed transitions of the atom and indicate their commonly used designation. For example, a $K \rightarrow L_3$ transition, K_{α_1} X-ray line, is one in which there is an initial vacancy in the K shell and a final vacancy in the L shell. In spite of the fact that the energy levels belong to the atom as a whole, the one-electron quantum numbers n, l, j are used in descriptions of atomic levels, as shown in Fig. 11.2. States with the same values of n and l but different values of $j (= l \pm 1/2)$ are well separated.

The main X-ray transitions for Pb are also shown in Fig. 11.2. The intensity ratios are indicated at the top of the diagram and are referenced to $K_{\alpha_1} = 100$ for K X-rays, $L_{\alpha_1} = 100$ for L X-rays, etc. The strong transitions obey the dipole selection rules. The intensity ratios and energy positions serve to establish the fingerprint pattern that identifies the element.

11.3 Dipole Selection Rules

The formula for the transition probability (Eq. 11.1) contains the matrix element $\langle \psi_f | \mathbf{r} | \psi_i \rangle$ known as the *dipole matrix element*. This quantity is identically zero—that is, transitions are forbidden—between certain initial, ψ_i, and final, ψ_f, states. Basically, a photon can be considered as a particle with unit angular momentum. Therefore, in order to conserve angular momentum as well as energy, transitions in which the angular momentum changes by one unit, $\Delta l = 1$, are allowed. These selection rules $(\Delta l = \pm 1, \Delta j = 0, \pm 1)$ result in a simplification of the observed X-ray spectrum. The allowed transitions are illustrated in Fig. 11.2, along with a few weak lines arising from transitions that are dipole forbidden but allowed for magnetic and electric quadrupole transitions.

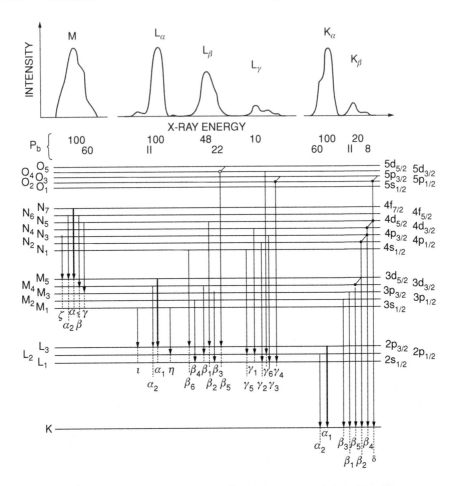

FIGURE 11.2. Principal X-ray transitions, including their common designations. The upper portion of the figure indicates the relative intensities of the various lines that form the characteristic X-ray spectrum and indicates the spectral distribution for Pb. [From F. Folkmann in Thomas and Cachard, 1978, with permission from Springer Science+Business Media.]

11.4   Electron Microprobe

The detection and measurement of the characteristic X-rays for materials excited by energetic electrons is the basis of electron microprobe analysis. The essential feature of the electron microprobe (Fig. 11.3) is the localized excitation of a small area of the sample surface with a finely focused electron beam. The volume of the sample material excited by the electrons has dimensions on the order of a micron, and hence the analytical technique is often referred to as *electron probe microanalysis* or *electron microprobe analysis* (EMA). The electron beam can be scanned across the surface to give an image of the lateral distribution of the material composition.

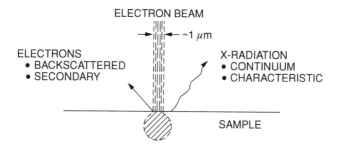

FIGURE 11.3. Electron beam interaction with a solid. Incident electrons create inner-shell vacancies within the first micron of the solid.

In materials analysis, there are only a few lines that are important, mainly K_α and K_β and the L series (L_α, L_β, and L_γ lines). The energies of the K series and L series are tabulated in the Appendices. The energies of the more important characteristic lines are shown in Fig. 11.4. In the analysis of X-ray spectra, it is common to refer to either energy or wavelength, depending on the type of detection system, either energy or wavelength dispersive. The most convenient form of analysis is the energy dispersive spectroscopy (EDS) mode using a Si(Li) detector whose basic operation is similar to that of a charged-particle solid-state detector as described in Chapter 3. An incoming X-ray creates a photoelectron that eventually dissipates its energy through the formation of electron–hole pairs. The number of pairs is proportional to the incident photon energy. Under bias, an electrical pulse is then formed with a magnitude proportional to the number of pairs or to the X-ray spectrum over a broad range of energies with an energy resolution of approximately 150 eV. The spectrum from Mn K X-rays taken with a

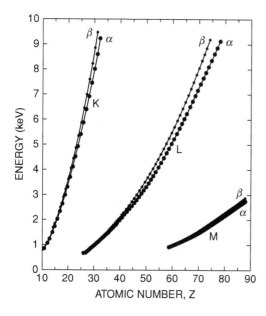

FIGURE 11.4. Energy of the K_α, K_β, L_α, L_β, and M_α lines of the elements as a function of atomic number.

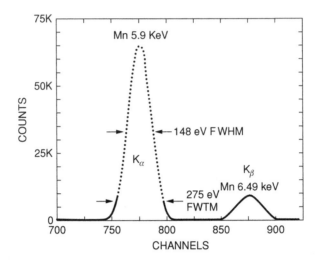

FIGURE 11.5. K X-ray spectra from Mn measured with an energy-dispersive Si(Li) solid-state detector. The $K\alpha$ line is at 5.89 keV, the K_β line at 6.49 keV, and the resolution of the detector, full width at half maximum, is 148 eV.

Si(Li) detector is shown in Fig. 11.5. The K_α and K_β lines are clearly resolvable. The full width at half maximum of the K_α line (148 eV) is determined by the resolution of the detector. The resolution is set by the statistical variations associated with the electron—hole pair creation process. Higher resolution is obtainable with wavelength-dispersive techniques, albeit at the expense of efficiency.

Wavelength-dispersive spectroscopy (WDS) involves X-ray diffraction from an analyzer crystal; only those X-rays that satisfy the Bragg relation ($n\lambda = 2d \sin \theta$) are constructively reflected into the detector. Higher-order reflections of a given wavelength can also be diffracted into the detector. The reflected wavelengths are $\lambda, \lambda/2, \lambda/3, \ldots$ corresponding to first-, second-, third-, reflections. In Fig. 11.6a,b, X-ray spectra are shown from a materials analysis of a nickel base alloy, using an energy-dispersive system and a wavelength-dispersive system. The energy resolution is sufficiently good in wavelength-dispersive systems (~ 5 eV) that closely spaced lines can easily be resolved.

With energy-dispersive systems, it is possible to have signal interference from different elements when the energies of X-rays are close together. The K_α line of element Z falls close to the K_β line of element $Z-1$ or $Z-2$; i.e., Br and Rb. As can be seen in Fig. 11.6b, an L transition from Ta falls close to the K transition of Ni; the Ta component of the material is, however, readily apparent from the wavelength-dispersive system. These interferences can complicate analysis of multielement samples.

The chemical binding energy shifts found in photoemission (XPS) are not as readily detectable in X-ray analysis, since the X-rays are due to a transition between two levels, both of which shift in the same direction due to bonding effects. Wavelength shifts to both longer and shorter wavelength are observed in very-high-resolution X-ray spectra for various elements upon chemical combination.

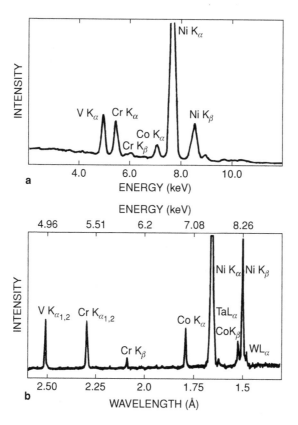

FIGURE 11.6. (a) Energy-dispersive and (b) wavelength-dispersive X-ray spectra from nickel base alloy. The energy-dispersive spectrum uses a Si(Li) solid-state detector, and the wavelength-dispersive spectrum uses a LiF diffracting crystal. [Adapted from Goldstein et al. 1981.]

In X-ray spectroscopy, as in most atomic transitions, there is an energy (line width) width or breadth associated with the lifetime of the core hole. This is a fundamental width set by the atomic processes involved and is the minimum energy width that could be observed in a high-resolution detector. The uncertainty principle states that

$$\Delta E \Delta t \geq \hbar. \tag{11.3}$$

For atomic energy levels, we may take $\Delta t = \tau$, where τ is equal to the mean life of the excited state. The mean life is determined by radiative and Auger transitions. Then,

$$\Delta E \cong \frac{\hbar}{\tau}. \tag{11.4}$$

The lifetime broadening can be as large as 10 eV for the L shells and more than 100 eV for the K shell of high Z elements. These widths are a consequence of the filling of an inner-electron-shell vacancy by outer electrons within a time interval on the order of 10^{-16} sec.

11.5   Transition Rate for Spontaneous Emission

The rate for a spontaneous transition from an electron in an initial state i to a final state f is

$$W = \frac{4}{3}\frac{e^2}{\hbar}\left(\frac{\omega_{fi}}{c}\right)^3 |\langle\psi_f|\mathbf{r}|\psi_i\rangle|^2, \tag{11.5}$$

where $e^2 = 1.44$ eV-nm, c is the speed of light, $\hbar\omega_{fi}$ is the energy of the emitted photon, and ψ_i and ψ_f are the wave functions of the initial and final states, respectively. This rate can be expressed as

$$W = 0.38 \times 10^{18}(\hbar\omega)^3 |\langle\psi_f|\mathbf{r}|\psi_i\rangle|^2,$$

in units of inverse seconds, where $\hbar\omega$ is in units of keV and the matrix element $(|\langle\psi_f|\mathbf{r}|\psi_i\rangle|^2)$ has dimensions of $(0.01$ nm$)^2$. A typical value of W is 10^{15}/s for K-shell transitions of elements in the middle of the periodic table. This formula is derived by considering the absorption transition probability in an electromagnetic field and the Einstein coefficients that describe the balance between photon absorption and induced and spontaneous transitions (Section 9.10).

11.6   Transition Rate for K_α Emission in Ni

We illustrate the use of Eq. 11.5 by explicitly calculating the transition rate for a particular case. Our starting point is the transition rate formula

$$W = \frac{4}{3}\frac{\omega^3 e^2}{\hbar c^3}\left|\left\langle\psi_f|\mathbf{r}|\psi_i\right\rangle\right|^2.$$

As a first estimate, we note that $\langle\psi_f|\mathbf{r}|\psi_i\rangle \sim a_0/Z$. The wave functions $\psi_f(\mathbf{r})$ and $\psi_i(\mathbf{r})$ have a finite intensity only for values of $|\mathbf{r}| \lesssim a_0/Z$, where a_0 is the Bohr radius, $a_0 = 0.053$ nm. Then

$$W = \frac{4}{3}\frac{\omega^3 e^2}{\hbar c^3}\frac{a_0^2}{Z^2}. \tag{11.6}$$

The energy of the transition, $\hbar\omega$, for 2p $\rightarrow$ 1s transitions in hydrogenic atoms is 10.2 Z^2 (eV) [i.e., $\hbar\omega = 13.6(1 - 1/n^2)Z^2$, where $n = 2$]. Therefore, the transition probability is proportional to Z^4:

$$W \propto Z^4. \tag{11.7}$$

This dependence will hold in more sophisticated treatments as well. We can give a simple order-of-magnitude estimate of W. The units of W are often given in terms of eV/$\hbar$. Thus we rewrite Eq. 11.6 as

$$W = \frac{4}{3}\frac{1}{\hbar}(\hbar\omega)^3\left(\frac{e^2}{\hbar c}\right)^3\left(\frac{a_0}{e^2}\right)^2\frac{1}{Z^2}, \tag{11.8}$$

(note that $e^2/\hbar c = 1/137$ and $e^2/a_0 = 27.2$ eV). Consider the K_α transition in Ni ($Z = 28$, $\hbar\omega = 7.5 \times 10^3$ eV). Then

$$W = 0.38 \, \text{eV}/\hbar.$$

Actually, the six 2p electrons in Ni can contribute to the process so that

$$W = 2.3 \, \text{eV}/\hbar.$$

The accepted result is $W = 0.551$ eV/$\hbar$. A factor of 4 is close for a crude estimate. In units of time and using the accepted value, $W = 8.3 \times 10^{14} \text{s}^{-1}$, or

$$\frac{1}{W} = 1.2 \times 10^{-15} \text{s}.$$

A detailed evaluation of the transition probability is given in Section 11.9.

The energy width Γ of an atomic state is related to the mean life τ of the state through the Heisenberg uncertainty principle:

$$\Gamma \tau = \hbar. \tag{11.9}$$

In the previous section, we estimated the radiative contribution to the width or, more precisely, the decay probability (per unit time) for a 2p to 1s transition, i.e.,

$$W_{\text{rad}} = \Gamma_{\text{rad}}/\hbar. \tag{11.10}$$

The width of the state is made up of all the processes that contribute to its finite lifetime:

$$\Gamma = \Gamma_{\text{rad}} + \Gamma_{\text{nonrad}}, \tag{11.11}$$

where Γ_{rad} represents all the radiative processes that contribute to the lifetime (i.e., X-ray emission) and Γ_{nonrad} contributes to all nonradiative contributions (i.e., Auger emission). The probability of a radiative decay is $\Gamma_{\text{rad}}/(\Gamma_{\text{rad}} + \Gamma_{\text{nonrad}})$ and is known as the *fluorescence yield* ω_X (see Section 11.3). For a state consisting of a K vacancy with atomic $Z > 40$, radiative processes dominate: $\Gamma_{\text{nonrad}} \ll \Gamma_{\text{rad}}$. In a hydrogenic approximation,

$$\Gamma = \frac{4}{3}(\hbar\omega)^3 \left(\frac{e^2}{\hbar c}\right)^3 \left(\frac{a_0}{e^2}\right)^2 \frac{0.74}{Z^2}(6), \tag{11.12}$$

where the extra factor of 6 allows for six 2p electrons. In the spirit of the hydrogenlike model, we note that the energy of the 2p–1s transition is given by

$$\hbar\omega = 13.6(1 - \frac{1}{4})Z^2, \tag{11.13}$$

so

$$\Gamma = 3.3 \times 10^{-8} Z^4 (\text{eV}). \tag{11.14}$$

This calculated result is only a factor of ~ 2 greater than the fit to experiment shown in Fig. 11.7. Note that the Z^4 dependence is close to that observed experimentally.

FIGURE 11.7. K level width as a function of atomic number. The dashed line is the calculated results (Eq. 11.14). The solid line is an empirical fit to the measured points. [From W. Bambynek et al., "X-Ray Fluorescence Yields, Auger, and Coster-Kronig Transitions Probabilities," *Rev. Mod. Phys.* 44, 716 (1972). Copyright 1972 by the American Physical Society.]

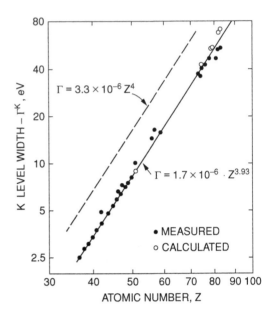

11.7   Electron Microprobe: Quantitative Analysis

The electron microprobe is used for identification of elements and the quantitative analysis of elemental composition. All elements with atomic number greater than that of Be can be analyzed in principle, but in practice the technique is mostly applied to Z approximately > 10. The detection limit for elements is 50–110 ppm except for low-Z elements (below Mg), which are detectable in concentrations greater than about 0.1 atomic percent.

Quantitative analysis of the concentration of a given element can be carried out with an accuracy of about 1% if suitable standards are available. The simplest procedure is to measure the yield Y_p of a given element at a wavelength λ_p, subtract the background yield Y_b, and determine the ratio K of the corrected yield in the sample to that in the standard $Y_p^s - Y_b^s$, where the background is due to bremsstrahlung (see Section 6.11). Using this definition,

$$K = \frac{Y_p - Y_b}{Y_p^s - Y_b^s}. \tag{11.15}$$

The concentration c_A of element A can be determined from the concentration c_A^s in the standard by

$$c_A = c_A^s\, K, \tag{11.16}$$

where the standard and sample are subject to identical electron beam impingement and X-ray detection conditions, and the standard has a composition near that of the sample to allow for equivalent X-ray absorption effects.

Often it is the concentration ratio c_A/c_B of two elements in a bulk or thin film sample that is desired. In this case, it is useful to establish a calibration curve that relates peak

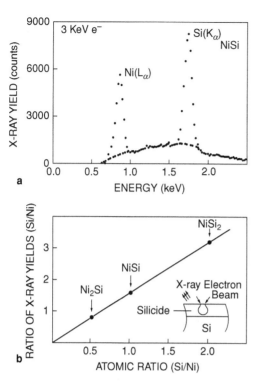

FIGURE 11.8. (a) Energy spectra from 3 keV electron bombardment of the nickel silicide. (b) Ratio of X-ray yields as a function of known atomic ratio for the different nickel silicides.

intensity ratios to atomic ratios. An example of this procedure is illustrated in Fig. 11.8, which shows the $Ni_{L\alpha}$ and $Si_{K\alpha}$ lines and the ratio of yields versus the measured Si/Ni atomic ratio in nickel silicide standards whose composition was determined by Rutherford backscattering analysis. The electron beam energy was sufficiently low (see electron ranges, Chapter 6) that the penetration of the electrons was confined to the silicide layer, and X-ray generation in the Si substrate was minimal.

11.7.1 Quantitative Analysis

The determination of an absolute concentration of an element in an unknown matrix represents a complicated problem. Consider first the yield of X-rays, Y_X, produced from a thin layer of width Δt and depth t into the sample:

$$Y_X(t) = N \Delta t \sigma_e(t) \omega_X \, e^{-\mu t / \cos \theta} I(t) \eta \frac{d\Omega}{4\pi}, \tag{11.17}$$

where

N is the number of atoms/volume;
$\sigma_e(t)$ is the ionization cross section at depth t, where the particle has energy E_t;
μ is the X-ray absorption coefficient;
ω_X is the fluorescence yield;

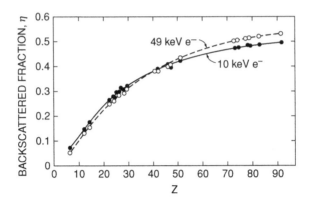

FIGURE 11.9. Variation of the backscattered fraction as a function of atomic number at $E_0 = 10$ keV and $E_0 = 49$ keV. [Goldstein et al., 1981, with permission from Springer Science+Business Media.]

θ is the detector angle;
$\eta d\Omega$ is the efficiency and solid angle of the detector;
$I(t)$ is the intensity of electron beam at depth t.

The total observed yield Y is given by

$$Y = \int_{t=0}^{R} Y(t)dt + \text{Secondary Fluorescence}, \qquad (11.18)$$

where R is the range of the electron, and the second term includes the effect of secondary fluorescence due to absorption of high-energy X-rays (generated by other heavy atoms within the matrix) and re-emission of the X-rays of interest.

Equation 11.17 takes into account (1) the change in cross section as a function of depth into the sample due to the change in electron energy as the beam penetrates the sample and (2) the attenuation of the beam, $I(t)$, as it penetrates the sample due to backscattering of the electrons.

The attenuation of the beam can be a surprisingly large factor. Figure 11.9 shows the backscattered fraction as a function of Z for two different incident energies. The backscattered fraction is almost energy independent.

11.7.2   Correction Factors

In practice, Eqs. 11.17 and 11.18 are usually not used explicitly in the evaluation of composition; rather, comparisons of X-ray yields from the unknown sample and a standard are used. Even under these conditions, however, corrections must be made, since many of the factors in Eq. 11.17 are matrix dependent. Extensive investigations in the field of microprobe analysis have generated an approach based on empirical correction factors that represent those matrix-dependent effects.

Some insight into the correction factors involved in quantitative analysis can be gained by considering the procedure for determining the concentration C_A of an element

A in an alloy from the ratio K of X-ray intensities from the sample and from a standard composed of element A. An expression of the form

$$C_A = KZAF \qquad (11.19)$$

is used (Birks, 1979; Goldstein et al., 1981), where Z is the atomic number correction factor, A is the absorption correction factor, and F is the fluorescence correction factor. These corrections relate to the three major effects that arise from the differing characteristics of the sample and the standard with respect to electron and X-ray interactions. The atomic number correction Z allows for the fact that the generation of primary X-rays in the samples does not increase linearly with concentration. The proportion of incident electrons that are backscattered and the volume of the sample in which X-rays are generated depend on sample composition. The absorption correction A is required because the absorption coefficients — i.e., the attenuation of the emerging X-radiation — will be different in the sample and the standard. The fluorescent correction F accounts for the generation of secondary X-rays from element A due to fluorescent excitation by X-rays emitted by another element (Fig. 11.10). The effect is strongest when the exciting lines have an energy slightly greater than the binding energy associated with the line that is being measured, i.e., near the maximum of the absorption cross section. For example, in a sample containing Fe and Ni, the Ni K X-rays could excite Fe K X-rays, and in a Cu–Au sample, the Au L lines can excite Cu K_α. The correction for fluorescence by characteristic lines depends on the atomic fluorescence yield ω_X and the fraction of the exciting element in the sample. For X-ray lines with energies below 3 keV, the values

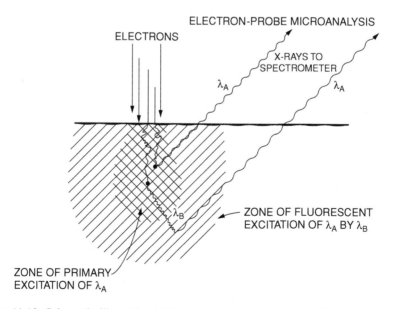

FIGURE 11.10. Schematic illustration of the generation of secondary radiation as a result of fluorescence excitation by primary radiation.

FIGURE 11.11. Scaled K-shell ionization cross sections by proton impact.
$E/E_B{}^K$ is a ratio of the incident energy
to the K-shell binding energy. M_1/M_p
is the ratio of the projectile mass to that
of the proton. [From J.D. Garcia, *Phys.
Rev. A1*, 1402 (1970). Copyright 1970
by American Physical Society.]

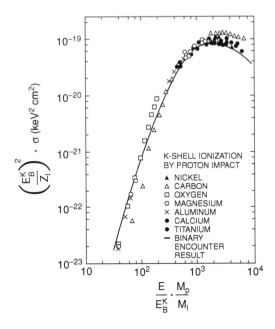

of ω_X are small and the fluorescence becomes negligible. The magnitude of each of the correction factors in Eq. 11.19 is between 2% and 10% in most cases.

11.8   Particle-Induced X-Ray Emission (PIXE)

Inner-shell ionization is caused by the time-dependent electric field created by the passage of a charge near an atom. In a classical sense, the field from a proton is precisely the same (aside from a sign) as that of an electron at the same velocity. In terms of kinetic energy, this occurs at an ion energy of $(M/m_e)E$, where M and m_e are the masses of the ion and electron, respectively, and E is the electron kinetic energy. Thus the velocity matching criteria require proton kinetic energies of $1836\,E$, which corresponds to the MeV ion range to match keV electrons.

Proton-induced K-shell ionization cross sections are shown in Fig. 11.11 in a reduced plot. The horizontal axis is $E/E^K{}_B$, the ratio of the ion energy to the K-shell binding energy. The vertical axis is $(E^K{}_B/Z_1)^2\sigma$, the product of the square of the K-shell binding energy and the reduced ionization cross section σ/Z_1^2. This scaling allows the formulation of a universal plot for the cross section. In Fig. 11.11, the ratio M_P/M_1 allows scaling for heavier projectiles. For protons, $M_P/M_1 = 1$ and $Z_1 = 1$; for He$^+$ ions, $M_P/M_1 = {}^1\!/_4$ and $Z_1 = 2$; etc. X-ray production cross sections are shown explicitly in Fig. 11.12. The X-ray production cross section $\sigma_{K\alpha}$ is related to the ionization cross section σ (Fig. 11.11) through the fluorescence yield ω_X; $\sigma_{K\alpha} = \omega_X\sigma$. The fluorescence yield is the probability of a radiative transition relative to all possible transitions (radiative and nonradiative). The maximum value of the cross section decreases with increasing atom number or binding energy. Further, the maximum of the

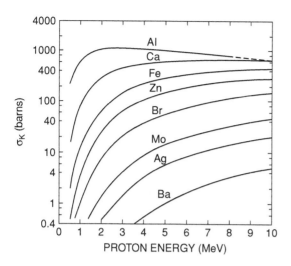

FIGURE 11.12. K_α X-ray production cross sections ($\sigma \omega_x$) versus proton energy. [From Cahill, 1980, reprinted with permission from *Annual Review of Nuclear and Particle Science*, Vol. 30. © 1980 by Annual Reviews, www.annualreviews.org.]

cross section is at higher energy for the heavier elements, corresponding to the idea of velocity matching, similar to that for electrons where the maximum occurs at energies of three to four times the binding energy E_B.

Particle-induced X-rays (PIXE) have been used for a variety of analytical problems with MeV accelerators. The major advantage of ion-induced X-ray analysis is a reduction in background relative to that of electrons, yielding a better sensitivity for trace element analysis. The background in the electron-microprobe-induced X-ray spectrum is due to electron bremsstrahlung (see Section 6.11). This is a continuous distribution associated with the deceleration of the electron as it traverses a solid. A quantum treatment of the bremsstrahlung process shows that the probability of photon emission decreases strongly with increasing mass of the charged particle projectile. Thus, at the same velocity, protons and electrons have approximately the same characteristic X-ray production probability but a large difference in bremsstrahlung background. Particle-induced X-rays can be combined with the formation of microbeams (of order 10 μm) to do high-sensitivity lateral mapping of trace element distributions. Such beams can also be brought into air for analysis of biological and vacuum-degradable samples (Cahill, 1980).

11.9   Evaluation of the Transition Probability for Radiative Transitions

In this section, we make a detailed evaluation of radiative transition probabilities and first concentrate on the dipole moment $\langle \psi_f | \mathbf{r} | \psi_i \rangle$. For ψ_i and ψ_f, we will use hydrogenic wave functions (see Table 9.1):

$$\psi_f = \psi_{1s},   \text{(11.20)}$$

$$\psi_{1s} = R_{10}(r) Y_{0,0}(\theta, \phi),   \text{(11.21)}$$

TABLE 11.1. Spherical harmonics.

l	m	Y_{lm}
0	0	$Y_{00} = \frac{1}{\sqrt{4\pi}}$
1	1	$Y_{11} = -\left(\frac{3}{8\pi}\right)^{1/2} \sin\theta \exp(i\phi)$
1	0	$Y_{10} = \left(\frac{3}{4\pi}\right)^{1/2} \cos\theta$
1	−1	$Y_{1-1} = \left(\frac{3}{8\pi}\right)^{1/2} \sin\theta \exp(-i\phi)$
2	2	$Y_{22} = \left(\frac{15}{32\pi}\right)^{1/2} \sin^2\theta \exp(2i\theta)$
2	1	$Y_{21} = -\left(\frac{15}{8\pi}\right)^{1/2} \cos\theta \sin\theta \exp(i\phi)$
2	0	$Y_{20} = -\left(\frac{5}{16\pi}\right)^{1/2} (3\cos^2\theta - 1)$
2	−1	$Y_{2-1} = \left(\frac{15}{8\pi}\right)^{1/2} \cos\theta \sin\theta \exp(-i\phi)$
2	−2	$Y_{2-2} = \left(\frac{15}{32\pi}\right)^{1/2} \sin^2\theta \exp(-2i\theta)$

where

$$R_{10} = \frac{2}{a^{3/2}} e^{-\rho}, \tag{11.22}$$

and $\rho = rZ/a_0 = r/a$;

$$\psi_i = \psi_{2p} = \frac{1}{\sqrt{3}} R_{21}(r)\{Y_{1,0}(\theta, \phi) + Y_{1,-1}(\theta, \phi) + Y_{1,1}(\theta, \phi)\}, \tag{11.23}$$

where $R_{21} = [1/4(2\pi)^{1/2}] \cdot (1/a^{3/2})\rho e^{-\rho/2}$ and the spherical harmonics, the Y's, are defined in Table 11-1. The factor $1/\sqrt{3}$ assures proper normalization, i.e., $\langle \psi_{2p} | \mathbf{r} | \psi_{2p} \rangle = 1$.

The matrix element of $\mathbf{r}$ is given by $\langle \psi_f | x | \psi_i \rangle + \langle \psi_f | y | \psi_i \rangle + \langle \psi_f | z | \psi_i \rangle$, or in spherical coordinates as

$$\langle \psi_f | x | \psi_i \rangle = \langle \psi_f | r \sin\theta \cos\phi | \psi_i \rangle, \tag{11.24}$$

$$\langle \psi_f | y | \psi_1 \rangle = \langle \psi_f | r \sin\theta \sin\phi | \psi_i \rangle, \tag{11.25}$$

and

$$\langle \psi_f | z | \psi_i \rangle = \langle \psi_f | r \cos\phi | \psi_i \rangle. \tag{11.26}$$

All the terms can be separated in the following sense:

$$\langle \psi_f | z | \psi_i \rangle = \frac{1}{3} \left\{ \int R_{21}(r) \cdot r \cdot R_{10}(r) r^2 dr \right.$$
$$\left. \times \int [(Y_{1,1} + Y_{1,-1} + Y_{1,0}) \cdot Y_{0,0}] \times \cos\theta \sin\theta \, d\theta \, d\phi \right\}. \tag{11.27}$$

Consider the ϕ dependence of integrals such as

$$\int Y_{1,-1}(\theta, \phi) Y_{0,0}(\theta, \phi) \sin\theta \, d\theta \, d\phi = \frac{1}{\sqrt{4\pi}} \left(\frac{3}{8\pi}\right)^{1/2} \times \int \sin^2\theta \, d\theta \int e^{-i\phi} d\phi,$$

where $Y_{1,-1} = (3/8\pi)^{1/2} \sin\theta e^{-i\phi}$. In general, any integral of the form

$$\int Y_{l,m}(\theta,\phi)Y_{l',m'}(\theta,\phi)\sin\theta\,d\theta\,d\phi$$

will result in a factor of the form

$$\int_0^{2\pi} e^{-i(m-m')\phi}\,d\phi.$$

This second integral is identically zero unless $m = m'$, thus yielding a selection rule. That is, the matrix element and hence the transition probability is identically zero unless $m = m'$. For the z component, the selection rule reduces to $m_f = m_i$. For the x and y components, the integral is of the form

$$\int \sin\phi\, e^{-i(m-m')\phi}\,d\phi$$

or

$$\int \cos\phi\, e^{-i(m-m')\phi}\,d\phi,$$

which yield a selection rule $m' = m \pm 1$ or $m_f = m_i \pm 1$. These *dipole selection rules* govern the observed spectra.

The z-matrix element for 2p to 1s hydrogenic wave functions reduces to

$$\frac{1}{3}\langle\psi_{100}|r\cos\theta|\psi_{210}\rangle = 0.248a. \tag{11.28}$$

In a similar manner, one can show that

$$\langle\psi_{1s}|x|\psi_{2p}\rangle = \langle R_{10}|r|R_{21}\rangle\delta_{l',l\pm1}\delta_{m',m\pm1}, \tag{11.29}$$

where

$$\delta_{l',l\pm1} = 1 \text{ if } \quad l' = l \pm 1$$
$$= 0 \qquad \text{otherwise.} \tag{11.30}$$

Then $\langle\psi_{1s}|x|\psi_{2p}\rangle$ reduces to

$$= \frac{2}{3\sqrt{24}}\frac{1}{a^3}\int e^{-3r/2a}\frac{r}{a}r^3\,dr, \tag{11.31}$$

or

$$\langle|x|\rangle = \left(\frac{81}{54}\right)^{1/2}\cdot\langle|z|\rangle, \tag{11.32}$$

and

$$\langle|y|\rangle = \left(\frac{81}{54}\right)^{1/2}\cdot\langle|z|\rangle.$$

Finally

$$\langle|x|\rangle + \langle|y|\rangle + \langle|z|\rangle = \langle|\mathbf{r}|\rangle, \tag{11.33}$$

and

$$\langle |\mathbf{r}| \rangle = 0.248a \left[1 + 2 \left(\tfrac{81}{54} \right)^{1/2} \right],$$

$$\langle |\mathbf{r}| \rangle^2 = 0.74a^2, \tag{11.34}$$

for 2p $\rightarrow$ 1s. Using Eq. (11.5), we find a value for the transition probability for Ni of

$$W = 1.2\,\mathrm{eV}/\hbar,$$

a factor of 2 greater than the accepted value of 0.55 eV/$\hbar$.

11.10   Calculation of the K_β / K_α Ratio

As a further example of the use of the transition probability formula, we calculate the K_β / K_α ratio for hydrogenic atoms. We use the following form for the radiative transition rate W:

$$W = \frac{4\omega^3 e^2}{3\hbar c^3} |\langle \psi_f | \mathbf{r} | \psi_i \rangle|^2 .$$

The ratio R is

$$R = \frac{W_{K_\beta}}{W_{K_\alpha}} = \frac{\omega_\beta^3 \, |\langle \psi_{100} | \mathbf{r} | \psi_{31} \rangle|^2}{\omega_\alpha^3 \, |\langle \psi_{100} | \mathbf{r} | \psi_{21} \rangle|^2}, \tag{11.35}$$

where

$$\psi_{3,1} = \frac{1}{\sqrt{3}} (\psi_{310} + \psi_{311} + \psi_{31-1})$$

and

$$\psi_{2,1} = \frac{1}{\sqrt{3}} (\psi_{210} + \psi_{211} + \psi_{21-1}).$$

These equations represent wave functions that are properly normalized and reflect the statistical weights of the degenerate sublevels, i.e.,

$$\langle \psi_{3,1} | \psi_{3,1} \rangle = 1/3(1 + 1 + 1) = 1.$$

We have also used the $\Delta l = \pm 1$ selection rule by considering p $\rightarrow$ s transitions only.
Consider the z component, $z = r \cos \theta$:

$$\begin{aligned}
\langle \psi_{100} | r \cos \theta | \psi_{3,1} \rangle = \frac{1}{3} \{ & \langle \psi_{100} | r \cos \theta | \psi_{310} \rangle \\
& + \langle \psi_{100} | r \cos \theta | \psi_{311} \rangle \\
& + \langle \psi_{100} | r \cos \theta | \psi_{31-1} \rangle \},
\end{aligned} \tag{11.36}$$

which reduces to $\sim 0.1a$.

For the $n = 2 \rightarrow 1$ transition: $\langle \psi_{100} | r \cos \theta | \psi_{2,1} \rangle$ reduces to $1/3 \langle \psi_{100} | r \cos \theta | \psi_{210} \rangle$, which is the matrix element in the K_α transition ($= 0.25a$) (Eq. 11.28) given in the

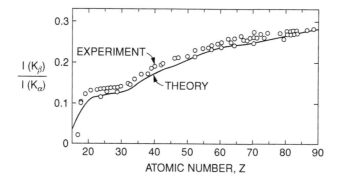

FIGURE 11.13. Measured K_β/K_α ratios as a function of atomic number. The theoretical curve represents a more sophisticated treatment that includes all electronic wave functions. The hydrogenic atom model result for $I(K_\beta)/I(K_\alpha)$ is 0.25. [From Bambynek et al., 1972, © 1972 by The American Physical Society.]

previous section. The ratio of K_β to K_α intensities is

$$R = \frac{\omega_{K_\beta}^3}{\omega_{K_\alpha}^3} \times \frac{(0.1)^2}{(0.25)^2}.$$

In hydrogenic atoms,

$$\omega_{K_\alpha} = \frac{3}{4} E_K,$$

$$\omega_{K_\beta} = \frac{8}{9} E_K,$$

so

$$R = \left(\frac{\frac{8}{9}}{\frac{3}{4}}\right)^3 \cdot \left(\frac{0.1}{0.25}\right)^2 = 0.25. \tag{11.37}$$

The measured values of the K_β/K_α intensity ratio are given in Fig. 11.13 and shown for a Mn spectrum in Fig. 11.5. Except for low Z values, the ratio is nearly independent of Z, and the calculation above is a reasonable estimate for $Z \geq 40$. At low Z, the calculation is influenced simply by the number of electrons available for the transitions (i.e., $R \cong 0$ for $Z \sim 10$, where there are no M-shell electrons).

Problems

11.1. Compare the values of L_α radiation given in Fig. 11.4 with values you would predict from the Bohr theory for elements with $Z = 30, 40, 50,$ and 60. Does the Bohr theory give a reasonable estimate (within 10%)?

11.2. Consider X-radiation emitted from a sample of CuTa irradiated by 15 keV electrons. List the appropriate bindings energies associated with $K_{\alpha 1}, K_{\beta 1}, L_{\alpha 1}, L_{\beta 1}$.

What energy resolution would be required to separate the signals for Cu K and Ta L radiation?

11.3. The energy width for the K vacancy state of Ca is about 1 eV. Estimate the ratio of radiative to nonradiative transitions.

11.4. Calculate the cross section for Ni K-shell production by 20 keV electrons and for 6.5 MeV protons. Compare the proton velocity with that of the Bohr velocity of Ni K-shell electrons.

11.5. You have thin films of $Hg_xCd_{1-x}Te$ whose composition is to be evaluated for a program on infrared detectors. You want to determine the Hg to Cd ratio and suspect that the sample may contain 1 at.% of Cu as an impurity. You mount a 100 nm thick film in an electron microscope (125 keV electrons) that has an energy-dispersive spectrometer for measurement of X-ray intensities with an energy resolution of 200 eV.

(a) For a given composition $x = 0.2$ of Hg_xCd_{1-x}, estimate the ratio of Hg to Cd K_α X-ray intensities from the ratio of cross sections and fluorescence yields (Appendix 11).

(b) Would you expect interference from K, L_α, or M X-ray lines from Hg, Cd, or Te?

(c) Estimate from Fig. 9.6 the mass absorption coefficient for Hg and Cd X-rays in Cd Te (assume values for Sn). Would X-ray absorption have a major influence in the intensity ratio?

(d) Estimate the Cu/Cd K X-ray intensity for 1 at.% Cu. Would interference or absorption have an influence?

11.6. Compare the Rutherford scattering cross section with the K_α X-ray production cross section for 4 MeV protons incident on Ag. Compare the two techniques on the basis of mass resolution and depth resolution.

11.7. Compare electron and X-ray emission processes for 4 keV X-rays and electrons incident on Al.

(a) What is the range, R_x, for the electrons in K-shell X-ray production, and what is the absorption depth where the incident X-ray flux decreases by 1/e (use Fig. 9.6)?

(b) What is the cross section for K-shell ionization by electrons and X-rays?

(c) What is the escape depth of K-shell photoelectrons and the absorption length (1/e attenuation) for the Al K_α X-rays?

(d) What is the ratio of photoelectron- to electron-induced X-ray emission yields for a 3 nm film and a 300 nm film?

References

1. W. Bambynek, B. Craseman, R. W. Fink, H.-U. Freund, H. Mark, C. D. Swift, R. E. Price, and Venugopala Rao, "X-ray Fluorescence Yields, Auger and Coster-Kronig Transition Probabilities," *Rev. Mod. Phys.* 44, 716 (1972).
2. L. S. Birks, *Electron Probe Microanalysis* (R. E. Krieger, New York, 1979).
3. T. A. Cahill, *Annu. Rev. Nucl. Part. Sci.* 30 (1980).
4. J. D. Garcia, *Phys. Rev. A* 1, 1402 (1970).

5. J. I. Goldstein, D. E. Newbury, P. Echlin, D. C. Joy, C. Fiori, and E. Lifshin, *Scanning Electron Microscopy and X-ray Microanalysis* (Plenum Press, New York, 1981).

6. S. A. E. Johansson and T. B. Johansson, *Nucl. Instrum. Meth.* 137, 473 (1976).

7. J. W. Mayer and E. Rimini, Eds., *Ion Beam Handbook for Material Analysis* (Academic Press, New York, 1977).

8. J. McGervey, *Introduction to Modern Physics* (Academic Press, New York, 1971).

9. F. H. Read, *Electromagnetic Radiation* (John Wiley and Sons, New York, 1980).

10. P. K. Richtmyer, E. H. Kennard, and J. N. Cooper, *Introduction to Modern Physics* (McGraw-Hill Book Co., New York, 1969).

11. L. Schiff, *Quantum Mechanics* (McGraw-Hill Book Co., New York., 1968).

12. J. P. Thomas and A. Cachard, Eds., *Material Characterization Using Ion Beams* (Plenum Press, New York, 1978).

13. D. B. Williams, *Practical Analytical Electron Microscopy in Materials Science* (Verlag Chemie International, Weinheim, 1984).

14. J. C. Willmott, *Atomic Physics* (John Wiley and Sons, New York, 1975).

15. R. Woldseth, *X-ray Energy Spectrometry* (Kevex Corp., 1973).

12
Nonradiative Transitions and Auger Electron Spectroscopy

12.1 Introduction

In previous chapters, we have discussed inner-shell vacancy formation by photon irradiation (the basis of X-ray photoelectron spectroscopy) or energetic electron and proton irradiation. The excited atoms can release their energy in radiative transitions (Chapter 11) with the emission of X-rays or in nonradiative transitions with the emission of electrons. The latter process forms the basis for Auger electron spectroscopy (AES), in which one determines composition by measuring the energy distribution of electrons emitted during irradiation with a beam of energetic electrons. As with other electron spectroscopies, the observation depth is about 1.0–3.0 nm and is determined by the escape depth (Chapter 7). The identification of atoms by core-level spectroscopies is based upon the values of the binding energies of the electrons. With Auger electron spectroscopy, the energy of the emergent electron is determined by the differences in binding energies associated with the deexcitation of an atom as it rearranges its electron shells and emits electrons (Auger electrons) with characteristic energies. Figure 12.1 shows the Auger radiationless deexcitation processes, in which the atom is left in the final state with two vacancies (or holes). If one of the final-state vacancies lies in the same shell as the primary vacancy (although not in the same subshell), the radiationless transition is referred to as a *Coster–Kronig transition*. This transition is significant because the Coster–Kronig transition rates are much higher than the normal Auger transitions and influence the relative intensities of the Auger lines. For example, in Fig. 12.1, if an L_1 shell has a vacancy, the L_2 to L_1 transition will be rapid (Coster–Kronig), therefore reducing vacancy transitions of the M electron to L_1.

12.2 Auger Transitions

12.2.1 Nomenclature

The nomenclature used to describe the Auger processes is shown in Fig. 12.1. For vacancies in the K shell, the Auger process is initiated when an outer electron such as an L_1 electron (dipole selection rules are not followed) fills the hole. The energy released can be given to another electron such as another L_1 or an L_3 electron, which is then ejected from the atom. The energy of the outgoing electron is $E_K - E_{L1} - E_{L1}$.

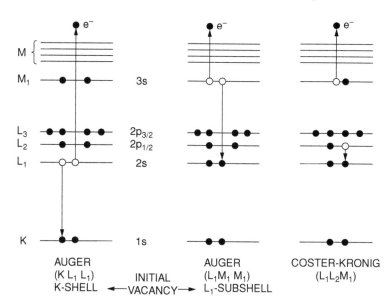

FIGURE 12.1. Schematic diagram of various two-electron deexcitation processes. The KL_1L_1 Auger transition corresponds to an initial K hole that is filled with an L_1 electron, while simultaneously the other L_1 electron is ejected to the vacuum. The LM_1M_1 Auger transition is the corresponding process with an initial 2s vacancy. The Coster–Kronig $L_1L_2M_1$ transition contains an initial L_1 hole that is filled with an L_2 electron accompanied by ejection of an M_1 electron.

The process described is called a *KLL* Auger transition in general terms and more specifically denoted as KL_1L_1 or KL_1L_3. If there are vacancies in the L shell, one can have Auger processes in which an electron from the M shell (M_1 electron) fills the L hole and another M-shell electron (for example, an M_1 electron) is ejected — an $L_1M_1M_1$ Auger transition. Since electron–electron interactions are strongest between electrons whose orbitals are closest together, the strongest Auger transitions are of the type *KLL* or *LMM*. For Coster–Kronig transitions, the vacancy is filled by electrons that come from the same shell, i.e., *LLM*. Auger transitions involving the outermost orbitals, the valence band, have an energy width of about twice that of the valence band. In Fig. 12.2, the Si $KL_1L_{2,3}$ and $L_{2,3}V_1V_2$ (or *LVV*) Auger transitions are indicated with V_1 and V_2 located at positions of maxima in the density of states in the valence band.

A complete nomenclature describing Auger transitions indicates the shells involved and the final state of the atom. The final state is usually described using the spectroscopic notation describing the orbitals. For example, a KL_1L_1 transition would leave the 2s shell empty (two vacancies) and the 2p shell with six electrons; the transition is KL_1L_1 $(2s^02p^6)$. A KL_2L_3 would leave the vacancies in the 2p shell and would be indicated as KL_2L_3 $(2s^22p^4)$. Even in the relatively simple *KLL* transition, there is a large variety of final states that can have slightly different energies and hence will correspond to slightly different Auger lines. In the following, we discuss these states in detail.

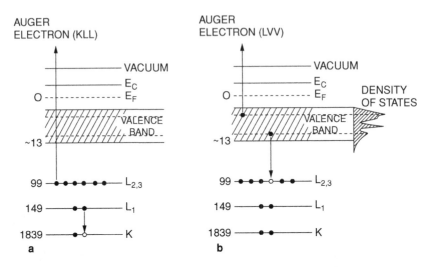

FIGURE 12.2. Schematic of the (a) KL_1 and (b) LVV Auger deexcitation processes in Si. Binding energies are indicated on the left. The energy of Auger electron in the KL_1L_{23} process is approximately 1591 eV, and the $L_{23}VV$ Auger electron has an energy of approximately 90 eV.

12.2.1.1 KL_1L_1

In the usual X-ray notation, this transition corresponds to an initial state of a single 1s hole and a final state of two 2s holes. We can consider electron holes as electrons to find the possible final configurations of the final states. The $n = 2$ shell is now considered as $2s^0 2p^6$ (where the filled shell is $2s^2 2p^6$) and has states given by the possible allowed quantum numbers consistent with the Pauli exclusion principle: $m_s = \pm 1$, $M_L = 0$, $M_s = 0$, where M_L and M_s are the total orbital and spin angular momenta, respectively. The notation ^1S indicates a state of total orbital momentum zero (S). This transition is properly written $KL_1L_1(^1S)$, although the final state (^1S) is the only one allowed, and therefore, in this case, the notation is slightly redundant.

12.2.1.2 KL_1L_2 or KL_1L_3

In this case, the final-state electron configuration is written $2s^1 2p^5$. The possible quantum states are ^1P and ^3P, where P denotes the total orbital angular momentum. This corresponds to two states coupling to a total angular momentum $L = 1$, i.e., a P state with the electron spins aligned ^3P and antialigned ^1P.

12.2.1.3 KL_2L_2, KL_2L_3, and KL_3L_3 Transitions

Here the final states can couple to total angular momenta states of $D(L = 2)$, $P(L = 1)$, and $S(L = 0)$, with different possible spin alignments to yield states ^1D, ^3P, and ^1S.

Thus, in KLL-type transitions, there is a total of six final states possible:

$$KL_1L_1 - 2s^0 2p^6(^1S),$$
$$KL_1L_{2,3} - 2s^1 2p^5(^1P, ^3P),$$

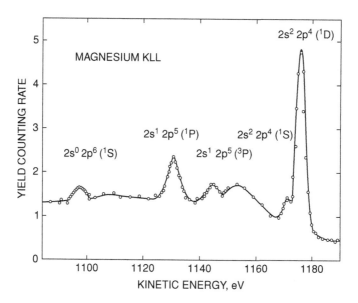

FIGURE 12.3. *KKL* spectrum of magnesium containing 5 of the 6 lines predicated in *L–S* coupling. [From Siegbahn et al., 1967.]

$$K L_{2,3} L_{2,3} - 2s^2 2p^4 (^1D, ^3P, ^1S)$$

These final states are shown experimentally in Fig. 12.3 for the case of magnesium. [Actually, the $2s^2 2p^4 (^3P)$ state is not observed due to lack of intensity.]

12.2.2   *Energies*

The energy of the Auger electrons can in principle be determined in the same way as that of X-rays: by the difference of the total energies before and after the transition. An empirical way of doing this, for example, is by

$$E_{\alpha\beta\gamma}^Z = E_\alpha^Z - E_\beta^Z - E_\gamma^Z - \frac{1}{2}(E_\gamma^{Z+1} - E_\gamma^Z + E_\beta^{Z+1} - E_\beta^Z), \qquad (12.1)$$

where $E_{\alpha\beta\gamma}^Z$ is the Auger energy of the transition $\alpha\beta\gamma$ of the element Z. The first three terms correspond to the difference in the binding energies of shells α, β, γ of the element Z. The correction term is small and involves the average of the increase in binding energy of the γ-electron when a β-electron is removed and of the β-electron when a γ-electron is removed. Measured values of Auger *KLL* transitions are given in Appendix 9 along with values of the binding energies (Appendix 6). A numerical test of the approximation of Eq. 12.1 is given in Table 12.1 for *KLL* transitions. The agreement is good. Figure 12.4 shows the dominant Auger energies versus atomic number. The strong Z dependence of the binding energies leads to a straightforward elemental identification using this technique.

TABLE 12.1. Tabulation of values used to calculate the energy of the KL_1L_2 Auger transition in Ni

(1) $E^{Ni}_{KL_1L_2} = E^{Ni}_K - E^{Ni}_{L_1} - E^{Ni}_{L_2} - \frac{1}{2}\left(E^{Cu}_{L_2} - E^{Ni}_{L_2} + E^{Cu}_{L_1} - E^{Ni}_{L_1}\right)$

(2) Electron binding energies in keV from Appendix 6

$E^{Ni}_K = 8.333$ $\qquad$ $E^{Cu}_{L_2} = 0.951$ $\qquad$ $E^{Cu}_{L_1} = 1.096$

$-E^{Ni}_{L_1} = 1.008$ $\qquad$ $-E^{Ni}_{L_2} = \dfrac{0.872}{0.079}$ $\qquad$ $-E^{Ni}_{L_1} = \dfrac{1.008}{0.088}$

$-E^{Ni}_{L_2} = \dfrac{0.872}{6.453}$ $\qquad$ $\dfrac{1}{2}(0.079 + 0.088) = 0.084$

$E^{Ni}_{KL_1L_2} = 6.453 - 0.084 = 6.369 \text{ keV}$

(3) Auger transition energy from Appendix 9

$E^{Ni}_{KL_1L_2} = 6.384 \text{ keV}$

12.2.3 Chemical Shifts

The chemical environment of an atom is reflected in changes in the valence-shell orbitals, which in turn influence the atomic potential and the binding energy of the core electrons. The binding energies of the inner-core K and L shells shift in unison with changes in the chemical environment. For this reason, the K_α X-ray emission lines, which are transitions between K and L shells, have only small shifts. For *KLL* Auger electron lines, both the K and L shells are involved, but unlike the K_α X-ray emission lines, the L shell is involved twice in the transition. The inner-shell electron that is ejected in *KLL* Auger processes therefore will display a chemical shift. Thus, one would expect chemical shifts in both AES and XPS spectra.

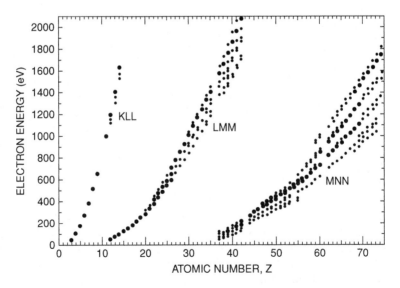

FIGURE 12.4. Principal Auger energies versus atomic number. The heavy points indicate the strong transitions for each element. [From Davis et al., 1976.]

Chemical shifts are evident in both AES and XPS spectra. However, the chemical shifts are more difficult to interpret in the two-electron Auger process than in the one-electron photoelectric process. Further, Auger line widths are broader than XPS lines. Consequently, the latter technique is typically used to explore changes in chemical binding.

12.2.4   Estimate of the Auger Transition Probability in a Hydrogenlike Atom: KLL Transition

In the usual manner, the transition probability for the Auger effect, W_A, can be written as

$$W_A = \frac{2\pi}{\hbar} \rho(k) \left| \phi_f(\mathbf{r}_1)\psi_f(\mathbf{r}_2)\frac{e^2}{|\mathbf{r}_1 - \mathbf{r}_2|}\phi_i(\mathbf{r}_1)\psi_i(\mathbf{r}_2)\,d\mathbf{r}_1 d\mathbf{r}_2 \right|^2, \tag{12.2}$$

where $\rho(k) = m(V/8\pi^3\hbar^2)k\sin\theta\,d\theta d\phi$ is the density of states associated with normalization in a box of volume V and for a *KLL* Auger transition from a hydrogenlike atom:

$$\phi_i\,(\mathbf{r}_1) = \frac{1}{\sqrt{6a^3}}\frac{\mathbf{r}_1}{a}e^{-\mathbf{r}_1/2a}Y_1^m(\theta_1, \phi_1), \tag{12.3}$$

$$\phi_f\,(\mathbf{r}_1) = \frac{2}{\sqrt{a^3}}e^{-\mathbf{r}_1/a}, \tag{12.4}$$

$$\psi_i(\mathbf{r}_2) = \frac{1}{\sqrt{6a^3}}\frac{\mathbf{r}_2}{a}e^{-\mathbf{r}_2/2a}Y_1^m(\theta_2, \phi_2), \tag{12.5}$$

$$\psi_f\,(\mathbf{r}_2) = \frac{1}{\sqrt{V}}e^{i\mathbf{k}.\mathbf{r}_2}. \tag{12.6}$$

These wave functions represent electrons in the 2p state, the 1s state, the 2p state, and a free electron, respectively, with $a = a_0/Z$, and a_0 the Bohr radius. For convenience, it is useful to write these equations as functions of r/a:

$$\phi_i\,(\mathbf{r}_1) = \frac{1}{\sqrt{6a^3}}\phi_i'\left(\frac{\mathbf{r}_1}{a}\right), \tag{12.3'}$$

$$\phi_f\,(\mathbf{r}_1) = \frac{2}{\sqrt{a^3}}\phi_f'\left(\frac{\mathbf{r}_1}{a}\right), \tag{12.4'}$$

$$\psi_i\,(\mathbf{r}_2) = \frac{1}{\sqrt{6a^3}}\psi_1'\left(\frac{\mathbf{r}_2}{a}\right), \tag{12.5'}$$

$$\psi_f\,(\mathbf{r}_2) = \frac{1}{\sqrt{V}}e^{ia\mathbf{k}\cdot\mathbf{r}_2/a}, \tag{12.6'}$$

and the potential as

$$\frac{e^2}{a}\left(\frac{1}{|\mathbf{r}_1/a - \mathbf{r}_2/a|}\right) = \frac{e^2}{a}V'(\mathbf{r}_1/a, \mathbf{r}_2/a). \tag{12.7}$$

In this calculation, we are considering a KL_2L_2 Auger transition, the transition of a 2p electron (L_2) to the 1s state (K), and the subsequent emission of another 2p electron (L_2) to a free electron. In this hydrogenic model, the energy of the Auger electron, E_A, is

$$E_A = E_K - E_L - E_L = E_K/2,$$

where E_K and E_L are the binding energies of the K and L shells, respectively, and $E_L = \frac{1}{4} E_K$ in the Bohr model. In the hydrogenic model,

$$E_K = e^2 Z^2/2a_0,$$

and

$$a = a_0/Z,$$

so

$$ak = (a_0/Z)\sqrt{mE_k/\hbar^2},$$
$$= 1/\sqrt{2},$$

where we have used the Bohr relation $a_0 = \hbar^2/me^2$. Then

$$\psi_f(\mathbf{r}_2) = \frac{1}{\sqrt{V}} e^{ir_2} \cos\theta_2/a\sqrt{2}.$$

In this form we can extract the basic dependence of the transition probability on atomic parameters such as the atomic number Z. Substituting Eqs. 12.3–12.7 (in modified form) into the formula for the transition probability, we find

$$W_A = \frac{2\pi}{\hbar} \frac{mk/d\Omega}{\hbar^2 8\pi^3} \frac{e^4 a}{9} F, \tag{12.8}$$

where $d\Omega = \sin\theta d\theta\, d\phi$ and

$$F = \left| \int\int \frac{d\mathbf{r}_1}{a^3} \cdot \frac{d\mathbf{r}_2}{a^3} \phi_f'\left(\frac{\mathbf{r}_1}{a}\right) \psi_f'\left(\frac{\mathbf{r}_2}{a}\right) \cdot V'\left(\frac{\mathbf{r}_1}{a}, \frac{\mathbf{r}_2}{a}\right) \phi_1'\left(\frac{\mathbf{r}_1}{a}\right) \psi_1'\left(\frac{\mathbf{r}_2}{a}\right) \right|^2.$$

F is a definite integration over all space for $\mathbf{r}_1/a$ and $\mathbf{r}_2/a$, resulting in a definite number that represents a matrix element of the potential factor $(|\mathbf{r}_2/a - \mathbf{r}_1/a|)^{-1}$.

Remembering that $ak = 1/\sqrt{2}$ and taking $d\Omega = 4\pi$, we can write the transition probability as

$$W_A = C\frac{e^4 m}{\hbar^3},$$

where C is a numerical constant dependent on the various factors in W_A and F. Noting that $a_0 = \hbar^2/me^2$ and $v_0 = e^2/\hbar$, we have the simple relationship

$$W_A = Cv_0/a_0, \tag{12.9}$$

where v_0 is the Bohr velocity, 2.2×10^8 cm/s; a_0 is the Bohr radius, 0.053 nm; and a_0/v_0 is a characteristic atomic time, 2.4×10^{-17} s. The integral F can be evaluated in the crude approximation $r_2 > r_1$ so that $1/|\mathbf{r}_2 - \mathbf{r}_1| \cong [1 + (r_1/r_2)\cos\theta_{1,2}]/r_2$, where

TABLE 12.2. Comparison of Auger transition
rates and K-level X-ray emission rates.

Atomic #	Element	Auger	K X-ray
10	Ne	0.23	0.005
11	Na	0.29	0.007
12	Mg	0.36	0.010
13	Al	0.40	0.014
14	Si	0.44	0.02
15	P	0.48	0.03
16	S	0.51	0.04
17	Cl	0.54	0.05
18	Ar	0.58	0.07
20	Ca	0.65	0.12
22	Ti	0.69	0.19
24	Cr	0.72	0.28
26	Fe	0.75	0.40
28	Ni	0.78	0.55
32	Ge	0.83	1.0
36	Kr	0.89	1.69
40	Zr	0.94	2.69
46	Pd	0.99	4.94
52	Te	1.04	8.40
58	Ce	1.07	11.6
65	Tb	1.10	21.8
70	Yb	1.13	29.6

$\theta_{1,2} = \theta_1 - \theta_2$. This approximation is based on the fact that the radial extension of the
1s wave function is small compared to the 2p function. The calculation of C is tedious
but straightforward, resulting in $C = 7 \times 10^{-3}$. The result for W_A (Eq. 12.9) is independent of Z, as suggested by Table 12.2. In a more complete calculation, one must
properly account for all the different equivalent pairs of electrons available for Auger
decay. The calculations can be done more precisely via numerical techniques, which
include more sophisticated wave functions and a better description of the interaction
potential (Bambynek et al., 1972). The main feature is that the Auger transition probability is roughly independent of Z in contrast to the strong Z dependence of radiative
transitions.

12.3   Yield of Auger Electrons and Fluorescence Yield

The lifetime of an excited state τ (a hole in a shell), is determined by the sum of
all possible decay processes. Radiative transitions occur with probability W_X. Auger
transitions have a probability W_A and Coster–Kronig (where the hole is filled by an
electron of the same shell) W_K. There are no other deexcitation mechanisms, so

$$1/\tau = W_X + W_A + W_K. \tag{12.10}$$

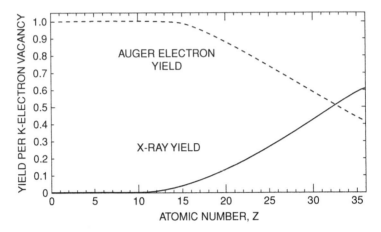

FIGURE 12.5. Auger electron and X-ray yields per K vacancy as a function of atomic number. The curves are from Equ 12.12. [Adapted from Siegbahn et al. 1967]

For transitions to vacancies in the K shell (as well as for holes in the L_3 and M_5 shells), Coster–Kronig transitions do not occur, and the probability for X-ray emission, ω_X, is given by

$$\omega_X = \frac{W_X}{W_A + W_X}, \tag{12.11}$$

ω_X is commonly called the *fluorescence yield*. For transitions to K-shell vacancies, the probability for radiative decay is proportional to Z^4 (Chapter 11), and the Auger probability is essentially independent of Z, as suggested by the semiempirical relation for ω_X of the form

$$\omega_X = \frac{W_X/W_A}{1 + W_X/W_A}, \tag{12.12}$$

where

$$\frac{W_X}{W_A} = (-a + bZ - cZ^3)^4, \tag{12.13}$$

with the numerical values $a = 6.4 \times 10^{-2}$, $b = 3.4 \times 10^{-2}$, and $c = 1.03 \times 10^{-6}$. This relationship yields the solid curve shown in Fig. 12.5. The *Auger electron yield* is $1 - \omega_X$. This figure shows the dominance of Auger transitions for low Z elements; in these cases, Auger emission is the important mechanism for relaxation of K vacancies. This curve does not imply that the Auger rate decreases at high Z, but emphasizes that the X-ray transition becomes the preferred method of deexcitation at high Z.

The fluorescence yield for K, L_3, and M_5 shells versus binding energy is shown in Fig. 12.6. The point of the figure is that the fluorescence yield is approximately the same for comparable transition energies independent of the electronic shell, in those cases where Coster–Kronig transitions do not occur. For K-shell transitions, the fluorescence yield is less than 0.1 for binding energies less than 2 keV, and the total Auger yield is

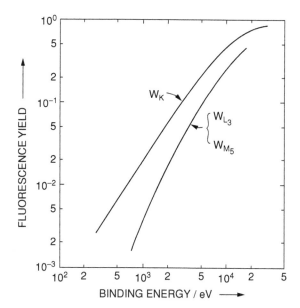

FIGURE 12.6. Fluorescence yield for K_1, L_2, and M_8 shells versus binding energy. [From J. Kirschner in Ibach, 1976, with permission from Springer Science+ Business Media.]

larger than 90% for low Z elements ($Z < 15$). Similarly, for L_3 transitions (Coster–Kronig transitions not allowed), Auger transitions dominate for $Z < 50$ where L-shell binding energies are less than 5 keV.

12.4 Atomic Level Width and Lifetimes

As pointed out in Chapter 11, the energy width ΔE, or more conventionally Γ, is related to the mean life τ of the state through the uncertainty principle, $\Gamma \tau = \hbar$. The decay probability per unit time is equal to the sum of the transition probabilities, so the total energy width of the state is given by

$$\Gamma = \Gamma_{\text{radiative}} + \Gamma_{\text{nonradiative}}. \tag{12.14}$$

There is a decay probability for each atomic process, but there is only a single lifetime for the hole. The natural line width for each process is given by the total lifetime. In the $Z < 30$ regime where Auger emission dominates, Table 12.2 shows that Auger rates vary from 0.23 to 0.80 eV/$\hbar$. The total width of the atomic transition then is 0.23–0.8 eV. For $Z > 30$, the K X-ray emission rates range up to 30 eV/$\hbar$, with a corresponding increase in atomic level width. The total lifetime $\tau = \hbar/\Gamma$, where $\hbar = 6.6 \times 10^{-16}$ s, will vary from about 10^{-17} to 10^{-15} s. Consequently, the measured X-ray spectrum will exhibit more line broadening at higher Z than at low Z, and hence Al or Mg is used as an X-ray source for XPS.

An X-ray spectrum measured with high resolution would be in the form of a Lorentzian centered about an energy E_X (Fig. 12.7):

$$Y(E) = \frac{A}{(E - E_X)^2 + \Gamma^2/4}. \tag{12.15}$$

FIGURE 12.7. Comparison of a Lorentzian line shape (Eq. 12.15) and a Gaussian line shape with the same full width half maximum. The Lorentzian is characterized by extended tails.

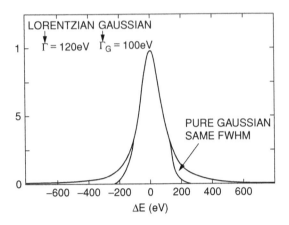

12.5 Auger Electron Spectroscopy

As with the other electron spectroscopies, Auger analysis is done under high vacuum conditions. Figure 12.8 shows schematically an experimental apparatus. The cylindrical mirror analyzer (CMA) has an internal electron gun whose beam is focused to a point on the specimen sample at the source point of the CMA. Electrons ejected from the sample pass through an aperture and then are directed through the exit aperture on the CMA to the electron multiplier. The pass energy E is proportional to the potential applied to the outer cylinder, and the range ΔE of transmitted electrons is determined by the resolution $R = \Delta E/E$, where R is typically 0.2–0.5%.

A schematic overall spectrum of electrons emitted from a solid irradiated by a 2 keV electron beam is shown in Fig. 12.9. The narrow peak on the right side is made up of elastically scattered electrons (no energy loss). Features at slightly lower energy correspond to electrons with characteristic energy losses due to electronic and plasma

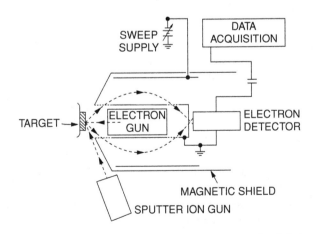

FIGURE 12.8. Experimental apparatus used in Auger spectroscopy. [After Palmberg in Czanderna et al., 1975.]

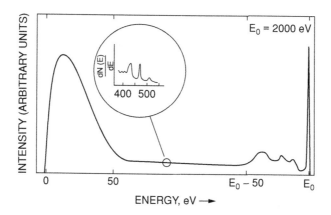

FIGURE 12.9. Spectrum of 2 keV electrons scattered from a solid. The inset shows the regime for Auger spectroscopy. The energy scale is nonlinear. [From Ibach, 1977, with permission from Springer Science+Business Media.]

excitations. Auger electron transitions generally appear as small features superimposed on the large background of secondary electrons. The usual practice is to use derivative techniques and generate a $dN(E)/dE$ function (inset to Fig. 12.9). Differential analysis of a hypothetical spectrum is shown in Fig. 12.10. The contribution from the slowly varying background is minimized by the derivative technique. The total backscattered background current with energy greater than 50 eV is typically 30% of the primary beam current. The noise level due to this current and the ratio of the analyzer ΔE to Auger line width generally establishes the signal-to-noise ratio and hence the detection

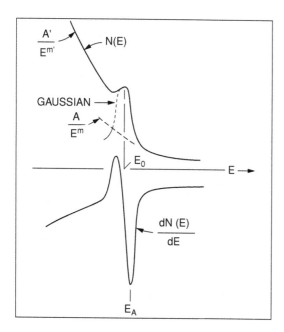

FIGURE 12.10. A hypothetical spectrum $N(E)$ containing a continuous background AE^{-m}, a Gaussian peak, and a low-energy step of form $A'E^{-m}$. The lower spectrum is the derivative spectrum. Note that the energy E_A of the most negative excursion of the derivative corresponds to the steepest slope of $N(E)$.

limit for impurities in the sample. A typical value for the detection limit is 1000 ppm, $\cong 0.1$ atomic %.

In practice, Auger spectroscopy is usually carried out in the derivative mode because of the small signal. The *differentiation* is conveniently done electronically by superimposing a small AC voltage on the outer-cylinder voltage and synchronously detecting the in-phase signal from the electron multiplier with a lock-in amplifier. The y-axis of the recorder is then proportional to $dN(E)/dE$ and the x-axis to the kinetic energy E of the electrons. The derivative spectrum is extracted directly. In this scheme, a perturbing voltage,

$$\Delta V = k \sin \omega t, \tag{12.16}$$

is superposed on the analyzer energy so that the collected electron current $I(V)$ is modulated. $I(V + \Delta V)$ can be written in a Taylor expansion:

$$I(V + k \sin \omega t) = I_0 + I'k \sin \omega t + k^2 \frac{\sin^2 \omega t}{2!} I'' \ldots, \tag{12.17}$$

where the prime denotes differentiation with respect to V. If we include higher-order terms in the expansion, then

$$I = I_0 + \left[kI' + \frac{k^3}{8} I''' \right] \sin \omega t - \left[\frac{k^2}{4} I'' + \frac{k^4}{48} I'''' \right] \cos 2\omega t, \tag{12.18}$$

where I_0 contains all non-time-dependent terms. In this calculation, we assumed $k \ll V$ so that terms of order k^3 and higher can be neglected in practice. Using a lock-in amplifier, for phase-sensitive detection, we select the component of the signal associated with the frequency ω, which is simply the desired quantity I' or dN/dE for a cylindrical mirror Auger analyzer. To satisfy this criterion, we require that k be less than the Auger width of ~ 5 eV.

An example of the use of derivative techniques is shown in Fig. 12.11 for 2 keV electrons incident on a Co sample. In the direct spectrum, $N(E)$, the main features are the peak of elastically scattered electrons and a nearly flat background. The arrows in Fig. 12.11a indicate the energies of oxygen and Co Auger transitions. The derivative spectrum (Fig. 12.11b) reveals the *LMM* Co and *KLL* carbon and oxygen signals.

For a free atom, the Auger yield Y_A is determined by the product of the electron impact ionization cross section (Chapter 6) and the probability for the emission of an Auger electron $(1 - \omega_x)$:

$$Y_A \propto \sigma_e \cdot (1 - \omega_X). \tag{12.19}$$

In a solid, the situation is more complicated even when considering the yield from a layer of the thickness of the electron escape depth λ. For example, primary electrons that penetrate the surface layer and then are backscattered can contribute to the Auger yield when the energy E_p of the primary electron is much greater than the binding energy. The yield is also strongly affected by the angles of incidence (diffraction effects influence the number of elastically scattered primaries) and of emission (geometric projection of the escape depth). Consequently, surface roughness plays a role; the escape probability of electrons from a rough surface is less than that from a smooth surface. In analyzing

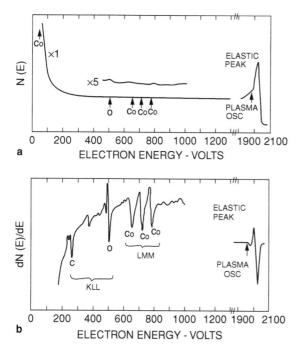

FIGURE 12.11. Comparison of (a) the spectrum $N(E)$ and (b) the derivative $dN(E)$ for 2 keV electrons incident on a Co sample. [From Tousset, in Thomas and Cachard, 1978, with permission of Springer Science+Business Media.]

solids, then, one must consider the modification of both the incident beam and the Auger electrons on passing through the solid.

Auger electron spectroscopy is a surface-sensitive technique. Figure 12.12 shows the oxygen signal corresponding to the absorption of 0.5 monolayers of oxygen atoms. In general, small amounts of the typical contaminants, C, N, and O, are easily detected.

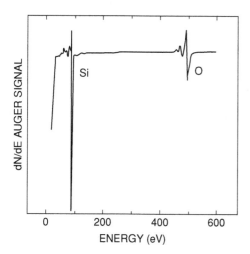

FIGURE 12.12. Auger spectra from single-crystal Si (111) after absorption of ~ 0.5 monolayers of oxygen atoms.

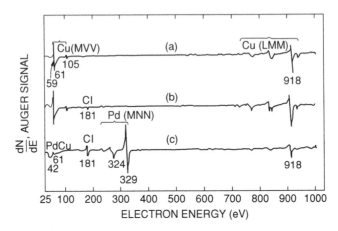

FIGURE 12.13. Auger signal of (a) freshly deposited Cu substrate. (b) Cu substrate just prior to Pd deposition, and (c) Pd/Cu bilayer, 1.35 nm of Pd.

Hydrogen cannot be detected in Auger measurements, since three electrons are needed in an Auger transition.

The Auger signal from a substrate is sensitive to the presence of surface layers. In Chapter 6, we noted that the substrate signal decreased as $e^{-x/\lambda}$, where $\lambda = 0.5$ nm for Si electrons penetrating Ge. Fig. 12.13 shows Auger spectra from a Cu substrate before and after the deposition of 1.35 nm of Pd. It is clear from the figure that the Cu signal is strongly attenuated by the Pd coverage. In particular, the low-energy Cu(MVV) line is completely attenuated due to the small escape length for 60 eV electrons; the high-energy line at 918 eV is only partly attenuated.

12.6   Quantitative Analysis

The determination of an absolute concentration of an element x in a matrix from the yield Y_A of Auger electrons is complicated by the influence of the matrix on the backscattered electrons and escape depth. For simplicity, let us consider $Y_A(t)$, the yield of *KLL* Auger electrons produced from a thin layer of width Δt at a depth t in the sample:

$$Y_A(t) = N_x \Delta t \cdot \sigma_e(t)[1 - \omega_X]e^{-(t\cos\theta/\lambda)} \cdot I(t) \cdot T \cdot d\Omega/4\pi, \qquad (12.20)$$

where

N_x	=	the number of x atoms/unit vol;
$\sigma_e(t)$	=	the ionization cross section at depth t;
ω_X	=	the fluorescence yield;
λ	=	the escape depth;
θ	=	the analyzer angle;
T	=	the transmission of the analyzer;
$d\Omega$	=	the solid angle of the analyzer;
$I(t)$	=	the electron excitation flux at depth t.

It is convenient to separate the excitation flux density into two components,

$$I(t) = I_p + I_B(t) = I_p(t)[1 + R_B(t)],$$

where I_p is the flux of primary electrons at depth t, I_B is the flux due to backscattered primary electrons, and R_B is the backscattering factor (Section 10.7).

When external standards are used with a known concentration N_X^S of element x in the standard, the concentration N_X^T in the test sample can be found from the ratio of Auger yields:

$$\frac{N_X^S}{N_X^T} = \frac{Y_X^S}{Y_X^T} \left(\frac{\lambda^T}{\lambda^S}\right) \left[\frac{1 + R_B^T}{1 + R_B^S}\right].$$

In this approach, the ionization cross section and the fluorescence yield are not required because the Auger yields from the same atom are measured. In addition, if the composition of the standard is close to that of the test sample, the element composition can be determined directly from the ratio of Auger yields if the measurements are made under identical experimental conditions. When the composition of the standard differs substantially from that of the test specimen, the influence of the matrix on electron backscattering and escape depth must also be considered.

Elemental sensitivities are acquired using pure element standards and are applied to unknown determinations in multielemental matrices. One must correct for the highly matrix-dependent parameters, which include the inelastic mean free path λ.

Even with corrections for escape depth and backscattering, the measured surface composition may not be related to the bulk composition of the sample because of the ion bombardment used in sputtering for sample cleaning and depth profiling (see Chapter 4).

12.7 Auger Depth Profiles

A major use of Auger electron spectroscopy is determining the composition as a function of depth in thin films and layered structures. The conventional apparatus is illustrated in Fig. 12.8, which consists of an electron gun and CMA assembly as well as a sputter ion gun. The Auger signal is generated in the near surface region of the sample (~ 3.0 nm), and ion sputtering provides the layer sectioning technique required for depth analysis. In routine laboratory use, the depth profiles are shown as Auger signal height versus sputter time. Further calibrations are required to convert sputter time to depth and signal height to atomic concentration. The combination of Rutherford backscattering spectrometry (RBS) and Auger electron spectroscopy (AES) is quite useful in such depth profile analyses because RBS gives quantitative information on depths and heavy mass constituents without the complications introduced by the intermixing due to sputtering. As discussed in Chapter 4, ion sputtering causes a change in the composition of the surface layers due to surface segregation and preferential sputtering. As compared to RBS, Auger depth profiling provides better depth resolution and is sensitive to both heavy and light elements.

In Fig. 12.14, we illustrate the data obtained from RBS and AES measurements on a sample prepared by depositing 100 nm of Ni on $\langle 100 \rangle$ InP (Fig. 12.14a) and annealing

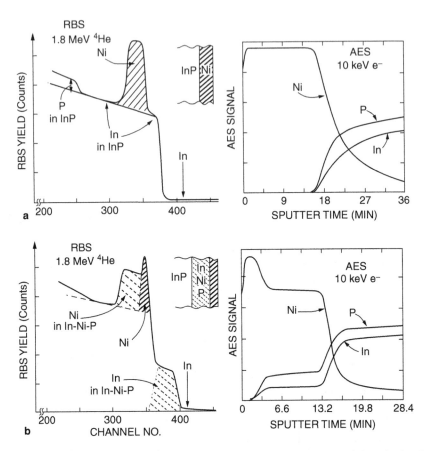

FIGURE 12.14. Comparison of RBS (left) and AES depth profiles of 100 nm Ni deposited on InP: (a) as deposited; (b) annealed at 250 °C for 30 min. [From Appelbaum, private communication.]

at 250°C for 30 min (Fig. 12.14b). In the RBS spectrum for the as-deposited case, the Ni signal is superimposed on the signal from the InP substrate. In the AES spectrum, both the In and P signals have comparable heights and can be clearly resolved. The long tail on the Ni signal, which extends well beyond the interface region, is clearly an artifact of the sputtering process because the Ni/InP interface is sharp, as can be inferred from the rear edge of Ni signal in the RBS spectrum. After annealing, the layer is partially reacted with an outer layer of Ni on a layer of $In_xP_yN_z$. The Ni layer and the reacted InPNi layer can be clearly seen in the AES spectrum, which has a P/In yield ratio of $\cong 2/1$. In the RBS spectrum, the heights of the Ni and In signals are nearly equal, which indicates that the ratio of Ni to In is about $3[\sigma_{In}/\sigma_{Ni} \cong 3.08]$. Analysis of the RBS spectra yields a P/In ratio of 0.5, a value quite different from the P-rich composition deduced from the AES data. The origin of the discrepancy possibly is due to preferential sputtering and segregation. The region of pure Ni in the reacted film is better resolved with AES due to its superior depth resolution. Further AES allowed a

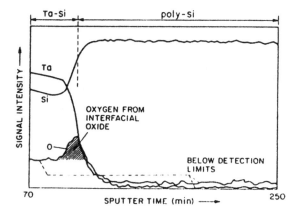

FIGURE 12.15. Sputter depth profiling with AES of the interface region of a Ta–Si film deposited on
polycrystalline Si. The shaded area
represents the oxygen signal from
the native oxide at the interface.
[From D. Pawlik, H. Oppolzer, and
T. Hillmer, *J. Vac. Sci. Technol.
B*, 3, 492 (1985). Copyright 1985
American Physical Society.]

determination of the carbon and oxygen at the interface region (not shown), which is
not possible with RBS.

One of the advantages of Auger electron spectroscopy is its sensitivity to low mass
impurities, such as carbon or oxygen, which are common contaminants at surfaces
and interfaces. The presence of these interfacial contaminants plays a disruptive role
in thin film reactions by retarding interdiffusion. The degradation of the planarity of
thin film structures following thermal processing is often directly correlated with these
contaminants. The presence of a native oxide of about 1.5 nm thicknesses is readily
apparent in the AES depth profile shown in Fig. 12.15. The removal of this native
oxygen layer is crucial for the formation of thin, uniform oxide layers on top of the
Ta–silicide layers during thermal oxidation. The presence of the native oxide layer
retards the release of Si from the poly-Si layer and leads to the oxidation of the whole
Ta–silicide layer rather than the formation of a SiO_2 layer on the surface. Auger electron
spectroscopy in conjunction with sputter depth profiling has the prerequisite sensitivity
to detect contaminant layers that impede thin film reactions.

Multilayer films are used in integrated circuits and optical structures as well as in
many other aspects of solid-state science. Auger electron spectroscopy with sputter
depth profiling has a natural application to the analysis of these structures.

Figure 12.16 shows sputter depth profiles of multilayer Cr/Ni thin film structures
deposited in a Si substrate. This impressive figure demonstrates the ability of Auger
spectroscopy, combined with sputtering, to profile a multilayer film of nearby elements
in the periodic chart in a semiquantitative manner. The rounding in the traces in the
upper portion of the figure reflects the irregularity in the surface topology that developed
during sputtering with a rastered beam of 5 keV Ar ions (see Chapter 4 and Carter et al.,
1983). In this example, the surface roughness could be minimized by rotating the sample
(lower portion of Fig. 12.16) during sputtering.

Modern analytical laboratories are now equipped with a variety of systems for depth
profiling of samples. When confronted with a layered or thin film sample containing unknown impurities or contaminants, the analyst will use all techniques at hand.
Sputter depth profiling with Auger analysis often represents the starting point for initial
analyses.

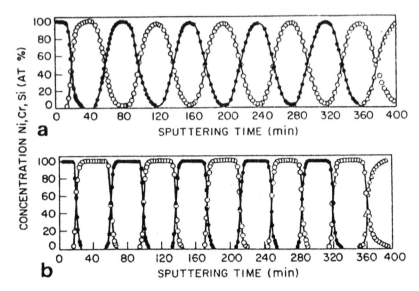

FIGURE 12.16. Auger electron spectroscopy sputter depth profiles of multilayer Cr/Ni thin film structures deposited on a Si substrate. The top Ni layer is about 25 nm thick and the other films are about 50 nm thick. Sputtering was carried out with a rastered beam of 5 keV Ar ions, with a stationary sample in the upper portion (a) and a rotating sample in the lower portion (b). The symbols represent: • nickel; ○ chromium; △ silicon. [From A. Zalar, *Thin Solid Films* 124, 223 (1985), with permission from Elsevier.]

Problems

12.1. You irradiate an AlP sample with 5 keV electrons and measure the *KLL* Auger electrons. Calculate the ratio of Al to P ionization cross sections, fluorescence yields ω_X, escape depths λ, and Auger yields.

12.2. Compare the Auger yields in Problem 12.1 with the electron microprobe K X-ray yields, ignoring X-ray absorption or electron backscattering corrections.

12.3. A Mg K_α X-ray creates a vacancy in the Cu L_1(2s) subshell. Estimate the energies of photoelectrons, LM_1M_1 Auger electrons, $L_1L_2M_1$ Auger electrons (Coster–Kronig transitions), and L X-rays. Would this L-shell vacancy preferentially be filled by radiative or nonradiative transitions? Make an estimate of the upper value of the fluorescence yield. Which Cu L level would not deexcite by Coster–Kronig transitions?

12.4. A beam of 10 keV electrons irradiates a 100 nm thick film of Ni on a Si substrate. Calculate the ratio of K X-ray and *KLL* Auger yields.

12.5. You are given a 20 nm thick layer of Ga_xAl_{1-x} as on an InP substrate about 1 mm thick and are asked to determine the Ga-to-Al ratio. You can carry out XPS, AES, or EMA analysis using 20 keV electrons and an Al K_α X-ray source. In order to compare the different techniques, you carry out the following calculations or comparisons.

(a) What is the cross section ratio σ_{Ga}/σ_{Al} for the K-shell electron impact ionization and L-shell photoeffect?

(b) What is the fluorescence yield ratio $\omega_X(Ga)/\omega_X(Al)$ for the K-shell hole?

(c) You measure the intensity of the K_α X-ray emission from Ga and Al with a detector system with 200 eV resolution. Would you expect interference from K, L_α, or M X-rays from As atoms or from the InP substrate? Would you expect electron backscattering from the InP substrate to influence the total or the ratio of X-ray yields from Ga and Al?

(d) In XPS measurements (neglecting work functions), what are the Ga and Al photoelectron energies and associated escape depths (λ)? What is the intensity ratio, assuming the same detector efficiency for both electron energies?

(e) In measurements of KL_1L_1 Ga and Al Auger electrons, what are the energies and associated escape depths? What is the ratio of Ga to Al transition rates?

(f) Compare the three techniques in terms of analysis depth, corrections or interferences, and yield ratios for values of x near 0.9.

12.6. Compare transitions for K-shell holes in $Z = 20$ and $Z = 36$ elements.

(a) What are the W_X/W_A ratios [Eq. 12.13]? Compare these values with the curve in Fig. 12.5.

(b) What are the atomic level widths and lifetimes? Compare the lifetime values for the two elements with the time for an electron to make a circular orbit in the Bohr model of the atom.

12.7. In an XPS analysis system with an Al K_α X-ray source, Auger electrons as well as photoelectrons are detected (see, for example, Fig. 10.7). For a vanadium target, what would be the energies and escape depths of the 2s photoelectrons and L_1MM Auger electrons? In comparison with 1.5 keV electrons, what is the ratio of electron to photon cross sections, σ_e/σ_{ph}, to form a 2s hole? The L shell fluorescence yield is small (Appendix 11 or Fig. 12.6), so estimate the ratio of photoelectron to Auger electrons, assuming Coster–Kronig transitions can be neglected. Is this a good assumption?

References

1. W. Bambynek et al., "X-ray Fluorescence Yields, Auger and Coster–Kronig Transitions," *Rev. Mod. Phys.* 44, 716 (1972).

2. T. A. Carlson, *Photoelectron and Auger Spectroscopy* (Plenum Press, New York, 1975).

3. G. Carter, B. Navinsek, and J. L. Whitton, "Heavy Ion Surface Topography Development," in *Sputtering by Particle Bombardment II*, R. Behrisch, Ed. (Springer-Verlag, Berlin, 1983).

4. C. C. Chang, "Analytical Auger Electron Spectroscopy," in *Characterization of Solid Surfaces*, P. F. Kane and G. R. Larrabee, Eds. (Plenum Press, New York, 1974), Chap. 20.

5. A. W. Czanderna, Ed., *Methods of Surface Analysis* (Elsevier, Amsterdam, 1975).

6. L. E. Davis, N. C. MacDonald, P. W. Palmberg, G. E. Riach, and R. E. Weber, *Handbook of Auger Electron Spectroscopy* (Physical Electronics Industries, Inc., Eden Prairie, MN, 1976).

7. G. Ertl and J. Kuppers, *Low Energy Electrons and Surface Chemistry* (Verlag Chemie International, Weinheim, 1974).

8. G. Herzberg, *Atomic Spectra and Atomic Structure* (Dover, New York, 1944).

9. H. Ibach, Ed., *Electron Spectroscopy for Surface Analysis, Topics in Current Physics*, Vol. 4 (Springer-Verlag, New York, 1977).

10. A. Joshi, L. E. Davis, and P. W. Palmberg, "Auger Electron Spectroscopy" in *Methods of Surface Analysis*, A. W. Czanderna, Ed. (Elsevier Science Publishing Co., New York, 1975), Chap. 5.

11. G. E. McGuire, *Auger Electron Spectroscopy Reference Manual* (Plenum Press, New York, 1979).

12. K. D. Sevier, *Low Energy Electron Spectroscopy* (Wiley-Interscience, New York, 1972).

13. K. Siegbahn, C. N. Nordling, A. Fahlman, R. Nordberg, K. Hamrin, J. Hedman, G. Johansson, T. Bergmark, S. E. Karlsson, I. Lindgren, and B. Lindberg, *ESCA, Atomic, Molecular and Solid State Structure Studied by Means of Electron Spectroscopy* (Almqvist and Wiksells, Uppsala, Sweden, 1967).

14. J. P. Thomas and A. Cachard, Eds., *Material Characterization Using Ion Beams* (Plenum Press, New York, 1978).

13
Nuclear Techniques: Activation Analysis and Prompt Radiation Analysis

13.1 Introduction

Analytical methods based on electronic interactions have long been used in the materials analysis laboratory, but determinations based on nuclear spectroscopy are much more recent. If radioactivity is produced by the irradiation and detected afterward, the method is called *activation analysis*; if radiations emitted instantaneously are detected, it is termed *prompt radiation analysis*. These two categories will be used for labeling major sections of the discussion that follows. For example, when a material containing carbon is irradiated with a beam of deuterons, one of the nuclear reactions with the carbon, ^{12}C, in the sample is the transformation to radioactive nitrogen, ^{13}N, by the prompt emission of a neutron, n. The carbon content of the sample can be determined either by measurement of the radiation emitted from the radioactive product nuclide, ^{13}N; by activation analysis; or by measuring the yield of neutrons, prompt radiation analysis. Radioactive nuclides that are used in analysis decay with half-lives ranging between milliseconds and thousands of years (the half-life of ^{13}N is 9.96 min); whereas, the prompt radiation from a nuclear reaction is emitted within times less than 10^{-12} sec after the nuclear reaction is initiated.

Several different kinds of nuclear interactions can be used for analysis:

1. A charged particle can elastically scatter from the charged target nucleus as in Rutherford scattering (Chapter 2) or nuclear elastic scattering.
2. Particles can excite the nucleus to a higher energy state (analogous to promoting an electron to a higher energy state in atomic spectroscopy); the nucleus can then be deexcited by γ-ray emission.
3. A different nucleus may be formed as a result of the nuclear reaction.

In most nuclear reactions, we have two particles or nuclei interacting to form two different nuclei. Thus,

$$a + b \rightarrow c + d$$
$$\text{Reactants} \quad \text{Products}$$

Any reaction must meet the requirement that the sum of the atomic numbers and mass numbers of reactants and products must balance. In other words

$$Z_a + Z_b = Z_c + Z_d$$
$$A_a + A_b = A_c + A_d.$$

Mass, however, *is* changed.

Although there is no theoretical limitation on what the nuclides a, b, c, and d can be, as a practical matter each side of the equation usually includes a very light nuclide. These are frequently termed *particles*. If we designate the particle (keep in mind that the distinction between particle and nucleus is a bit ambiguous) by the lower case, we can write a nuclear reaction,

$$a + X \rightarrow b + Y.$$

In a common shorthand notation, one would write

$$X(a,b)Y.$$

A nucleus with four neutrons and three protons is designated

$$^{7}_{3}X,$$

where the subscript is Z, the number of protons, and the superscript is the total nucleons, which we term the mass number, A. More generally, therefore, a nucleus is designated

$$^{A}_{Z}X.$$

A nuclear reaction of interest in determining boron depth profiles is

$$^{1}_{0}n + ^{10}_{5}B \rightarrow ^{4}_{2}He^{+} + ^{7}_{3}Li$$

In shorthand,

$$^{10}B(n, \alpha)^{7}Li$$

The reactant and product light particles are placed in parentheses and separated by a comma.

As in any specialized field, a certain nomenclature has developed based on convenience and tradition. The terms most frequently used are given below.

Nucleon	Either a proton or a neutron.
Nuclide	A specific nuclear species with a given proton number Z and neutron number N.
Isotopes	Nuclides of same Z and different N.
Isobars	Nuclides of same mass number A, where $A = Z + N$.
Isotones	Nuclides with same N but different Z.
Isomer	Nuclide in an excited state with a measurable half-life.
Proton	$^{1}_{1}H$
Deuteron (d)	$^{2}_{1}H$, one proton and one neutron.
Tritron (t)	$^{3}_{1}H$, one proton and two neutrons.
Alpha (α)	$^{4}_{2}He$, two protons and two neutrons.

Let us consider the irradiation of nuclei in a flux of protons, specifically proton irradiation of ^{12}C (Fig. 13.1). Some of the incident protons can be scattered due to long-range Coulomb interaction with the nuclei, known as *elastic scattering* (as described in

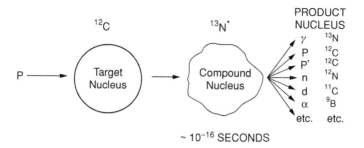

FIGURE 13.1. Schematic representation of the formation and decay of the compound nucleus ^{13}N during a nuclear reaction between protons and ^{13}C.

Chapter 2). Charged particles cannot effectively react through the nuclear force unless they have an energy comparable to the Coulomb barrier $Z_1 Z_2 e^2 / R \cong Z_1 Z_2 / A^{1/3}$ (in MeV) of the target atoms, which sets the lower limit of usable energy. If the protons have sufficient energy to overcome the Coulomb barrier, they may actually be captured by the nucleus to form a *compound nucleus*. The compound nucleus is now in a highly excited state, and the kinetic energy of the incident particle also adds to the excitation energy. In the compound nucleus model, it is assumed that the excitation energy is randomly distributed among all the nucleons in the resultant nucleus so that none of them has enough energy to escape immediately, and thus the compound nucleus has a lifetime which is long (10^{-14}–10^{-18} sec) compared with the time it takes for a nucleon to traverse a nucleus (10^{-21}–10^{-22} sec). The highly excited compound nucleus can now deexcite in many different ways by emitting γ-rays, protons, neutrons, alpha particles, etc. The incident protons can however also transfer sufficient energy to single nucleons or groupings of nucleons (such as deuterons and alphas) so that they may be directly ejected from the nucleus. Examples of such direct interactions are (p,n), (p,α), (α,p), and (α,n) reactions. Compound nucleus reactions are more likely at relatively low energies; whereas, the probability for a direct interaction increases with energy. Some of the nuclear reactions that can occur during proton irradiation are

(p,p)	Elastic scattering (Rutherford)
(p,p)	Compound nucleus elastic scattering
(p,p')	Inelastic scattering
(p,γ)	Prompt γ-ray emission
(p,n)	Prompt neutron emission
(p,α)	Prompt alpha emission

The probability of reaction between an incident particle and a target nucleus can be approximated by the geometrical cross section presented by the target nucleus to a point-size projectile. The radius of a nucleus is given rather accurately by the empirical formula

$$R = R_0 A^{1/3},$$

where A is the mass number and R_0 is a constant equal to 1.4×10^{-13} cm. For a medium weight nucleus such as ^{66}Zn, we can therefore calculate the geometrical cross section

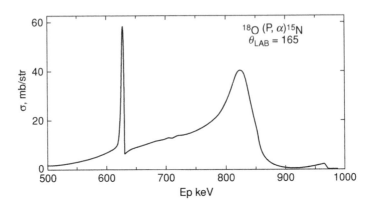

FIGURE 13.2. Cross section σ in millibarns per steradian versus incident proton energy E_p at a detection angle $\theta = 165°$ for $^{18}O(p,\alpha)^{18}N$ reaction. [From Amsel and Samuel, *Anal. Chem.* 39, 1689 (1967). Copyright 1967 American Chemical Society.]

as follows:

$$\sigma_{geo} = \pi(1.4 \times 10^{-13} \times 66^{1/3})^2 = 1.006 \times 10^{-24} \, cm^2.$$

Since most cross sections are of the order of 10^{-24} cm^2, it has become convenient to express cross sections in units of the barn, where

$$1 \, barn = 10^{-24} \, cm^2.$$

The reaction cross sections in general cannot be given by simple analytical functions. For example, Fig. 13.2 shows that the $^{18}O(p,\alpha)^{15}N$ reaction has a cross section that varies smoothly with energy below and above the resonance at 0.629 MeV.

The Breit–Wigner treatment deals with resonant cross sections in a quantitative way. The probability of the reaction

$$X(a,b)Y$$

may be denoted by the cross section $\sigma(a, b)$. According to the two-step compound nucleus view of nuclear reactions,

$$\sigma(a,b) = \sigma_c(a) \times (\text{relative probability of emission b}),$$

where $\sigma_c(a)$ is the cross section for the formation of the compound nucleus. The relative probability for the emission of b is just Γ_b/Γ where Γ_b is the transition rate for emission of b, also called the partial level width for b, and Γ is the total level width $\Gamma = \hbar\tau$, where τ is the mean life for a state so that

$$\sigma(a,b) = \sigma_c(a) \cdot \Gamma_b/\Gamma.$$

In general, the values of the cross sections and level widths depend on the energy of the incident particle, and on the charge and mass of the target nucleus. In its simplest form, the Breit–Wigner formula gives the value of the cross section in the neighborhood of a single resonance level in the compound nucleus, formed by an incident particle with zero angular momentum. Under these conditions, the formula is

$$\sigma(a,b) = \frac{\lambda^2}{4\pi} \frac{\Gamma_a \Gamma_b}{(E - E_0)^2 + (\Gamma/2)^2}, \tag{13.1}$$

where λ is the de Broglie wavelength of the incident particle ($\lambda = h/mv$), E_0 the energy at the peak of the resonance, E the energy of the incident particle, and Γ_a the partial level width for the emission of a in the inverse reaction. It is clear from this equation that the cross section will be at a maximum for $E = E_0$.

13.2   *Q* Values and Kinetic Energies

Nuclear reactions obey the following conservation laws:

1. Conservation of nucleons (A)
2. Conservation of charge (Z)
3. Conservation of mass-energy (E)
4. Conservation of momentum (p)

If the exact rest masses of the reactants and of the products of a nuclear reaction are totaled, there is likely to be a difference between the two because mass and energy may be exchanged according to the equation

$$E = mc^2,$$

where E is the energy, m is the mass, and c is the speed of light with 1 mass unit equal to 931.4 MeV.

The mass difference will correspond to either an emission or an absorption of energy. Thus a complete nuclear reaction should be written in the form

$$X + a \rightarrow Y + b \pm Q,$$

where Q is the energy balance, usually given in MeV. If energy is released by the reaction, the Q value will be positive. If the Q value is negative, energy must be supplied, and there will be a definite threshold below which these endoergic reactions will not occur.

Once a nuclear reaction has occurred, the radiations emitted are characteristic of the excited nuclei, in much the same way that optical radiation is characteristic of an excited emitting atom. It is the existence of a unique set of well-defined energy levels in the atom or the nucleus that permits the use of the emitted radiation as an identification of the source.

Symbolically, a nuclear reaction may be written

$$M_1 + M_2 \rightarrow M_3 + M_4 + Q,$$

where M_1 is the incident nucleus, M_2 is the target, M_3 is the emitted radiation, which may be either a nuclear particle or a gamma ray, M_4 is the residual nucleus, and Q is the energy released (absorbed) in the reaction. Q is simply the difference between the total energy, at rest, of the interacting system before the reaction takes place and after the reaction has occurred. If the M's are taken to be masses,

$$Q = (M_1 + M_2)c^2 - (M_3 + M_4)c^2. \qquad (13.2)$$

Let us consider the reaction

$$^{35}_{17}\text{Cl} + ^{1}_{0}\text{n} \rightarrow ^{32}_{15}\text{P} + ^{4}_{2}\text{He}^+ + Q,$$

which can be written in an abbreviated form as

$$^{35}\text{Cl}(n, \alpha)^{32}\text{P}.$$

The reaction is balanced with respect to nucleons and charge in that the reactants and products have the same total number of nucleons (36) and protons (17). The reaction energy Q is equivalent to the difference in mass between the reactants and the products:

$$\begin{aligned}
\Delta M &= M(^{35}\text{Cl}) + M(n) - M(^{32}\text{P}) - M(^{4}\text{He}^+) \\
&= 34.96885 + 1.00867 - 31.97391 - 4.00260 \\
&= +0.00101 \text{ amu}, \\
Q &= 0.00101 \times 931.4 = +0.94 \text{ MeV}.
\end{aligned}$$

Similarly, the Q value can be calculated for the following reaction:

$$^{14}\text{N}(p, n)^{14}\text{O}, \quad Q = -5.931 \text{ MeV}.$$

Q values can thus be positive or negative. Positive Q values denote exoergic nuclear reactions. Negative Q values denote endoergic nuclear reactions.

If the residual nucleus M_4 is left in an excited energy state, the Q for the reaction will be reduced, relative to the value that would be obtained if the residual nucleus were left in its ground state. The reduction is just the amount of the excitation energy. For a well-defined beam energy, the energy spectrum of M_3 will be *characteristic* of the Q values possible in the reaction or, equivalently, to the excited states of the residual nucleus. Even if the emitted particles M_3 are not observed, the prompt gamma rays emitted in the decay of the excited M_4 nucleus will be characteristic of that nucleus.

In the case of activation analysis, the identifying characteristics for the radioactive M_4 nucleus can be the half-life for decay, the types of radiation emitted, and the characteristic gamma rays emitted from the daughter nuclei of M_4.

For the prompt radiation analysis (PRA) case, if M_3 is a gamma ray, the nuclear reaction is called a *direct capture reaction*. The case $M_1 = M_3$ and $Q = 0$ is just the elastic scattering reaction. When $M_1 = M_3$ but $Q \neq 0$, the reaction is called *inelastic scattering*, and, finally, when $M_1 \neq M_3$, it is commonly termed a *rearrangement collision*.

In contrast to the case of atoms, nuclear characteristics usually differ markedly between two isotopes of the same chemical element. The emitted radiations or reaction products are specific not only to the chemical element but also to a particular isotope of that element. It is this property that provides the basis for the many important

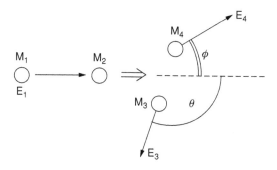

FIGURE 13.3. Notation used in nuclear reactions where the incident particle is dentoted by mass M_1 and energy E_1 and the emitted particle M_3 is detected at an angle θ relative to direction of the incident particle.

applications of stable and radioisotope tracers. Nuclear reactions, with only a few exceptions, are not affected by the state of atomic electrons and so do not give information directly about the chemical bonds or the chemical compound form of the elements in a sample.

For a reaction induced by an incident particle M_1 of energy E_1 (Fig. 13.3), the energy E_3 of the emitted particle M_3 in the direction θ (relative to the incident direction in the laboratory) is determined by the conservation of total energy and momentum, and in the nonrelativistic case is given by

$$E_3^{1/2} = A \pm (A^2 + B)^{1/2}, \tag{13.3}$$

where

$$A = \frac{(M_1 M_2 E_1)^{1/2}}{M_3 + M_4} \cos\theta \quad \text{and} \quad B = \frac{M_4 Q + E_1(M_4 - M_1)}{M_3 + M_4}, \tag{13.4}$$

using $M_1 + M_2 = M_3 + M_4$.

Equations 13.3 and 13.4 show that E_3 is characteristic of the reaction for a given E_1 and θ. In fact, the residual nucleus can be left in the ground state or in excited states, each state corresponding to a different Q value for the same reaction, and hence to a different value of E_3. The energy spectrum of the emitted particles will exhibit a series of peaks that are specific to the reaction and lead to the detection of a given nucleus M_2. The energy of a peak allows the identification of the reaction (and hence of the nucleus M_2), and from the intensity of the peak, the amount of the M_2 species can be determined.

Equation 13.3 can be approximated within a wide energy range by

$$E_3 = \alpha E_1 + \beta, \tag{13.5}$$

where α and β (as A and B in Eq. 13.3) are specific to the reaction under study and depend on the detection angle θ.

The kinematics of deuteron-induced reactions are shown in Fig. 13.4 for some specific reactions. The relation between the emitted particle energy E_3 and incident energy E_1 roughly follows Eq. 13.5, with different values of α and β for each reaction. The dashed line $E_3 = E_1$ denotes the maximum energy of elastically scattered particles and

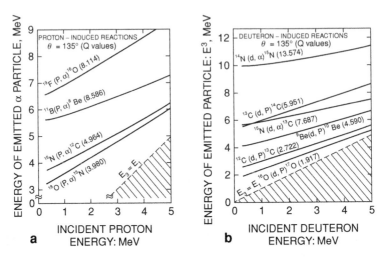

FIGURE 13.4. The energy of the emitted particle at $\theta = 135°$ versus incident particle energy E_1 for (a) proton-induced reactions and (b) deuteron-induced reactions. The Q values for the reaction are given in parentheses, and the dashed lines $E_3 = E_1$ show the maximum energy of a particle scattered from a heavy mass element in the target. [Adapted from Feldman and Picraux, 1977.]

may be considered the high-energy limit for the more abundant elastic scattering. The detection of light species in a heavy mass substrate can be carried out in many cases without interference from incident particles elastically scattered from the substrate.

13.3   Radioactive Decay

A nuclear reaction can take place through the formation of a compound nucleus in two distinct stages: (a) the incident particle is absorbed by the target nucleus to form a compound nucleus, and (b) the compound nucleus disintegrates by ejecting a particle or emitting a γ-ray. We designate the compound nucleus $_{Z}^{A}X$ and assume that the nuclear reaction produces the compound nucleus in the excited state E^*, as shown in the left side of Fig. 13.5.

The level E^* can decay either by the emission of the *radiative capture* γ-*rays* $\gamma_1, \gamma_2, \gamma_3$, and γ_4, to reach the ground state of $_{Z}^{A}X$, or (as in this case) by ejection of the protons p_0, p_1, and p_2 of three distinct energies. The proton groups feed excited states of the residual nucleus $_{Z-1}^{A-1}Y$, which can deexcite by γ-ray emission (γ_5, γ_6, and γ_7) to yield the ground state of $_{Z-1}^{A-1}Y$. This nucleus is itself unstable and decays by ejection of a β-particle to the excited or ground state of $_{Z}^{A-1}X$. The transitions γ_1–γ_7 and proton emission are most likely to occur very rapidly after the formation of the compound nucleus, i.e., within $<10^{-12}$ sec, but the half-life for the β-decay and hence for the emission of γ_8 will be very much longer. Thus two types of activation techniques can be distinguished: prompt techniques, where the samples are measured while the irradiation is in progress; and delayed methods, which depend on the measurement of

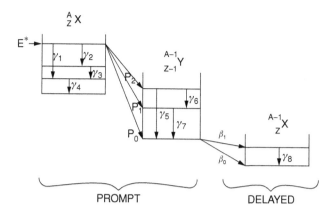

FIGURE 13.5. Energy-level representation of an excited compound nucleus $_Z^A X$ is an excited state E^+ that promptly decays by γ-ray emission to its ground state or by proton emission to the states of the residual nucleus.

a radionuclide with a half-life sufficiently long for the sample to be removed from the place of irradiation before the radioactivity is counted.

13.3.1   Beta Decay

As the atomic number increases, there is an excess of neutrons needed for nuclear stability, and nuclides with neutron-to-proton ratios, N/Z ratios, that differ from the stability line undergo radioactive decay. This decay occurs by emission of beta particles: either electrons, β^-, or positrons, β^+. When the N/Z ratio of a radioactive nucleus is greater than that of a stable nucleus with the same mass number, a neutron is converted into a proton, with the emission of an electron and an antineutrino ($\bar{\nu}$):

$$N/Z \text{ ratio too large: } n \to p^+ + \beta^- + \bar{\nu}.$$

If, however, the N/Z ratio is too small, the nucleus can become stable by converting protons into neutrons within the nucleus by the following processes: the emission of a positron, or the capture of an atomic orbital electron [*electron capture* (EC)];

$$N/Z \text{ ratio too small: } \begin{cases} p^+ \to n + \beta^+ + \nu \\ p^+ + e^- \to n + \nu \end{cases}.$$

The neutrino ν in β-decay shares the decay energy with the β-particle. β-particles thus have a continuous energy spectrum, with an average energy of about one-third of the maximum β-decay energy. Energy spectra for ^{64}Cu (half-life = 12.9 hr) emission are shown in Fig. 13.6.

If a nucleus decays by electron capture (EC), the resultant hole in the electron orbital (usually in the K shell) can be filled by an electron from an outer shell. Decay by electron capture is thus associated with X-ray emission, which may also be measured for analytical purposes.

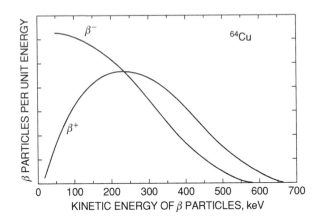

FIGURE 13.6. Energy spectra of the positive (β^+) and negative (β^-) electrons emitted by ^{64}Cu. The pronounced difference between the two spectral shapes results largely from the Coulomb effect.

Coulomb instability of the nucleus becomes very large for the heavier nuclides. Since the helium nucleus is very stable, α-decay takes place for $Z > 83$. Delayed neutron and proton emission also occurs for nuclides far from the stability line. These modes of decay are, however, not very common and as such are not very important for nuclear analysis.

13.3.2 Gamma Decay

During β-decay, the product nucleus may be left in an excited state. Deexcitation usually occurs by the emission of a gamma ray. The decay schemes for ^{27}Mg and ^{64}Cu are shown in Fig. 13.7. Because deexcitation by γ-ray emission is much more probable than β-decay, the γ-decay rate will be the same as the rate of the β-decay with which it is associated.

From the decay scheme, we can see that ^{64}Cu decays by both β^+ (19%) and β^- (39%) emission, while the other 42% of its decay is by electron capture, which will lead to ^{64}Ni X-ray emission.

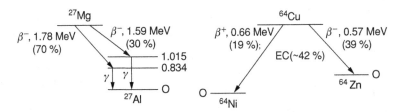

FIGURE 13.7. The principal decay schemes for ^{27}Mg (half-life = 9.5 min) and ^{64}Cu (half-life = 12.9 hr). The decay of ^{27}Mg shows the gamma rays associated with β-decay and ^{64}Cu the $\beta^+ - \beta^-$ branching.

When positrons pass through matter, they can annihilate with electrons (their antiparticle). If such a positron comes to rest and then annihilates with a free electron, conservation of linear momentum requires that two gamma rays be emitted, in opposite directions (180°), each having an energy equal to m_0c^2 (0.51 MeV). Positron decay is thus associated with 0.51 MeV γ-rays, which is referred to as *annihilation radiation*. If the electron is bound in an atom, annihilation with the production of a single photon can occur, because the atom can garner some momentum. The probability for such a process is, however, extremely small.

13.4   Radioactive Decay Law

The rate of radioactive decay, A, is proportional to the number of radioactive nuclei, N, present:

$$A = \frac{dN}{dt} = -\lambda N, \tag{13.6}$$

where λ is the decay constant. If at some particular time there are N_0 radioactive nuclei in a sample, then we can find the number of radioactive nuclei N_t, remaining at a later time t, by integration of the above equation:

$$\int_{N_0}^{N_t} \frac{dN}{N} = -\int_0^t \lambda \, dt,$$

$$N_t = N_0 e^{-\lambda t}. \tag{13.7}$$

It is convenient to express the decay constant λ as a half-life, which is defined as the time $T_{1/2}$ required for any number of radioactive nuclei to decay to half their initial value:

$$\frac{N_t}{N_0} = \frac{1}{2} = e^{-\lambda T_{1/2}} \tag{13.8}$$

and

$$\lambda = \frac{\ln 2}{T_{1/2}} = 0.693/T_{1/2}.$$

From Eq. 13.6, we now have

$$A_t = -\lambda N_0 e^{-0.693t/T_{1/2}}$$
$$= A_0 e^{-0.693t/T_{1/2}}. \tag{13.9}$$

Equation 13.9 is known as the *radioactive decay law*. If the decay rate (A_t) is plotted as a function of decay time (t) on a semilog plot, the decay rate will follow a straight line with intercept A_0. The half-life of the radionuclide may be obtained from the slope.

Information regarding the half-lives, types of decay, and decay energies can be obtained from the nuclide chart. More details about decay-level schemes can be obtained from compilations such as the *Table of Isotopes* (Lederer et al., 1967).

13.5   Radionuclide Production

In activation analysis, we are interested in the amount of radioactivity formed as a function of irradiation time. Let us consider the nuclear reaction

$$X(a,b)Y,$$

with Y the radioactive product nucleus, which we determine as a measure of X in the sample. The rate at which nuclide Y is formed may be expressed as

$$\left(\frac{dN_Y}{dt}\right)_{growth} = N_X \sigma \phi,$$

where N_X is the number of target nuclides and N_Y the number of product nuclides formed, σ the reaction cross section (cm^2), and ϕ the flux of incident particles (particles cm^{-2}- sec^{-1}). If the product nucleus is stable, the total number of nuclides Y formed during an irradiation time of t seconds is

$$N_Y = N_X \sigma \phi t. \tag{13.10}$$

If Y is radioactive, however, and decays with a decay constant λ_Y, we have

$$\left(\frac{dN_Y}{dt}\right)_{decay} = -N_Y \lambda_Y.$$

The total rate for forming the radioactive product nuclide Y is then given by

$$\frac{dN_Y}{dt} = N_X \sigma \phi - N_Y \lambda_Y. \tag{13.11}$$

If the target number N_X is a constant, as it usually is, one obtains by integration

$$N_Y = \frac{N_X \sigma \phi}{\lambda_Y}(1 - e^{-\lambda_Y t}).$$

The activity A_t (in disintegrations per second) of the radioactive nuclide Y at time t is given by $A_t = N_Y/\lambda_Y$ so that

$$A_t = N_X \sigma \phi (1 - e^{-\lambda_Y t}). \tag{13.12}$$

For $t \to \infty$, the term in brackets (growth factor) equals unity and $A_\infty = N_X \sigma \phi$. The activity A_∞ is referred to as the *saturation* A_{sat}, and Eq. 13.12 may be written as

$$A_t = A_{sat}(1 - e^{-\lambda t}) \tag{13.13}$$

The saturation factor A_t/A_{sat} is plotted in Fig. 13.8 as a function of irradiation time t expressed in number of half-lives, where $\lambda = 0.693/T_{1/2}$.

13.6   Activation Analysis

Activation analysis is a highly sensitive, nondestructive (if chemical separations are not used) technique for qualitative and quantitative determination of trace amounts of elements in a sample. It has been particularly useful for the simultaneous determination of many elements in complex samples because it provides a simple alternative to much more difficult, tedious, and destructive analytical techniques.

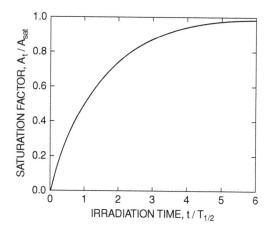

FIGURE 13.8. Growth of radioactive nuclide activity A_i/A_{sat} as a function of irradiation time $t/T_{1/2}$. After an irradiation time of four half-lives, about 94% of saturation activity A_{sat} has formed.

In activation analysis, the element to be determined is uniquely identified by the half-life and energy of the radiation emitted by its radioactive product nuclide. For optimum analytical sensitivity, it is desirable to form the maximum amount of activity in the sample from the elements to be determined and to measure such activities with maximum efficiency in the presence of other interfering activities. Fortunately, the irradiation and measurement parameters can be changed over wide ranges in order to select the optimum conditions for analysis.

Figure 13.8 shows that 50% of the maximum activity is obtained after an irradiation time equal to one half-life and more than 90% after four half-lives. Irradiations longer than a few half-lives will thus only increase the amount of unwanted background activity from other longer-lived activities also found in the sample.

The measurement of activities with short half-lives should be carried out as soon as possible after the irradiation. For longer-lived activities, better sensitivity is usually obtained if measurement takes place after time is allowed for the shorter-lived activities in the sample to decay. Figure 13.9 shows the decay curve for two activities. It is clear that the best time to measure the 24-hour activity would be after allowing 20 hours of decay time. The different components contributing to a decay curve are usually determined by decay curve stripping using a graphic approach or a computer.

Activation analysis is primarily a technique used for detection of impurities in bulk materials. The activation processes, involving neutrons, are deeply penetrating and thus sample many atoms throughout the material. The detection limits for thermal neutron activation analysis are generally in the range of 10^{-8}–10^{-10} grams. In terms of monolayers, 5×10^{-8} g of Ni would correspond to one monolayer on a 1 cm^2 sample. There is no intrinsic depth sensitivity in neutron activation analysis, so depth profiles must be generated by the use of sample thinning techniques such as sputtering or chemical etching.

13.7 Prompt Radiation Analysis

In prompt radiation analysis, the presence of an element in a sample is detected through the nuclear radiations emitted instantaneously from nuclear reactions produced

FIGURE 13.9. Decay diagram of a mixture of two independently decaying nuclides with half-lives of 4 and 24 hrs.

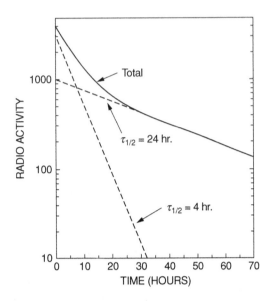

in the target by the irradiating beam. Detection limits can be quite good, but typically not as good as those that can be achieved under ideal conditions with activation analysis.

One of the important advantages of prompt analysis and the backscattering techniques discussed in Chapter 3 is that they can be used to measure the depth distribution of elements in the surface or near-surface regions of the sample. The dependence of the characteristics of the emitted radiations on depth is due to the energy loss suffered by the incident ions as they penetrate into the sample and also to the energy losses suffered by charged particles emitted from the reaction as they emerge from within the sample.

Since nuclear-reaction analysis can provide essentially background-free detection of light elements ($Z \leq 15$), depth distributions of trace amounts within the near-surface region can be measured. The primary emphasis in our discussion of prompt radiation analysis is the determination of concentration depth profiles of trace element impurities. In the use of prompt analysis for depth profiles, two different methods are applied, namely, the energy-analysis method and the resonance method. The former is used when the nuclear-reaction cross section is a smoothly varying function of energy. The latter method is used when a sharp peak (resonance) (see Fig. 13.2) in the cross section as a function of energy is present, and the depth profile is derived from a measurement of the nuclear-reaction yield as a function of the energy of the analyzing beam.

13.7.1   Energy-Analysis Method

13.7.1.1   Thermal Neutron-Induced Reactions

Elemental depth distributions of certain trace elements can be determined by a thermal neutron beam to produce reactions with certain elements that yield monoenergetic

TABLE 13.1. Energies and cross sections for thermal neutron reactions.

Element	Reaction	Energy of emitted particles in keV	Cross section (barns)
Li	^{8}Li(n,α)T	2056	940
B	^{10}B(n,α)^{7}Li	1472	3836
Be	^{7}Be(n,p)^{7}Li	1439	48,000
Na	^{22}Na(n,p)^{22}Ne	2248	29,000

charged particles. These isotropically emitted particles lose energy in passing through matter, and their residual energy on leaving the sample surface is primarily dependent on the amount of matter through which the particle has passed. For a given sample with a known atomic density, the energy of a detectable particle is determined by the depth at which the initial reaction took place. A quantitative image of the elemental distribution with depth—a depth profile—is generated directly from the charged particle spectrum. For impurity distributions close to the surface so that the energy loss rate is nearly constant, the energy difference ΔE between detected energies of particles emitted from atoms at the surface or at depth t is determined by the rate of energy loss dE/dx along the outgoing path

$$\Delta E = t(dE/dx); \tag{13.14}$$

whereas, in Rutherford backscattering (Chapter 3), the energy-to-depth conversion is determined by the energy loss along both the inward and outward paths.

Thermal neutron reaction cross sections can be substantially greater than geometric cross sections ($\cong 1$ barn), as indicated in Table 13.1. For a total thermal neutron flux of 10^8 n/cm^2, the sensitivity for impurity detection is about 10^{14} atoms/cm^2 for boron.

Figure 13.10 shows the charged particle spectrum of a thin film of boron (10 nm) on Ni. The four peaks in the spectrum correspond to the secondary and primary α-particles and lithium ions from the ^{10}B(n,α)^{7}Li reaction. The measured energies of the four charged particles are highest when the reaction occurs at the surface of the sample. When the reaction occurs within the sample, the particles must pass through overlying matter, and the entire charged particle spectrum is shifted to lower energies. The maximum distance the particles can travel and still exit the surface—the range—varies with sample composition, but is typically 1–10 μm for solids. Figure 13.10b shows the spectrum shift of the primary alpha peak that occurs when the thin boron film is covered with 50 nm of Cu. Besides the spectral shift, the covered B film also shows indications of diffusion into the nickel substrate. The left side of the peak has broadened, and the height has decreased, indicating movement of the boron.

13.7.1.2 Charged-Particle-Induced Reactions

In charged-particle irradiation of targets at energies sufficiently high so that the incident particle can penetrate the Coulomb barrier, a variety of nuclear reactions can occur, as shown in Fig. 13.11 for deuteron irradiation of a thin aluminum nitride target.

The yield depends, as for Rutherford backscattering, on the differential cross section of the reaction; however, unlike backscattering, there is no simple analytical formula for the cross section. The cross-section curves can be obtained from the nuclear physics

FIGURE 13.10. (a) Charged particle spectrum from ^{10}B for a thermal neutron-irradiated sample of Ni covered with 10 nm of B. The portion of the spectra associated with 1471 keV α is expanded in (b) for samples covered (dashed line) and samples not covered (solid line) with a 50 nm thick Cu film. [From J.E. Riley, Jr., and R.F. Fleming, private communication.]

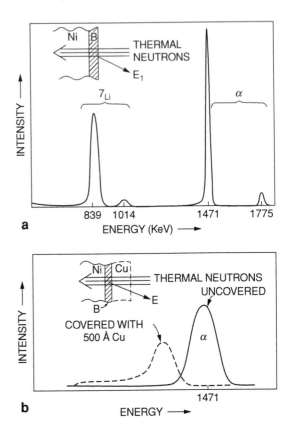

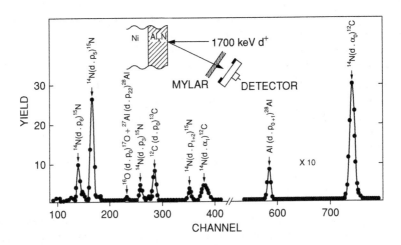

FIGURE 13.11. Energy spectrum at 8.23 keV/channel from deuteron bombardment of a 170 nm thick Al nitride film on a Ni backing for a deuteron incident energy of 1.7 MeV and a detection angle of 160°. The mylar film is used to block the scattered deuterons from the detector. [From Cachard and Thomas, 1978, with permission of Springer Science+Business Media.]

literature (see Feldman and Picraux, 1977). As a general feature, medium and high Z nuclei do not undergo nuclear reactions (in the MeV range) owing to Coulomb barrier repulsion. This feature, combined with the fact that the emitted particles have energies well above that of the incident particle energy (due to the high Q values of most of the reactions), allows background-free detection of light elements on heavier substrates. The abundant elastically scattered particles are stopped in a thin absorber to prevent count-rate saturation of the detector and electronic systems.

The number of detected particles, Q_D, is proportional to the total number N_S of atoms/cm^2,

$$Q_D = N_S \, \sigma(\theta) Q \Omega, \qquad (13.15)$$

where $\sigma(\theta)$ is the differential cross section, Ω the detection solid angle, and Q the number of incident particles (Eq. 2.9).

For incident energy E_0, the energy E_3 of the detected species for a surface reaction is $E_3 \cong \alpha E_0 + \beta$ as given in Eq. 13.5. Then the energy difference ΔE between the detected particles originating from the surface and from depth t depends both on the energy loss dE/dx of the incident particle on its inward path and the energy loss of the reaction particle on its outward path,

$$\Delta E = t \left[\alpha \frac{dE}{dx} \bigg|_{\text{in}} + \frac{dE}{dx} \bigg|_{\text{out}} \right]. \qquad (13.16)$$

Again we use the approximation that the energy losses are constant in the near-surface region. The reaction factor α weights the energy loss in the inward path in the same fashion as the backscattering kinematic factor weights $dE/dx|_{\text{in}}$ (Eq. 3.20a). Eq. 13.16 defines the correspondence between the depth scale and the energy scale. If the cross section is known, the concentration profile can be deduced from the shape of the experimental spectrum.

The $^{16}O(d,\alpha)^{14}N$ reaction can be used for oxygen depth-profile measurements. This reaction, at low deuteron energy, emits only a ground state α-group, α_0. For α-particles corresponding to the first excited state of the ^{14}N nucleus, the reaction has a negative Q value of 0.829 MeV (and therefore a threshold energy) and will not occur for a deuteron energy below 933 keV. At low deuteron bombarding energies the α_0 energy at large angles is low, and the stopping power or energy loss per unit length therefore is relatively high and provides improved depth resolution. Figure 13.12 shows the energy spectrum that was observed for a 600 nm thick SiO$_2$ layer with the beam at normal incidence to the target ($\phi = 0°$) and the detector at 145°. In order to avoid interference from the $^{16}O(d,p_0)^{17}O$ reaction, the detector depletion depth was not allowed to exceed 26 μm. While the α-particles were stopped in this thickness and deposited their full energy in the detector, the protons deposited only a portion of their energy and were displaced thereby to lower energies in the particle energy spectrum.

Figure 13.12 shows clearly the advantage of making measurements with particles having a high stopping power. Thus, while the proton groups from the $^{18}O(d,p)^{17}O$ are quite narrow and cannot be used for depth profile measurements, the α_0 group is quite wide and can be so used.

FIGURE 13.12. Portion of the energy spectrum at 6.75 keV/channel obtained at $\theta = 145°$ from the irradiation of a 600 nm SiO_2 layer of 900 keV deuterons. The energy width ΔE_0 of the oxygen signal is directly related to the oxide thickness. [From Turos et al., *Nucl. Inst. Meth.* 111, 605 (1973). Copyright 1973, with permission from Elsevier.]

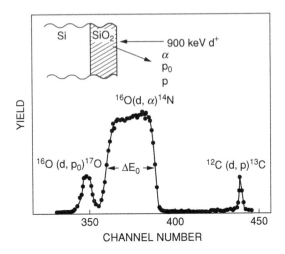

13.7.2   The Resonance Method

Many nuclear reactions have the property that the reaction yield exhibits one or more sharp peaks or *resonances* as a function of bombarding energy. Such a resonance is measured experimentally by varying the incident beam energy in small increments and measuring the quantity of radiation emitted per unit beam influence at each energy. The use of the resonance method in depth profiling of trace elements takes advantage of the sharp peak (see, for example, Fig. 13.2) in the nuclear reaction cross section as a function of energy. Consider the ideal case shown in Fig. 13.13, where only one resonance exists in the cross-section curve and where off-resonance cross-section values can be neglected. The method consists of measuring the reaction yield (most often γ-rays) due to the interaction between the incident beam and the impurity atoms as a function of incident beam energy. Incident ions having an energy E_0 (i.e., larger than E_R, the resonance energy) are slowed down until E_R is reached at depth x, where the nuclear reaction will then occur at a rate proportional to the impurity concentration. The depth x and the incident beam energy E_0 are related through the equation

$$E_o = E_R + \left(\frac{dE}{dx}\right)_{in} \frac{x}{\cos\theta_1}, \qquad (13.17)$$

where θ_1 is the angle between the incident beam and the surface normal. The stopping power $(dE/dx)_{in}$ for the incident beam is assumed to be a constant. A more elaborate analysis can be done by taking into account the detailed cross section function, energy straggling, and other factors (Russell et al., 1996).

Neglecting the finite experimental depth resolution, it is seen that the yield curve in Fig. 13.13 can be converted into the desired concentration profile by simply changing scales of yield and energy to corresponding scales of concentration and depth, respectively. An example of the use of nuclear resonance is shown in Fig. 13.14, which gives the gamma yield as a function of beam energy for a hydrogen-implanted target. The reaction between fluorine and hydrogen has a strong resonance at about 16.4 MeV, so the hydrogen concentration profile can be obtained directly. The extracted

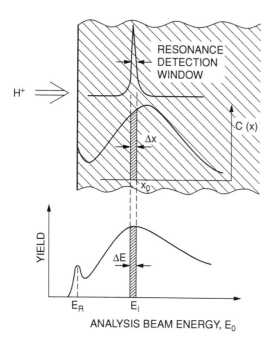

FIGURE 13.13. Principle of concentration profile measurements using resonant reactions.

hydrogen concentration profile indicates only the hydrogen within the sample and does not include the surface hydrogen present due to contamination.

Nuclear reaction analysis (NRA) is a method of determining the absolute concentration (atoms/cm^2) of light impurities in and on a solid. It thus provides an absolute calibration for other surface-sensitive techniques, particularly Auger analysis and secondary ion mass spectroscopy (SIMS). In a typical application, a light particle of interest is implanted into a heavier substrate. SIMS provides a sensitive depth profile, while NRA determines an absolute concentration. Reaction analysis is particularly useful for hydrogen detection and absolute hydrogen surface coverages. Table 13.2 lists the most used charged particle reactions for light atom detection.

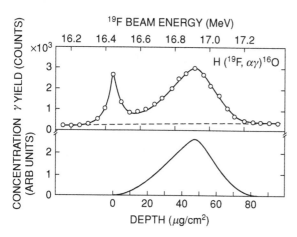

FIGURE 13.14. Range profile of 12 keV H ions implanted into Al$_2$O$_2$ to a fluence of 4×10^{16}/cm^2 measured by use of the nuclear reaction ^{4}H(^{19}F,$\alpha\gamma$) ^{18}O. The upper part of the figure is the raw experimental data and the lower part is the extracted depth. [Adapted from Feldman & Mayer, 1986.]

TABLE 13.2. Most used charged particle reactions for light atom detection. [From Feldman & Mayer, 1986.]

Nucleus	Reaction	Incident energy (E_α) (MeV)	Emitted energy (MeV)	Approximately $\sigma_{LAB}(E_0)$ (mb/sr)	Yield* (counts/μC)
^{2}H	^{2}H(d,p)^{2}H	1.0	2.3	5.2	30
^{2}H	^{2}H(^{3}He, p)^{4}He	0.7	13.0	61	380
^{3}He	^{3}He(d,p)^{4}He	0.45	13.6	64	400
^{6}Li	^{6}Li(d,α)^{4}He$^+$	0.7	9.7	6	35
^{7}Li	^{7}Li(p,α)^{4}He$^+$	1.5	7.7	1.5	9
^{9}Be	^{9}Be(d,α)^{7}Li	0.6	4.1	~1	6
^{11}B	^{11}B(p,α)^{8}Be	0.65	5.57(α_0)	0.12(α_0)	0.7
		0.65	3.70(α_1)	90(α_1)	550
^{12}C	^{12}C(d,p)^{13}C	1.20	3.1	35	210
^{13}C	^{13}C(d,p)^{14}C	0.64	5.8	0.4	2
^{14}N	^{14}C(d,α)^{12}C	1.5	9.9(α_0)	0.6(α_0)	3.6
		1.2	6.7(α_1)	1.3(α_1)	7.0
^{15}N	^{15}N^{16}(N,α)^{12}C	0.8	3.9	~15	90
^{16}O	^{16}O(p,α)^{17}O	0.90	2.4(p$_0$)	0.74(p$_0$)	5
		0.90	1.6(p$_1$)	4.5(p$_1$)	28
^{18}O	^{18}O(p,α)^{15}N	0.730	3.4	15	90
^{19}F	^{19}F(p,α)^{16}O	1.25	6.9	0.5	3
^{23}Na	^{23}Na(p,α)^{20}Ne	0.592	2.238	4	25
^{31}P	^{31}P(p,α)^{28}Si	1.514	2.734	16	100

* For a 1×10^{16} cm^{-2} surface layer and a solid angle of 0.1sr at $150°$ C

Problems

13.1. A magnesium foil 0.1 mm thick is irradiated with a beam of 22 MeV deuterons (beam current = 100 μA) that has a cross-sectional area less than that of the foil. Sodium-24 (half-life = 15.0 hr) is formed by the ^{26}Mg(d,α)^{24} Na reaction, which has an average cross section σ = 25 mbarns, throughout the thickness of the foil. What is the activity of ^{24}Na (disintegrations/sec) in the foil during a 2-hr irradiation?

13.2. A steel sample weighing 2.5 g is irradiated for 30 min in a reactor with a thermal neutron flux $\phi = 4.2 \times 10^{13}$ neutrons/cm^2-sec. The ^{27}Mg activity (half-life = 9.5 min) formed by the ^{26}Mg(n, γ)^{27}Mg reaction is measured 10 min after the irradiation, and the 0.834 MeV γ-ray gives a count rate of 625 cpm. For a counting efficiency of 3% and σ(n, γ) = 30 mbarns, calculate the wt% Mg concentration in the sample.

13.3. How would you dope silicon with phosphorous using a nuclear reactor? Calculate the dopant concentration of a silicon sample irradiated in a thermal neutron flux of 2×10^{14} neutrons s^{-1} cm^{-2} for 6 hr. [^{30}Si(n, γ)^{31}Si; σ = 0.12 barns.]

13.4. Find the threshold kinetic energy (MeV) for the following incident particles to disintegrate the deuteron into a proton and a neutron.
    (a) Electrons
    (b) Protons
    (c) Alpha particles

13.5. A thin 8.0 mg foil of ^{113}Cd is exposed to a flux of thermal neutrons. How many ^{114}Cd nuclei are formed? $\phi_n = 1.6 \times 10^{13}$ n/cm^2-s, the irradiation time is 2 hr, and $\sigma(n, \gamma) = 2 \times 10^4$ barns.

13.6. Compare proton backscattering and ^{15}N(p,α)^{12}C reactions from one monolayer of nitrogen atoms (10^{15} N/cm^2) on a carbon substrate.

    (a) For backscattering with 0.8 MeV protons through 180° with a detector solid angle of 0.01 str, calculate the yield (assuming pure Rutherford scattering, Eq. 2.17).

    (b) Compare the RBS yield with the nuclear reaction yield (Table 13.2) for the same detector.

    (c) What are the relative advantages of the two techniques?

13.7. The geometrical cross section for a reaction is given by $\sigma_{geom} = \pi R^2$, where R is the nuclear radius and the distance of closest approach d equals $Z_1 Z_2 e^2 / E$.

    (a) Calculate these values for 1 MeV p,d and α-particles incident on ^{14}N.

    (b) Compare the values of the geometrical cross section with the Rutherford scattering cross sections for 1 MeV particles ($\theta = 180°$) and with the ^{15}N(p, α) and ^{14}N(d, α) cross sections deduced from Table 13.2.

    (c) Do these values give support to the rule of thumb that the onset of nuclear reactions occurs when the particles penetrate the Coulomb barrier?

13.8. Compare the depth scales in eV/nm for detection of F in an Al thin film with RBS for 3 MeV ^{4}He ions incident ($\theta = 180°$) and for the ^{19}F(p,α) reaction with 1.25 MeV protons.

13.9. For X(a,b)Y resonant nuclear reactions, there is an inverse reaction a(X,Y)b. If the energy of the resonance at E_a is known, what is E_X in terms of E_a and the masses of the reactants? Evaluate your answer using energy values for ^{19}F(p,α) in Table 13.2 compared with those for ^{1}H(F,α) in Fig. 13.14.

References

1. A. Cachard and J. P. Thomas, "Microanalysis by Direct Observation of Nuclear Reactions," in *Material Characterization Using Ion Beams*, J. P. Thomas and A. Cachard, Eds. (Plenum Press, New York, 1978).

2. G. R. Choppin and J. Ryaberg, *Nuclear Chemistry* (Pergamon Press, Oxford, 1980).

3. R. D. Evans, *The Atomic Nucleus* (McGraw-Hill Book Co., New York, 1955).

4. F. Everling, L. A. Koenig, J. H. E. Mattauch, and A. H. Wapstra, "Consistent Set of Energies Liberated in Nuclear Reaction," *1960 Nuclear Data Tables* (U.S. Government Printing Office, Washington, DC, 1961).

5. L. C. Feldman and J. W. Mayer, *Fundamentals of Surface and Thin Film Analysis* (Prentice-Hall, New Jersey, 1986)

6. L. C. Feldman and S. T. Picraux, "Selected Low Energy Nuclear Reaction Data," in *Ion Beam Handbook for Material Analysis*, J. W. Mayer and E. Rimini, Eds. (Academic Press, New York, 1977).

7. G. Friedlander, J. W. Kennedy, and J. M. Miller, *Nuclear and Radiochemistry* (John Wiley and Sons, New York, 1964).

8. G. B. Harvey, *Nuclear Chemistry* (Prentice-Hall, Englewood Cliffs, NJ, 1965).

9. I. Kaplan, *Nuclear Physics* (Addison-Wesley, Reading, MA, 1964).

10. M. Lederer, J. M. Hollander, and I. Perlman, *Table of Isotopes*, 6th ed. (John Wiley and Sons, New York, 1967).
11. W. E. Meyerhof, *Elements of Nuclear Physics* (McGraw-Hill Book Co., New York, 1967).
12. P. A. Tipler, *Modern Physics* (Worth Publishers, New York, 1978).
13. E. A. Wolicki, "References to Activation and Prompt Radiation Analysis: Material Analysis by Means of Nuclear Reactions," in *New Uses of Ion Accelerators*, J. F. Ziegler, Ed. (Plenum Press, New York, 1975), Chap. 3.
14. S. W. Russell, T. E. Levine, A. E. Bair, and T. L. Alford, "Guidelines to the Application of Nuclear Resonance to Quantitative Thin Film Analysis," *Nuclear Instruments and Methods B*, vol 118, 201–205 (1996).

14
Scanning Probe Microscopy

14.1   Introduction

Since the conception of scanning probe microscopy (SPM) in the 1990s, the technique has evolved from a novelty to a standard analytical tool in both academic and industrial settings. The number of variations and applications has dramatically escalated over the last ten years. SPM is a fundamentally simple and inexpensive technology that is capable of imaging and measuring surfaces on a fine scale and of altering surfaces at the atomic level. There are three elements common to all probe microscopes. Firstly, a small, sharp probe comes within a few tenths of nanometers of the sample's surface, and the interactions between the surface and the probe are used to interrogate the surface. Secondly, a detection system monitors the product of the probe–surface interaction (e.g., a force, tunneling current, change in capacitance, etc.). Thirdly, either the probe or sample is raster-scanned with nanoscale precision. By monitoring of the interaction intensity, any surface variation translates to topographical information from the surface and generates a three-dimensional image of the surface.

Over twenty different variations of the SPM currently exist; however, the most commonly used are Atomic Force Microscopy (AFM) and Scanning Tunneling Microscopy (STM). In AFM, the probe tip is affixed to a cantilevered beam (Fig. 14.1). The probe interacts with the surface and the resulting force deflects the beam in a repulsive manner, as described by Hooke's Law. In the same manner that a spring changes dimensions under the influence of forces, the attractive and repulsive forces between atoms of the probe and the surface can also be monitored when brought extremely close to each other. Hence, the net forces acting on the probe tip deflect the cantilever, and the tip displacement is proportional to the force between the surface and the tip. As the probe tip is scanned across the surface, a laser beam reflects off the cantilever. By monitoring the net (x, y, and z) deflection of the cantilever, a three-dimensional image of the surface is constructed. In STM, a sharp metallic probe and a conducting sample are brought together until their electronic wave functions overlap (Fig. 14.2). By applying a potential bias between them, a tunneling current is produced. The probe is mounted on a piezoelectric drive that scans the surface. Combination of the piezoelectric drive with a feedback loop allows imaging of the surface in either a constant-current or a constant-height mode.

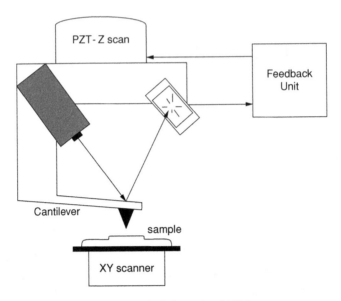

FIGURE 14.1. Schematic of AFM.

Other methods are also utilized to detect the deflection of the cantilever. A plate is placed above the AFM cantilever, which acts as the other plate of a capacitor. The capacitance between the two plates reflects the deflection of the cantilever. Another mode uses laser interferometry, where a beam is split with one part reaching the detector directly while the other part is focused on the back of the cantilever and is reflected back

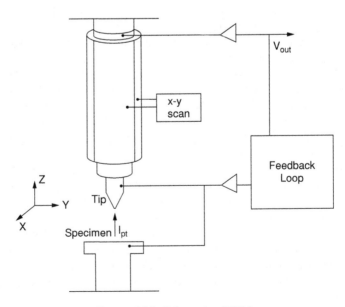

FIGURE 14.2. Schematic of STM.

to the detector. The coherent beams travel different paths and produce an interference pattern, which changes as the cantilever moves up and down. This pattern allows direct imaging of surfaces and defects in real space, with subnanoscale resolution in contrast to diffraction-based analyses. Diffraction analysis samples a macroscopic volume of materials; whereas, SPM can sample small areas ($<1 \ \mu m^2$). Operating in the near field, probe–sample spacing is on the order of typical wavelengths used in electron microscopy. Hence, imaging is not diffraction limited and the spatial resolution is not a function of the wavelengths. Ideally, analysis of nanoscale features can be done in vacuum or ambient. In comparison to electron microscopy, the SPM has much lower capital and maintenance costs and essentially no sample preparation.

14.2   Scanning Tunneling Microscopy

14.2.1   Theory

Scanning tunneling microscopy (STM) has the potential to image the surface of materials. Under carefully controlled conditions, STM has subatomic resolution and is capable of imaging individual atoms and electronic structure. However, analyses using STM are typically limited to electrically conducting materials, since the technique measures a current between the probe tip and the sample surface. The tip of the probe consists ideally of a single atom that comes in close proximity to the surface (Fig. 14.3). In the same manner as a profilometer, the tip is scanned over the surface. However, it does not touch the surface. The separation distance is normally a few tenths of nanometers. This distance allows the wave functions to overlap and results in a finite probability that the electron can surmount the barrier between the probe tip and the sample surface.

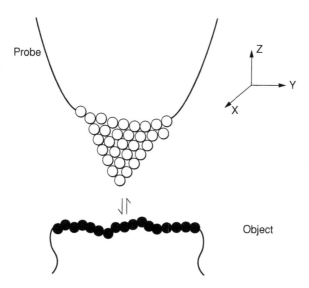

FIGURE 14.3. Schematic showing the interaction between the atoms of the probe tip and the atoms of the sample under interrogation.

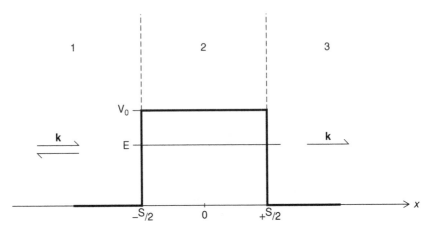

FIGURE 14.4. Schematic of an electron with energy E impinging upon a one-dimensional rectangular potential barrier of height V_O and width s.

Typically, the probe tip is grounded and the sample is biased in the range of tens of millivolts, resulting in a tunneling current. In normal electrical conduction, the two metallic surfaces conduct electricity when they are touching in a contiguous matter. There is, however, a case when the two metallic surfaces (the tip and the sample) are not actually touching and yet a current flows. The current that flows is referred to as a *tunneling current*.

According to quantum mechanics, there is a finite probability that the electron can *tunnel* through the barrier, without ever acquiring the full energy (kinetic plus potential energy) necessary to surmount the barrier. Assuming elastic tunneling (the electron does not lose nor gain energy), we consider an electron with energy E and mass m impinging upon a one-dimensional barrier with an energy height of V_O (Fig. 14.4). The electron can be reflected by the barrier (region 1), or tunnel (region 2), or complete the tunneling process (region 3). Starting with Schrödinger's time-independent equation for region 1,

$$-\frac{\hbar^2}{2m}\frac{d^2\psi_1}{d^2x} = E\psi_1,$$
$$\psi_1 = e^{ikx} + Ae^{-ikx}, \tag{14.1}$$

where the wave vector k is equal to $[2mE/\hbar^2]^{1/2}$ and $\hbar$ is Plank's constant divided by 2π. In region 2, the wave function relationship is described as

$$-\frac{\hbar^2}{2m}\frac{d^2\psi_2}{d^2x} + V_O = E\psi_2,$$
$$\psi_2 = B'e^{ikx} + C'e^{-ikx} = Be^{-\xi x} + Ce^{-i\xi x}, \tag{14.2}$$

where, in this case, ξ is equal to $[-k'^2]^{1/2} = [2m(V_O - E)/\hbar^2]^{1/2}$. Finally in region 3,

$$-\frac{\hbar^2}{2m}\frac{d^2\psi_3}{d^2x} = E\psi_3,$$
$$\psi_3 = De^{ikx}. \tag{14.3}$$

The incident current density j_i and transmitted current density j_t are given (Wiesendagner, 1994) as

$$j_i = \frac{-i\hbar}{m}\left(\psi_3^*(x)\frac{d\psi_3}{dx} - \psi_3(x)\frac{d\psi_3^*}{dx}\right),$$

$$j_i = \frac{-i\hbar}{m}|D|^2,$$

$$j_t = \frac{\hbar k}{m}. \tag{14.4}$$

By matching the wave functions ψ_j and the corresponding first derivatives at the edges of the barriers, $x = 0$ and $x = s$ (discontinuities in the potential), the transmission coefficient T, the ratio of current density j_t and incident current j_i are determined:

$$T = {j_t}\big/{j_i} = |D|^2$$

$$= \frac{1}{1 + (k^2 + \xi^2)^2/(4k^2\xi^2)\sinh^2(\xi s)}$$

$$\approx \frac{16k^2\xi^2}{(K^2 + \xi^2)^2}\exp(-2\xi s), \tag{14.5}$$

where the term ξ equals $[2m(V_O - E)]^{1/2}/\hbar$ and is referred to as the *decay rate*. Hence, the effective barrier height $\phi(= V_O - E)$ and the barrier width s dictate the tunneling current. For the case of the tunneling microscope, where the gap between the sample and probe is 0.1 nm, any small bias applied between the probe tip and sample will generate a large electrostatic field. An approximation of the magnitude of the tunneling current (I) is an exponential function of the separation distance between the probe tip and the sample:

$$I = C\rho_s\rho_t\exp\left(s\phi^{1/2}\right) \tag{14.6}$$

where ρ_s and ρ_t are the electron densities of the sample surface and probe tip, respectively. C is proportionatly constant, the probe tip scans across the surface using a piezoelectric crystal that changes its volume when a voltage is applied to it. As the tip moves in the x- or y-direction along the sample's surface, the current varies according to Eq. 14.6. The output current differs when the probe tip is right on top of an atom (smaller distance) as compared to when the probe tip is above a space between atoms (larger distance). Hence, the relative electrostatic potential of an individual atom is detected as an increase in the tunneling current as a function of spatial position in the x–y scan across the sample's surface. For case where the ϕ equals 5 eV, a variation in s from 0.1 to 1.0 nm results in a variation in tunneling current by a factor of 7.5.

In Fig. 14.5a, the actual probe-tip-surface displacement, s, is held constant, and is the constant-height mode of operation. Consequently, the output current varies with the electron density. Monitoring of the probe current as a function of the x–y displacement yields a topographical representation of the surface morphology. On the basis of Eq. 14.6, this mode of operation is sensitive to small fluctuations in s and results in an exponential increase or decrease in output current.

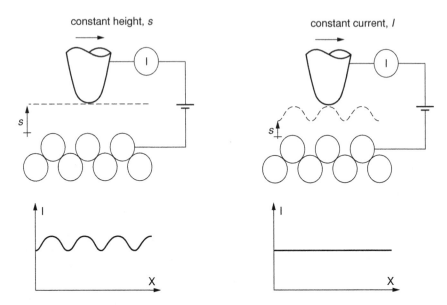

constant height, s   constant current, I

FIGURE 14.5. Schematic of STM probe tip scanning in the x-direction when operating in (a) constant-height mode and (b) constant-current mode.

Often it is desired that the output signal vary linearly with the probe-tip surface displacement, s. In this case, a feedback loop is employed in *constant-current mode*, which controls the height of the probe-tip–specimen separation, s (Fig. 14.5b). The height of the probe tip is controlled by a piezoelectric crystal, a material which expands linearly when a voltage is placed across it. The expanding crystal pushes the tip closer to the sample. Hence, the voltage needed to expand the piezoelectric crystal to keep the current constant varies linearly with the actual height of the atoms on the sample. By montoring the feedback voltage, we can directly measure the displacement of the tip.

Typical STM analyses are conducted in ultrahigh vacuum to minimize contamination. Fig. 14.6a displays a negative-bias (-1.06 V) STM image of a Si(111) 7×7 surface. The area of the image is approximately 24 nm × 24 nm. Note that the terrace at the kinked edge is clearly visible with atomic resolution. In the accompanying Fig. 14.6b, the high-resolution STM image is taken with a negative bias of (-0.12 V). The line denotes the asymmetry of the faulted and unfaulted halves of the 7×7 unit cell. With low-bias voltage (-0.12 V), the adatoms in the faulted region gives rise to more tunneling current than the matrix atoms in the unfaulted region. Figure 14.7 shows UHV-STM of Quasi-1D gold wires grown on a Si(557) surface taken using a 1.66 V bias. The Si(557) surface can be regarded as a combination of Si(111) terraces and single height steps; hence, the Au atoms absorb to the (111) terraces, forming quasi–one-dimensional gold wires.

Another application of STM is the manipulation of atoms. Figure 14.8 shows the formation of a *quantum dot corral*. In this case, Fe atoms are adsorbed onto a Cu(111) surface at a temperature of approximately 4K. The STM probe tip descends directly on

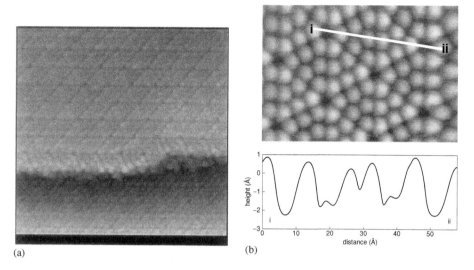

(a)

(b)

FIGURE 14.6. (a) Constant-height STM image of a Si(111) 7 × 7 surface from an area of 24 × 24 nm. (b) The high-resolution STM image is taken with a negative bias (−0.12 V). The line scan denotes the asymmetry between the faulted and unfaulted regions. [From J.M. Macleod et al., *Review of Scientific Instruments*, Vol. 74, pp. 2429–2437]

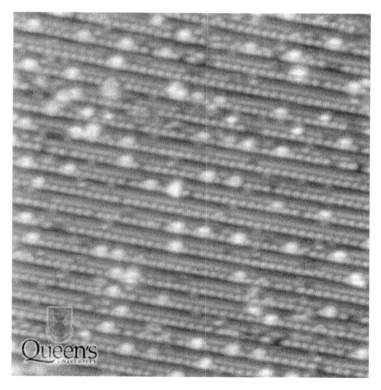

FIGURE 14.7. UHV-STM of Quasi-1D gold wires grown on a Si(557) surface taken using a 1.66 V bias. Au atoms absorb to the (111) terrace. (With permission from A. McLean, J. Macleod, and J. Lipton-Duffin.)

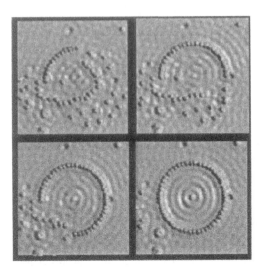

FIGURE 14.8. Series of images displaying the formation of a quantum dot corral using the STM. Fe atoms are adsorbed onto a Cu(111) surface and positioned using the STM. [Image reproduced by permission of IBM Research, Almaden Research Center. Unauthorized use not permitted.]

top of a Fe atom and increases the attractive force by increasing the tunneling current. The probe tip translates across the surface to the appropriate locale with the tethered Fe atom. Once the appropriate position is determined, the Fe atom is untethered by reducing the tunneling current.

14.3   Atomic Force Microscopy

14.3.1   Theory

When the probe–surface spacing is relatively large (∼1 nm or greater), the interactions are dominated by long-range van der Waals forces (Fig. 14.9). These attractive forces depend exponentially on distance and are extremely sensitive to probe-tip shape. Other attractive forces include metallic adhesion forces and charge accumulation between the probe tip and the nearest surface atom. At small spatial separations (∼0.1 nm or less), the wave function of the probe–surface overlap and short-range quantum-mechanical exchange-correlation forces dominate due to the Pauli exclusion principle. These forces decay exponentially with increasing distance:

$$F = -\gamma(\Delta\varepsilon_C)\chi\left(\sigma\frac{H}{e^2}\right)^{1/2},\tag{14.7}$$

where $\Delta\varepsilon_C$ is the width of the conduction band, γ is a dimensionless factor approximately equal to one, χ is the decay rate of the force, and σ is the conductance.

The slope of the van der Waals curve is very steep in the repulsive or contact region (Fig. 14.9). As a result, the repulsive van der Waals force balances almost any force that attempts to push the atoms closer together. For example, in AFM, when the cantilever pushes the tip against the sample, the cantilever deforms as opposed to pressing the probe tip closer to the surface. Only negligible reduction in the interatomic separation between the probe-tip atom and the surface atoms is probable, even with the use of

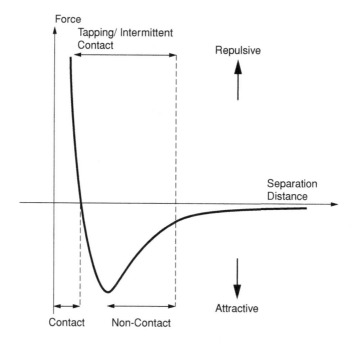

FIGURE 14.9. Schematic of van der Waals forces as a function of probe-tip–surface spacing.

a rigid cantilever. These cantilevers will apply substantial forces onto the sample's surface and will more than likely result in surface deformation. The point of balance between attractive and repulsive forces is defined as the *mechanical point contact*. Van der Waals forces also dominate the interaction between nonmagnetic and electrically natural solids that are separated by a distance of several nanometers.

In addition to the repulsive van der Waals force described above, two other forces are generally present during contact AFM operation: a capillary force exerted by the thin water layer often present in an ambient environment, and the force exerted by the cantilever itself. The capillary force arises when water wicks its way around the tip, applying a strong attractive force (about 10^{-8}N) that holds the tip in contact with the surface. As long as the tip is in contact with the sample, the capillary force should be constant because the distance between the tip and the sample is virtually incompressible. It is also assumed that the water layer is reasonably homogeneous. The variable force in contact AFM is the force exerted by the cantilever. The total force that the tip exerts on the sample is the sum of the capillary plus cantilever forces, and must be balanced by the repulsive van der Waals force for contact AFM. The magnitude of the total force exerted on the sample varies from 10^{-8}N (with the cantilever pulling away from the sample almost as hard as the water is pulling down the tip) to the more typical operating range of 10^{-7} to 10^{-6}N.

The atomic force microscope (AFM) or scanning force microscope (SFM), like all other scanning probe microscopes, utilizes a sharp probe moving over the surface of a sample in a raster scan (see Fig. 14.1). In the case of the AFM, the probe is a tip on the end of a cantilever that bends in response to the force between the tip and the sample.

An optical detector monitors the extent of bending of the lever. The cantilever beam obeys Hooke's Law for small displacements, and the interaction force between the tip and the sample can be found. The movement of the tip or sample is performed by an extremely precise positioning device made from piezoelectric ceramics. The scanner is capable of subnanometer resolution in the x-, y- and z-directions.

14.3.2   Modes of Operation

The AFM typically operates in either of two principal modes: *constant-force mode* (with feedback control) and *constant-height mode* (without feedback control). If the electronic feedback is engaged, then the probe-tip positioning is controlled by the piezoelectric device, which moves the sample (or tip) up and down and responds to any changes in force that are detected, altering the tip–sample separation to restore the force to a predetermined value (similar to Fig. 14.5a). This mode of operation is known as *constant force* and usually results in accurate topographical images. In constant-force mode, the speed of scanning is limited by the response time of the feedback circuit, but the total force exerted on the sample by the tip is well controlled (similar to Fig. 14.5b). Constant-force mode is generally preferred for most applications.

Constant-height or *deflection* mode operates with the feedback electronics minimized. This mode is particularly practical for imaging very flat samples at high resolution. The minimized feedback eliminates issues of thermal drift or the possibility of a rough sample damaging the tip and/or cantilever. Constant-height mode is often used for taking atomic-scale images of atomically flat surfaces, where the cantilever deflections and thus variations in applied force are small. Constant-height mode is also essential for recording real-time images of changing surfaces, where high scan speed is essential.

Once the AFM has detected the cantilever deflection, it can generate the topographic data set by operating in one of the two modes—constant-height or constant-force mode. In constant-height mode, the spatial variation of the cantilever deflection can be used directly to generate the topographic data set because the height of the scanner is fixed as it scans. In constant-force mode, the deflection of the cantilever can be used as input to a feedback circuit that moves the scanner up and down in the z-axis, responding to the topography by keeping the cantilever deflection constant. In this case, the image is generated from the scanner's motion. With the cantilever deflection held constant, the total force applied to the sample is constant.

14.3.3   Probe–Sample Interaction

The way in which image contrast is obtained can be achieved in many ways. The three main classes of interaction are *contact mode, tapping mode*, and *noncontact mode*. Contact mode is the most common method of operation of the AFM. In this case, the tip and sample reside in the repulsive region of Fig. 14.5 during the scan. A consequence of contact-mode operation is that large lateral forces on the sample surface have a tendency to drag the probe tip.

In the noncontact mode, the probe tip resonates at a distance above the sample surface of the sample such that it is no longer in the repulsive region of Fig. 14.5. If noncontact

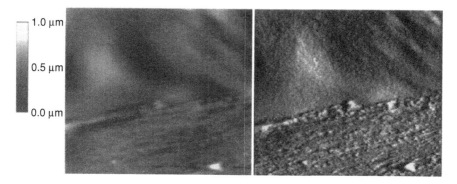

FIGURE 14.10. Image of a piece of mica using AFM (a) in noncontact mode and (b) in contact mode. Note that in this case the contact mode provides more details. [From Le Grimellec et al., *Biophys. J.*, 75, 695–703 (1998). With permission of The Biophysical Society.]

mode is used, it is advantageous to conduct the analysis in vacuum. Conducting the analysis in ambient conditions usually results in a thin layer of water contamination between the probe tip and the surface. Fig. 14.10 (a and b) shows a comparison of noncontact- and contact-mode imaging for a piece of mica. In this case, the contact mode provides more details; however, the noncontact mode minimizes risk of damage to the probe tip. Recent advancements have led to the development of a high-resolution,

FIGURE 14.11. High resolution, noncontact AFM image taken of the Ge/Si(105) surface. The size of image is 4.2 nm × 4.2 nm. The frequency shift was set at −60 Hz. The oscillation amplitude and resonant frequency of the cantilever were 3.8 nm and 280 482 Hz, respectively (with permission from Yukio Hasegawa).

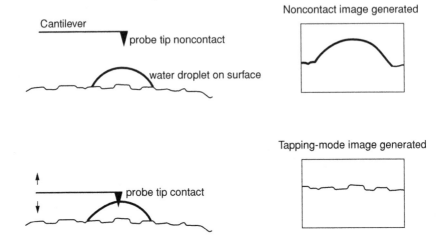

FIGURE 14.12. (a) Schematic displaying the consequence of adsorbed moisture from the ambient during AFM analysis. (b) Tapping-mode AFM is used to avoid the false image of the surface.

atomically resolved electrostatic potential profile, with use of an AFM equipped with a Kelvin probe. Figure 14.11 shows a noncontact AFM micrograph of Ge atoms residing on a Si(105) substrate. The image reveals the electrostatic potential variations among the dangling bond states and all dangling bonds of the surface, regardless of their electronic configuration. These results clearly demonstrate that high-resolution noncontact AFM with a Kelvin-probe method is a tool for imaging atomic structures and determining electronic properties of surfaces. The resolution and results are on a par with or better than STM, whose images strongly deviate from the atomic structure by the electronic states involved.

Figure 14.12a shows the consequence of water contamination on the resulting image during AFM. The presence of the moisture from the ambient adsorbs on the surface and can give a false image. To avoid this problem, tapping mode is the commonly employed mode when operating in air or other gases (Figure 14.12b). In this case, the cantilever resonates at its resonant frequency on the order of 100–400 kilohertz. The probe tip contacts the surface only during a fraction of the resonant period as a means to reduce the influence of the lateral forces on the probe tip. The feedback controls adjust such that the amplitude of the cantilever resonates at a constant value. An image can be formed from this amplitude signal, as there will be small variations in this oscillation amplitude due to the control electronics not responding instantaneously to changes on the specimen surface. This mode eliminates lateral forces between the tip and the sample and has become an important SPM technique, since it overcomes some of the limitations of both contact and noncontact AFM. In the example below, tapping-mode AFM was used to compare the surface morphology of as-deposited and plasma-treated parylene surfaces (Fig. 14.13). AFM results, taken in ambient, reveal the increased surface roughness associated with the oxygen plasma treatment and correlated with mechanical adhesion analysis.

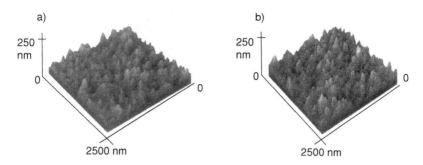

FIGURE 14.13. AFM images of Pa-n on Si: (a) untreated, as-deposited Pa-n on Si, and (b) oxygen plasma-treated Pa-n. Both images were taken in ambient using tapping mode.

14.3.4 Tip Effects

Image contrast using the AFM results from interaction forces between the probe tip and sample. Understanding of the probe-tip feature then becomes important for interpretation of the image. During analyses at high magnification and on surfaces with substantial surface morphology, the probe condition influences the image resolution. A given feature may be alleged to be topographical that is actually a direct result of the probe-tip shape. For example, for the probe to interact with individual atoms, these interactions must be limited to the fewest number of atoms on the probe tip. Hence, the resolution of the AFM probe tip depends on the sharpness of the tip. Commercially fabricated probe tips have a radius of curvature on the order of 10 nm. Sharper (i.e., better) tips are also available. Only a tip with sufficient sharpness can properly image a given Z-gradient. Some gradients will be steeper or sharper than any tip can be expected to image without artifact. Figure 14.14 shows AFM images using a 40 nm diameter probe tip (a) and a 5 nm diameter probe tip (b). Another effect is tip broadening, which arises when the tip's radius of curvature is on the order of or larger that the feature under investigation. When the probe tip scans over the specimen, the sides of the tip

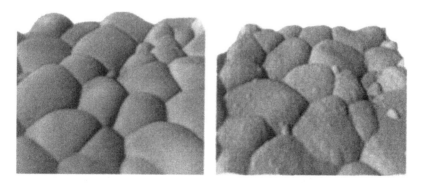

FIGURE 14.14. AFM images using a 40 nm diameter probe tip (a) and a 5 nm diameter probe tip (b). [Courtesy of Paul E. West, Ph.D, CTO, Pacific NanoTechnology.]

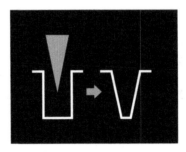

FIGURE 14.15. Schematic of the apparent change in the image of a square hole to a slope-edged hole because of the shape of the probe.

make contact before the base, and the microscope produces a self-image of the probe tip (a false image), rather than of the object surface (see Fig. 14.15).

References

1. R. Wiesendanger, *Scanning Probe Microscopy and Spectroscopy* (Cambridge University Press, New York, 1994).
2. D. A. Bonnell, Ed., *Scanning Tunneling Microscopy and Spectroscopy* (VCH Publishers, New York, 1993).
3. C. Kittel, *Introduction to Solid State Physics*, 5th ed. (John Wiley and Sons, New York, 1976).

Appendix 1
K_M for $^4\text{He}^+$ as Projectile and Integer Target Mass

ATOMIC MASS (amu)	180°	170°	160°	145°	90°	60°	45°	30°
4					0.000	0.250	0.500	0.750
6	0.040	0.041	0.043	0.051	0.200	0.476	0.659	0.832
8	0.111	0.113	0.118	0.134	0.333	0.589	0.738	0.873
10	0.184	0.186	0.193	0.213	0.429	0.661	0.787	0.897
12	0.250	0.253	0.260	0.282	0.500	0.711	0.820	0.914
14	0.309	0.311	0.319	0.342	0.556	0.748	0.844	0.926
16	0.360	0.363	0.371	0.394	0.600	0.776	0.863	0.935
18	0.405	0.408	0.416	0.439	0.636	0.799	0.877	0.942
20	0.444	0.447	0.455	0.478	0.660	0.817	0.889	0.948
22	0.479	0.482	0.490	0.512	0.692	0.833	0.899	0.952
24	0.510	0.513	0.521	0.542	0.714	0.846	0.907	0.956
26	0.538	0.540	0.548	0.569	0.733	0.857	0.914	0.960
28	0.563	0.565	0.572	0.592	0.750	0.866	0.920	0.962
30	0.585	0.587	0.594	0.614	0.765	0.875	0.925	0.965
32	0.605	0.607	0.614	0.633	0.778	0.882	0.929	0.967
34	0.623	0.626	0.632	0.650	0.789	0.889	0.933	0.969
36	0.640	0.642	0.649	0.666	0.800	0.895	0.937	0.971
38	0.655	0.657	0.664	0.681	0.810	0.900	0.940	0.972
40	0.669	0.671	0.678	0.694	0.818	0.905	0.943	0.974
42	0.682	0.684	0.690	0.706	0.826	0.909	0.946	0.975
44	0.694	0.696	0.702	0.718	0.833	0.913	0.948	0.976
46	0.706	0.707	0.713	0.728	0.840	0.917	0.950	0.977
48	0.716	0.718	0.723	0.738	0.846	0.920	0.952	0.978
50	0.726	0.727	0.733	0.747	0.852	0.923	0.954	0.979
52	0.735	0.736	0.742	0.755	0.857	0.926	0.956	0.980
54	0.743	0.745	0.750	0.763	0.862	0.929	0.958	0.980
56	0.751	0.753	0.758	0.771	0.867	0.931	0.959	0.981
58	0.759	0.760	0.765	0.778	0.871	0.933	0.960	0.982
60	0.766	0.767	0.772	0.784	0.875	0.935	0.962	0.982
62	0.772	0.774	0.778	0.791	0.879	0.937	0.963	0.983
64	0.779	0.780	0.784	0.796	0.882	0.939	0.964	0.983
66	0.784	0.786	0.790	0.802	0.886	0.941	0.965	0.984
68	0.790	0.792	0.796	0.807	0.889	0.943	0.966	0.984
70	0.795	0.797	0.801	0.812	0.892	0.944	0.967	0.985

(*continued*)

ATOMIC MASS (amu)	180°	170°	160°	145°	90°	60°	45°	30°
72	0.801	0.802	0.806	0.817	0.895	0.946	0.968	0.985
74	0.805	0.807	0.811	0.821	0.897	0.947	0.969	0.986
76	0.810	0.811	0.815	0.826	0.900	0.949	0.970	0.986
78	0.814	0.816	0.819	0.830	0.902	0.950	0.970	0.986
80	0.819	0.820	0.824	0.834	0.905	0.951	0.971	0.987
82	0.823	0.824	0.827	0.837	0.907	0.952	0.972	0.987
84	0.826	0.828	0.831	0.841	0.909	0.953	0.972	0.987
86	0.830	0.831	0.835	0.844	0.911	0.955	0.973	0.988
88	0.834	0.835	0.838	0.847	0.913	0.956	0.974	0.988
90	0.837	0.838	0.842	0.851	0.915	0.957	0.974	0.988
92	0.840	0.841	0.845	0.854	0.917	0.957	0.975	0.988
94	0.843	0.844	0.848	0.856	0.918	0.958	0.975	0.989
96	0.846	0.847	0.851	0.859	0.920	0.959	0.976	0.989
98	0.849	0.850	0.853	0.862	0.922	0.960	0.976	0.989
100	0.852	0.853	0.856	0.864	0.923	0.961	0.977	0.989
102	0.855	0.856	0.859	0.867	0.925	0.962	0.977	0.990
104	0.857	0.858	0.861	0.869	0.926	0.962	0.978	0.990
106	0.860	0.861	0.864	0.872	0.927	0.963	0.978	0.990
108	0.862	0.863	0.866	0.874	0.929	0.964	0.979	0.990
110	0.865	0.866	0.868	0.876	0.930	0.964	0.979	0.990
112	0.867	0.868	0.871	0.878	0.931	0.965	0.979	0.990
114	0.869	0.870	0.873	0.880	0.932	0.966	0.980	0.991
116	0.871	0.872	0.875	0.882	0.933	0.966	0.980	0.991
118	0.873	0.874	0.877	0.884	0.934	0.967	0.980	0.991
120	0.875	0.876	0.879	0.886	0.935	0.967	0.981	0.991
122	0.877	0.878	0.881	0.888	0.937	0.968	0.981	0.991
124	0.879	0.880	0.882	0.889	0.938	0.968	0.981	0.991
126	0.881	0.882	0.884	0.891	0.938	0.969	0.982	0.992
128	0.882	0.883	0.886	0.892	0.939	0.969	0.982	0.992
130	0.884	0.885	0.887	0.894	0.940	0.970	0.982	0.992
132	0.886	0.887	0.889	0.896	0.941	0.970	0.982	0.992
134	0.887	0.888	0.891	0.897	0.942	0.971	0.983	0.992
136	0.889	0.890	0.892	0.898	0.943	0.971	0.983	0.992
138	0.890	0.891	0.894	0.900	0.944	0.971	0.983	0.992
140	0.892	0.893	0.895	0.901	0.944	0.972	0.983	0.992
142	0.893	0.894	0.896	0.903	0.945	0.972	0.984	0.992
144	0.895	0.896	0.898	0.904	0.946	0.973	0.984	0.993
146	0.896	0.897	0.899	0.905	0.947	0.973	0.984	0.993
148	0.898	0.898	0.900	0.906	0.947	0.973	0.984	0.993
150	0.899	0.900	0.902	0.908	0.948	0.974	0.984	0.993
152	0.900	0.901	0.903	0.909	0.949	0.974	0.985	0.993
154	0.901	0.902	0.904	0.910	0.949	0.974	0.985	0.993
156	0.902	0.903	0.905	0.911	0.950	0.975	0.985	0.993
158	0.904	0.904	0.906	0.912	0.951	0.975	0.985	0.993
160	0.905	0.906	0.908	0.913	0.951	0.975	0.985	0.993
162	0.906	0.907	0.909	0.914	0.952	0.976	0.986	0.993
164	0.907	0.908	0.910	0.915	0.952	0.976	0.986	0.993
166	0.908	0.909	0.911	0.916	0.953	0.976	0.986	0.994
168	0.909	0.910	0.912	0.917	0.953	0.976	0.986	0.994
170	0.910	0.911	0.913	0.918	0.954	0.977	0.986	0.994
172	0.911	0.912	0.914	0.919	0.955	0.977	0.986	0.994

(continued)

ATOMIC MASS (amu)	180°	170°	160°	145°	90°	60°	45°	30°
174	0.912	0.913	0.915	0.920	0.955	0.977	0.987	0.994
176	0.913	0.914	0.916	0.921	0.956	0.978	0.987	0.994
178	0.914	0.915	0.917	0.921	0.956	0.978	0.987	0.994
180	0.915	0.916	0.917	0.922	0.957	0.978	0.987	0.994
182	0.916	0.916	0.918	0.923	0.957	0.978	0.987	0.994
184	0.917	0.917	0.919	0.924	0.957	0.978	0.987	0.994
186	0.918	0.918	0.920	0.925	0.958	0.979	0.987	0.994
188	0.918	0.919	0.921	0.925	0.958	0.979	0.988	0.994
190	0.919	0.920	0.922	0.926	0.959	0.979	0.988	0.994
192	0.920	0.921	0.922	0.927	0.959	0.979	0.988	0.994
194	0.921	0.921	0.923	0.928	0.960	0.980	0.988	0.994
196	0.922	0.922	0.924	0.928	0.960	0.980	0.988	0.995
198	0.922	0.923	0.925	0.929	0.960	0.980	0.988	0.995
200	0.923	0.924	0.925	0.930	0.961	0.980	0.988	0.995

The Kinematic Factor, K_{M_2}, defined by Eq. 2.5 for a $^4\text{He}^+$ atom as projectile and integral atomic masses for the target atom M_2. The angle indicates the scattering angle as measured in the laboratory frame of reference.

Appendix 2
Rutherford Scattering Cross Section of Elements for 1 MeV^4 He

ELEMENT	AT. # (Z_2)	AVG. MASS (amu)	$d\sigma/d\Omega$ in 10^{-24} cm^2/Steradian							
			179.5°	170°	160°	145°	90°	60°	45°	30°
Be	4	9.01	0.053	0.055	0.058	0.069	0.297	1.294	3.836	18.454
B	5	10.81	0.097	0.098	0.104	0.122	0.482	2.038	6.008	28.848
C	6	12.01	0.147	0.150	0.159	0.185	0.704	2.944	8.661	41.550
N	7	14.01	0.214	0.218	0.230	0.266	0.974	4.023	11.803	56.568
O	8	16.00	0.291	0.297	0.312	0.360	1.285	5.267	15.429	73.896
F	9	19.00	0.384	0.390	0.410	0.471	1.642	6.681	19.542	93.540
Ne	10	20.18	0.478	0.486	0.511	0.586	2.032	8.254	24.131	115.486
Na	11	22.99	0.590	0.599	0.629	0.720	2.471	9.998	29.210	139.749
Mg	12	24.31	0.707	0.718	0.754	0.862	2.945	11.903	34.767	166.318
Al	13	26.98	0.838	0.851	0.893	1.021	3.466	13.979	40.812	195.201
Si	14	28.09	0.974	0.991	1.039	1.187	4.023	16.216	47.335	226.391
P	15	30.97	1.129	1.146	1.201	1.371	4.626	18.623	54.348	259.896
S	16	32.06	1.285	1.306	1.370	1.563	5.267	21.192	61.838	295.707
Cl	17	35.45	1.460	1.483	1.555	1.773	5.954	23.933	69.818	333.833
Ar	18	39.95	1.647	1.672	1.752	1.997	6.685	26.840	78.283	374.272
K	19	39.10	1.835	1.861	1.951	2.223	7.446	29.904	87.221	417.011
Ca	20	40.08	2.033	2.064	2.163	2.465	8.253	33.136	96.646	462.065
Sc	21	44.96	2.249	2.285	2.394	2.727	9.108	36.542	106.561	509.435
Ti	22	47.90	2.476	2.513	2.633	2.998	10.001	40.110	116.956	559.113
V	23	50.94	2.706	2.751	2.882	3.281	10.935	43.844	127.835	611.101
Cr	24	52.00	2.947	2.997	3.139	3.574	11.909	47.740	139.194	665.397
Mn	25	54.94	3.208	3.255	3.410	3.882	12.926	51.806	151.039	722.006
Fe	26	55.85	3.469	3.522	3.690	4.200	13.982	56.034	163.365	780.923
Co	27	58.93	3.741	3.802	3.983	4.533	15.082	60.431	176.177	842.153
Ni	28	58.71	4.024	4.089	4.283	4.875	16.219	64.990	189.469	905.690
Cu	29	63.54	4.330	4.392	4.601	5.235	17.404	69.721	203.250	971.543
Zn	30	65.37	4.633	4.702	4.925	5.604	18.627	74.615	217.511	1039.703
Ga	31	69.72	4.947	5.026	5.264	5.989	19.894	79.677	232.258	1110.176
Ge	32	72.59	5.272	5.358	5.611	6.384	21.201	84.903	247.486	1182.958
As	33	74.92	5.606	5.700	5.970	6.791	22.549	90.294	263.198	1258.051
Se	34	78.96	5.970	6.054	6.340	7.213	23.940	95.853	279.394	1335.455
Br	35	79.91	6.326	6.416	6.720	7.644	25.370	101.575	296.072	1415.167
Kr	36	83.80	6.693	6.791	7.112	8.090	26.843	107.465	313.235	1497.192
Rb	37	85.47	7.070	7.175	7.514	8.547	28.357	113.520	330.880	1581.526
Sr	38	87.62	7.457	7.569	7.927	9.017	29.912	119.740	349.009	1668.171
Y	39	88.91	7.855	7.974	8.351	9.499	31.508	126.126	367.620	1757.125
Zr	40	91.22	8.263	8.390	8.786	9.994	33.146	132.679	386.716	1848.391
Nb	41	92.91	8.681	8.816	9.232	10.501	34.825	139.397	406.294	1941.967

(continued)

ELEMENT	AT. # (Z_2)	AVG. MASS (amu)	$d\sigma/d\Omega$ in 10^{-24} cm²/Steradian							
			179.5°	170°	160°	145°	90°	60°	45°	30°
Mo	42	95.94	9.110	9.253	9.690	11.021	36.547	146.282	426.358	2037.855
Tc	43	99.000	9.549	9.701	10.159	11.554	38.310	153.333	446.904	2136.053
Ru	44	101.000	9.998	10.639	12.099	40.114	40.114	160.549	467.933	2236.560
Rh	45	102.900	10.458	11.129	12.657	41.959	41.959	167.930	489.446	2339.379
Pd	46	106.400	10.928	11.631	13.228	43.846	43.846	175.479	511.443	2444.508
Ag	47	107.900	11.408	12.143	13.810	45.774	45.774	183.192	533.922	2551.948
Cd	48	112.400	11.899	12.668	14.406	47.746	47.746	191.073	556.886	2661.699
In	49	114.800	12.400	13.203	15.014	49.757	49.757	199.119	580.333	2773.760
Sn	50	118.700	12.951	13.749	15.635	51.811	51.811	207.331	604.264	2888.132
Sb	51	121.800	13.475	14.306	16.269	53.905	53.905	215.709	628.678	3004.814
Te	52	127.600	14.008	14.875	16.916	56.043	56.043	224.254	653.576	3123.808
I	53	126.900	14.552	15.452	17.572	58.219	58.219	232.961	678.955	3245.109
Xe	54	131.300	15.107	16.043	18.243	60.438	60.438	241.837	704.820	3368.723
Cs	55	132.900	15.671	16.643	18.926	62.698	62.698	250.878	731.167	3494.647
Ba	56	137.300	16.246	17.256	19.622	65.001	65.001	260.085	757.998	3622.882
La	57	138.900	16.832	17.878	20.330	67.343	67.343	269.458	785.312	3753.426
Ce	58	140.100	17.427	18.512	21.050	69.728	69.728	278.996	813.108	3886.281
Pr	59	140.900	18.034	19.156	21.783	72.153	72.153	288.700	841.389	4021.446
Nd	60	144.200	18.650	19.812	22.529	74.621	74.621	298.570	870.154	4158.923
Pm	61	147.000	19.277	20.479	23.287	77.130	77.130	308.607	899.401	4298.709
Sm	62	150.400	19.914	21.157	24.058	79.681	79.681	318.809	929.133	4440.808
Eu	63	152.000	20.562	21.846	24.841	82.273	82.273	329.177	959.347	4585.216
Gd	64	157.300	21.220	22.547	25.638	84.907	84.907	339.712	990.047	4731.935
Tb	65	158.900	21.888	23.258	26.446	87.582	87.582	350.412	1021.228	4880.964
Dy	66	162.500	22.567	23.980	27.267	90.299	90.299	361.278	1052.893	5032.304
Ho	67	164.900	23.256	24.713	28.100	93.057	93.057	372.309	1085.042	5185.955
Er	68	167.300	23.955	25.457	28.946	95.856	95.856	383.507	1117.673	5341.915
Tm	69	168.900	24.665	26.212	29.805	98.697	98.697	394.870	1150.788	5500.186
Yb	70	173.000	25.385	26.979	30.676	101.579	101.579	406.400	1184.388	5660.769
Lu	71	175.000	26.115	27.755	31.559	104.503	104.503	418.095	1218.470	5823.660
Hf	72	178.500	26.865	28.544	32.456	107.469	107.469	429.956	1253.035	5988.864
Ta	73	181.000	27.607	29.343	33.364	110.475	110.475	441.983	1288.084	6156.376
W	74	183.900	28.369	30.153	34.285	113.524	113.524	454.176	1323.617	6326.200
Re	75	186.200	29.141	30.975	35.219	116.613	116.613	466.534	1359.633	6498.335
Os	76	190.200	29.923	31.807	36.166	119.745	119.745	479.059	1396.133	6672.781
Ir	77	192.200	30.716	32.650	37.124	122.917	122.917	491.750	1433.115	6849.534
Pt	78	195.100	31.519	33.505	38.096	126.132	126.132	504.605	1470.581	7028.602
Au	79	197.000	32.332	34.370	39.079	129.387	129.387	571.628	1508.531	7209.979
Hg	80	200.600	33.156	35.246	40.076	132.684	132.684	530.817	1546.964	7393.666
Tl	81	204.400	34.096	36.134	41.085	136.023	136.023	544.171	1585.881	7579.665
Pb	82	207.200	34.943	37.032	42.106	139.403	139.403	557.691	1625.281	7767.972
Bi	83	209.000	35.801	37.942	43.140	142.824	142.824	571.377	1665.164	7958.590
Po	84	210.000	36.669	38.862	44.186	146.287	146.287	585.228	1705.531	8151.519
At	85	210.000	37.547	39.792	45.244	149.791	149.791	599.245	1746.380	8346.758
Rn	86	222.000	38.435	40.737	46.318	153.339	153.339	613.431	1787.716	8544.311
Fr	87	223.000	39.335	41.690	47.402	156.926	156.926	627.780	1829.533	8744.169
Ra	88	226.000	40.244	42.655	48.498	160.555	160.555	642.295	1871.833	8946.343
Ac	89	227.000	41.164	43.630	49.607	164.225	164.225	656.976	1914.617	9150.824
Th	90	232.000	42.094	44.617	50.729	167.937	167.937	671.823	1957.885	9357.617
Pa	91	231.000	43.035	45.614	51.863	171.689	171.689	686.836	2001.635	9566.719
U	92	238.000	43.986	46.623	53.010	175.485	175.485	702.015	2045.870	9778.132

Appendix 3
^{4}He$^+$ Stopping Cross Sections

Atom	No.	400	600	800	1000	1200	1400	1600	1800	2000	2400	2800	3200	3600	4000
H	1	14.02	14.11	13.5	12.49	11.34	10.19	9.154	8.289	7.606	6.75	6.081	5.534	5.108	4.683
He	2	16.72	17.88	18.03	17.52	16.63	15.56	14.46	13.44	12.52	11.12	10.02	9.117	8.416	7.714
Li	3	22.28	21.99	21.46	20.64	19.6	18.42	17.28	16.24	15.35	13.63	12.3	11.22	10.33	9.587
Be	4	27.09	26.76	25.89	24.71	23.4	22.06	20.8	19.65	18.64	16.55	14.93	13.61	12.61	11.68
B	5	32.6	33.49	32.67	31.27	29.48	27.59	25.74	24.1	22.7	20.09	18.09	16.51	15.22	14.14
C	6	33.32	36.58	37.21	36.19	34.27	31.99	29.72	27.68	25.97	23.1	20.8	18.99	17.59	16.36
N	7	46.23	48.45	48.12	46.24	43.54	40.58	37.71	35.15	32.98	29.39	26.68	24.45	22.6	21.05
O	8	44.34	47.72	48.39	47.34	45.29	42.81	40.27	37.91	35.84	32.39	29.54	27.21	25.18	23.55
F	9	40.07	43.99	45.66	45.73	44.76	43.15	41.24	39.24	37.31	34.44	31.79	29.46	27.45	25.75
Ne	10	39.32	43.59	45.54	45.86	45.1	43.68	41.92	40.06	38.24	35.79	33.56	31.55	29.66	27.99
Na	11	42.02	44.08	44.95	44.88	44.24	43.14	41.96	40.73	39.56	36.57	34.04	31.86	29.95	28.24
Mg	12	56.04	57.26	56.78	55.26	53.21	50.99	48.82	46.85	45.11	41.57	38.67	36.2	34.16	32.33
Al	13	55.39	54.86	53.81	52.43	50.85	49.18	47.5	45.85	44.25	40.38	37.38	34.96	32.92	31.08
Si	14	70.15	71.09	69.44	66.3	62.5	58.62	55.02	51.88	49.26	44.71	41.27	38.44	36.11	34.09
P	15	64.66	68.45	67.57	65.13	62.06	58.88	55.87	53.08	50.67	45.88	42.17	39.14	36.66	34.55
S	16	62.12	68.61	69.72	67.75	64.72	61.48	58.37	55.41	52.89	47.81	43.88	40.68	38.06	35.83
Cl	17	83.26	86.5	84.96	80.68	75.18	69.52	64.36	60.04	56.65	51.15	46.98	43.53	40.68	38.34
Ar	18	83.61	88.7	87.82	83.47	77.52	71.27	65.56	60.81	57.13	51.75	47.56	44.07	41.18	38.79
K	19	83.16	89.08	90.61	88.88	85.56	80.68	75.61	70.84	66.64	60.13	54.97	50.8	47.18	44.45
Ca	20	93.78	97.3	97.14	94.47	90.5	85.61	80.58	75.64	71.19	64.19	58.63	54.1	50.37	47.21
Sc	21	92.58	96.27	96.3	93.86	90.42	85.73	81.12	76.65	72.53	65.71	60.17	55.6	51.8	48.58
Ti	22	91.07	95.41	95.76	93.54	89.87	85.55	81.14	77.0	73.31	67.61	62.56	58.04	54.17	50.84
V	23	86.19	90.13	90.55	88.7	85.58	81.89	78.13	74.6	71.45	66.28	61.75	57.64	53.95	50.79
Cr	24	79.42	84.81	86.62	85.97	83.76	80.68	77.24	73.78	70.51	66.16	62.13	58.54	55.17	52.12
Mn	25	77.08	82.69	84.4	83.6	81.35	78.41	75.3	72.36	69.72	65.29	61.39	58.01	54.85	52

Atom	No.	400	600	800	1000	1200	1400	1600	1800	2000	2400	2800	3200	3600	4000
Fe	26	80.15	86.9	89.26	88.64	86.13	82.59	78.65	74.71	71.05	66.57	62.86	59.59	56.64	54.02
Co	27	72.11	79.07	82.04	82.29	80.82	78.38	75.5	72.55	69.75	65.5	62.01	58.96	56.23	53.73
Ni	28	68.29	74.6	77.74	78.66	78.07	76.56	74.54	72.3	70.04	66.0	61.0	56.55	52.64	49.94
Cu	29	62.41	68.2	71.77	73.58	74.05	73.5	72.24	70.51	68.48	64.9	62.0	59.34	56.91	54.72
Zn	30	65.53	70.47	72.85	73.47	72.98	71.71	69.97	68.12	66.04	62.25	59.06	56.35	53.91	51.77
Ga	31	74.23	78.12	79.41	79.12	77.79	75.86	73.6	71.25	68.82	64.46	60.89	57.91	55.29	53
Ge	32	77.76	81.9	82.76	82.1	80.26	77.76	75.18	72.46	69.8	65.08	61.25	58.1	55.38	53.02
As	33	81.41	87.03	87.98	87.02	84.84	82.01	79.14	75.97	73.05	67.85	63.66	60.24	57.32	54.8
Se	34	83.2	89.4	89.8	87.8	84.9	81.6	78.4	75.3	72.4	67.06	62.88	59.43	56.5	53.99
Br	35	95.55	101.1	101.1	97.91	93.04	87.7	82.65	78.29	74.75	69.06	64.64	50.96	57.9	55.17
Kr	36	102.2	108.2	108	104.2	98.67	92.74	87.26	82.66	79.04	63.04	68.14	64.21	60.94	58
Rb	37	98.18	108.3	110.1	107.4	102.6	97.34	92.4	87.67	83.47	76.51	71.01	66.59	62.85	59.7
Sr	38	109	117	117.4	114.2	109	103.1	97.75	92.56	87.93	80.3	74.2	69.51	65.46	62.04
Y	39	110	120.4	121.1	117.3	111.6	105.5	99.6	93.9	88.8	84.35	77.65	72.11	67.92	64.37
Zr	40	115.4	126	126.8	123.2	117.9	112	106.2	100.7	97.5	87.64	81.02	75.72	71.2	67.44
Nb	41	118.1	128.2	128.7	125.1	119.8	114	108.4	103	98.1	90.26	83.66	78.21	73.66	69.88
Mo	42	109.8	120.5	122.2	119.6	115.1	110	104.8	99.88	95.21	88.09	81.86	76.49	72.01	68.18
Tc	43	116	126.8	128.9	126.3	121.2	115.4	109.2	103.3	97.85	90.14	83.58	78.09	73.43	69.48
Ru	44	104.1	116.8	120.5	119.5	116	111.3	105.8	100.7	95.86	89.05	82.95	77.71	73.21	69.35
Rh	45	100.9	113.6	117.7	117.2	113.9	109.5	104.6	99.85	95.42	89.21	83.44	78.33	73.87	70.03
Pd	46	89.09	104.9	111.9	112.9	110.3	105.8	100.5	95.32	90.65	86.26	81.88	77.72	63.77	70.04
Ag	47	88.63	101.9	108.4	110.2	108.8	105.4	100.9	96.02	91.22	86.66	82.54	78.53	73.98	68.91
Cd	48	96.33	107	112	113	111.4	108.1	103.8	99.24	94.71	89.47	84.81	80.43	76.31	72.53
In	49	104.3	110.1	113.7	115.2	114.8	112.7	109.3	105	100.0	94.46	89.9	85.69	81.81	78.17
Sn	50	108.2	115.8	118.6	118.3	115.8	112.1	107.9	103.6	99.49	93.82	89.04	84.81	81.14	77.7
Sb	51	116.2	122.2	122.2	119.9	116.8	113.3	110	106.8	103.7	97.58	92.45	88.01	84.26	80.61
Te	52	121.3	127.2	126.5	123.4	119.4	115.3	111.2	107.4	103.9	97.5	92.23	87.74	83.82	80.34
I	53	135	141.7	141	135.8	128.5	120.6	113.2	106.7	101.5	95.02	89.84	85.31	81.53	78.07
Xe	54	144.7	149.7	148.2	143	136	128.7	122	116.2	111.5	104.1	98.28	93.26	89.05	85.31
Cs	55	129.7	141.5	143.1	139.7	134.4	128.7	123.2	117.9	113.1	105	98.36	92.91	88.36	84.29
Ba	56	141.2	150.7	151.4	147.4	141.3	134.9	128.7	122.9	117.6	108.7	101.6	95.71	90.86	86.53
La	57	144.7	156.5	156.9	152.3	145.7	138.6	131.7	125.1	119.0	109.9	102.6	96.71	91.72	87.5
Ce	58	136.4	146.1	147.7	144.5	139.1	133.2	127.4	121.8	116.7	108	101	95.22	90.56	86.27

(continued)

Atom	No.	400	600	800	1000	1200	1400	1600	1800	2000	2400	2800	3200	3600	4000
Pr	59	134.1	143.8	145.7	142.9	137.8	132.2	126.5	121.0	116.0	107.5	100.5	94.84	90.27	86.02
Nd	60	131.9	141.6	143.5	141	136.5	131	125.5	120.2	115.3	106.9	99.99	94.4	89.8	85.71
Pm	61	129.7	139.4	141.4	139.2	135.1	129.8	124.4	119.2	114.5	106.2	99.41	93.91	89.35	85.35
Sm	62	127.7	137.7	139.4	137.4	133.2	128.4	123.3	118.3	113.6	105.5	89.8	93.37	88.86	84.96
Eu	63	125.8	135.6	137.4	135.6	131.7	127.1	122.1	117.3	112.7	104.7	98.15	92.81	88.34	84.53
Gd	64	130.1	139.9	141.7	139.9	135.8	131.1	125.8	120.8	116.0	107.7	100.8	95.23	90.57	86.59
Tb	65	122.2	131.7	133.6	132.2	128.7	124.5	119.7	115.2	110.8	103.1	96.8	91.62	87.24	83.63
Dy	66	111.5	123.9	128.1	127.9	125.4	121.8	117.7	113.5	109.5	102.3	96.4	91.36	87.11	83.41
Ho	67	107.5	118.4	122.4	122.3	120.1	117	113.2	109.5	105.7	98.56	92.71	87.83	83.66	80.21
Er	68	106.1	116.8	120.8	120.7	118.6	115.7	112	108.4	104.8	97.7	91.98	87.17	83.07	79.65
Tm	69	104.7	115.2	119.2	119.2	117.2	114.1	110.8	107.2	103.7	96.82	81.22	86.46	82.43	79.06
Yb	70	103.5	113.8	117.7	117.8	115.9	113	109.7	106.2	102.9	96.06	90.57	85.87	81.89	78.54
Lu	71	106.3	116.9	120.4	120.2	118.1	115	111.6	108.0	104.5	97.42	91.75	86.89	82.79	79.34
Hf	72	109.7	120.8	124.5	124.3	122.2	118.9	115.4	111.6	108.0	100.6	94.7	89.62	85.32	81.72
Ta	73	105.8	117.5	121.7	121.8	119.8	116.7	113.1	109.4	105.6	98.96	93.14	88.4	84.19	80.53
W	74	103.4	114.2	118	118.2	116.5	113.9	110.9	107.6	104.4	97.87	92.17	87.42	83.2	79.61
Re	75	114.4	125.8	129.8	129.8	127.3	124.1	120.4	116.5	112.7	105.2	98.97	93.58	88.98	85.12
Os	76	112.5	124.5	129	129.5	127.3	124.3	120.7	117.0	113.3	106	99.75	94.32	89.69	85.79
Ir	77	110.7	123.2	128.2	129.3	127.3	124.5	121.1	117.7	114.0	106.8	100.7	95.29	90.62	86.67
Pt	78	103.1	117.6	124.2	126.2	125.6	123.7	121.2	118.2	115.0	108.4	102.5	97.21	92.6	88.64
Au	79	109.9	122.7	128	129.1	127.9	125.3	122.3	118.9	115.5	110	104.9	99.92	95.42	91.39
Hg	80	103.5	116.9	122.7	124.2	123.5	121.6	119.2	116.3	113.3	107.1	101.5	96.45	91.9	87.95
Tl	81	113.4	125	129.5	130	128.4	125.8	122.9	119.6	116.4	109.8	104.0	98.82	94.11	89.99
Pb	82	126.4	138.1	141.9	141.6	139.1	135.9	132.4	128.7	125.0	117.7	111.5	106.0	100.9	96.47
Bi	83	124.6	136	139.2	138.3	135.4	131.9	128.3	124.5	120.8	113.6	107.6	102.2	97.4	93.07
Po	84	127.3	140	143	141.8	138.4	134.6	130.6	126.5	122.6	115.2	109.0	103.6	98.76	94.38
At	85	128.2	142.7	146.1	145.1	141.1	137	132.7	128.5	124.4	116.7	110.4	104.9	100.0	95.6
Rn	86	127.7	144.4	148.7	147.7	143.5	139.2	134.7	130.3	126.0	118.1	111.7	106.0	101.2	96.72
Fr	87	143.7	158.2	160.9	158.5	153	147.6	142.2	137.0	132.1	123.2	116.0	109.9	104.7	99.92
Ra	88	155.2	167.8	169.8	166.1	160.1	154	147.9	142.1	136.8	127.1	119.4	112.9	107.4	102.5
Ac	89	158.1	171.3	173.8	170	164.1	157.6	151	144.9	139.4	129.2	121.3	114.5	108.9	103.9
Th	90	159.4	173.4	176.6	173.2	167.2	160.5	153.7	147.4	141.7	131.2	123.0	116.0	110.2	105.2
Pa	91	153.1	166.8	170.4	168	162.8	156.8	150.8	145.0	139.5	129.6	121.8	115.1	109.5	104.7
U	92	150.7	164.4	168.4	166.6	161.9	156.1	150.3	144.6	139.3	129.5	121.7	115.2	109.7	105.0

Appendix 4
Electron Configurations and Ionization Potentials of Atoms

Z	Element	1s	2s	2p	3s	3p	3d	4s	4p	4d	4f	5s	5p	5d	6s	6p	6d	7s	Ionization potential (eV)
1	H	1																	13.595
2	He	2																	24.580
3	Li	2	1																5.390
4	Be	2	2																9.320
5	B	2	2	1															8.296
6	C	2	2	2															11.260
7	N	2	2	3															14.532
8	O	2	2	4															13.614
9	F	2	2	5															17.422
10	Ne	2	2	6															21.564
11	Na	2	2	6	1														5.138
12	Mg	2	2	6	2														7.644
13	Al	2	2	6	2	1													5.984
14	Si	2	2	6	2	2													8.149
15	P	2	2	6	2	3													10.486
16	S	2	2	6	2	4													10.357
17	Cl	2	2	6	2	5													12.967
18	Ar	2	2	6	2	6													15.759
19	K	2	2	6	2	6		1											4.339
20	Ca	2	2	6	2	6		2											6.111
21	Sc	2	2	6	2	6	1	2											6.540
22	Ti	2	2	6	2	6	2	2											6.280
23	V	2	2	6	2	6	3	2											6.740
24	Cr	2	2	6	2	6	5	1											6.764
25	Mn	2	2	6	2	6	5	2											7.432
26	Fe	2	2	6	2	6	6	2											7.870
27	Co	2	2	6	2	6	7	2											7.864
28	Ni	2	2	6	2	6	8	2											7.633
29	Cu	2	2	6	2	6	10	1											7.724
30	Zn	2	2	6	2	6	10	2											9.391
31	Ga	2	2	6	2	6	10	2	1										6.00
32	Ge	2	2	6	2	6	10	2	2										7.88
33	As	2	2	6	2	6	10	2	3										9.81
34	Se	2	2	6	2	6	10	2	4										9.75
35	Br	2	2	6	2	6	10	2	5										11.84

(continued)

Z	Element	1s	2s	2p	3s	3p	3d	4s	4p	4d	4f	5s	5p	5d	6s	6p	6d	7s	Ionization potential (eV)
36	Kr	2	2	6	2	6	10	2	6										14.00
37	Rb	2	2	6	2	6	10	2	6		1								4.176
38	Sr	2	2	6	2	6	10	2	6		2								5.692
39	Y	2	2	6	2	6	10	2	6	1	2								6.377
40	Zr	2	2	6	2	6	10	2	6	2	2								6.835
41	Nb	2	2	6	2	6	10	2	6	4	1								6.881
42	Mo	2	2	6	2	6	10	2	6	5	1								7.131
43	Tc	2	2	6	2	6	10	2	6	(5)	(2)?								7.23
44	Ru	2	2	6	2	6	10	2	6	7	1								7.365
45	Rh	2	2	6	2	6	10	2	6	g	1								7.461
46	Pd	2	2	6	2	6	10	2	6	10									8.33
47	Ag	2	2	6	2	6	10	2	6	10		1							7.574
48	Cd	2	2	6	2	6	10	2	6	10		2							8.991
49	In	2	2	6	2	6	10	2	6	10		2	1						5.785
50	Sn	2	2	6	2	6	10	2	6	10		2	2						7.33
51	Sb	2	2	6	2	6	10	2	6	10		2	3						8.639
52	Te	2	2	6	2	6	10	2	6	10		2	4						9.01
51	I	2	2	6	2	6	10	2	6	10		2	5						10.44
54	Xc	2	2	6	2	6	10	2	6	10		2	6						12.127
55	Cs	2	2	6	2	6	10	2	6	10		2	6		1				3.893
56	Ba	2	2	6	2	6	10	2	6	10		2	6		2				5.210
57	La	2	2	6	2	6	10	2	6	10		2	6	1	2				5.61
58	Ce	2	2	6	2	6	10	2	6	10	1	2	6	1	2				6.91
59	Pr	2	2	6	2	6	10	2	6	10	3	2	6		2				5.70
60	Nd	2	2	6	2	6	10	2	6	10	4	2	6		2				6.31
61	Pm	2	2	6	2	6	10	2	6	10	5	2	6		2				
62	Sm	2	2	6	2	6	10	2	6	10	6	2	6		2				5.6
63	Eu	2	2	6	2	6	10	2	6	10	7	2	6		2				5.67
64	Gd	2	2	6	2	6	10	2	6	10	7	2	6	1	2				6.16
65	Tb	2	2	6	2	6	10	2	6	10	8	2	6	1	2				6.74
66	Dy	2	2	6	2	6	10	2	6	10	9	2	6	1	2	?			6.32
67	Ho	2	2	6	2	6	10	2	6	10	10	2	6	1	2	?			
68	Er	2	2	6	2	6	10	2	6	10	11	2	6	1	2	?			
69	Tm	2	2	6	2	6	10	2	6	10	13	2	6		2				
70	Yb	2	2	6	2	6	10	2	6	10	14	2	6		2				6.22
71	Lu	2	2	6	2	6	10	2	6	10	14	2	6	1	2				6.15
72	Hf	2	2	6	2	6	10	2	6	10	14	2	6	2	2				5.5
73	Ta	2	2	6	2	6	10	2	6	10	14	2	6	3	2				7.7
74	W	2	2	6	2	6	10	2	6	10	14	2	6	4	2				7.98
75	Re	2	2	6	2	6	10	2	6	10	14	2	6	5	2				7.87
76	Os	2	2	6	2	6	10	2	6	10	14	2	6	6	2				8.70
77	Ir	2	2	6	2	6	10	2	6	10	14	2	6	7	2				9.20
78	Pt	2	2	6	2	6	10	2	6	10	14	2	6	9	1	?			9.0
79	Au	2	2	6	2	6	10	2	6	10	14	2	6	10	1				9.22
80	Hg	2	2	6	2	6	10	2	6	10	14	2	6	10	2				10.434
81	Tl	2	2	6	2	6	10	2	6	10	14	2	6	10	2	1			6.106
82	Pb	2	2	6	2	6	10	2	6	10	14	2	6	10	2	2			7.415
83	Bi	2	2	6	2	6	10	2	6	10	14	2	6	10	2	3			7.287
84	Po	2	2	6	2	6	10	2	6	10	14	2	6	10	2	4			8.43
85	At	2	2	6	2	6	10	2	6	10	14	2	6	10	2	5			9.2
86	Rn	2	2	6	2	6	10	2	6	10	14	2	6	10	2	6			10.745

(*continued*)

Z	Element	1s	2s	2p	3s	3p	3d	4s	4p	4d	4f	5s	5p	5d	6s	6p	6d	7s	Ionization potential (eV)
87	Fr	2	2	6	2	6	10	2	6	10	14	2	6	10	2	6		1	4.0
88	Ra	2	2	6	2	6	10	2	6	10	14	2	6	10	2	6		2	5.277
89	Ac	2	2	6	2	6	10	2	6	10	14	2	6	10	2	6	1	2?	6.9
90	Th	2	2	6	2	6	10	2	6	10	14	2	6	10	2	6	2	2?	
91	Pa	2	2	6	2	6	10	2	6	10	14	2	6	10	2	6	1	2?	
92	U	2	2	6	2	6	10	2	6	10	14	2	6	10	2	6	1	2	4
93	Np	2	2	6	2	6	10	2	0	10	14	2	6	10	2	6	1	2?	
94	Pu	2	2	6	2	6	10	2	6	10	14	2	6	10	2	6	1	2?	
95	Am	2	2	6	2	6	10	2	6	10	14	2	6	10	2	6		2	
96	Cm	2	2	6	2	6	10	2	6	10	14	2	6	10	2	6	1	2?	
97	Bk	2	2	6	2	6	10	2	6	10	14	2	6	10	2	6	1	2?	
98	Cf	2	2	6	2	6	10	2	6	10	14	2	6	10	2	6	1	2?	

Appendix 5
Atomic Scattering Factor

$\frac{\sin\theta}{\lambda}$ (nm⁻¹)	0.0	1.0	2.0	3.0	4.0	5.0	6.0	7.0	8.0	9.0	10.0	11.0	12.0
H	1.0	0.8	0.5	0.3	0.1	0.1	0.0	0.0	0.0	0.0	0.0	0.0	
He	2.0	1.9	1.5	1.1	0.8	0.5	0.4	0.2	0.2	0.1	0.1	0.1	
Li⁺	2.0	2.0	1.8	1.5	1.3	1.0	0.8	0.6	0.5	0.4	0.3	0.3	
Li	3.0	2.2	1.8	1.5	1.3	1.0	0.8	0.6	0.5	0.4	0.3	0.3	
Be⁺²	2.0	2.0	1.9	1.7	1.6	1.4	1.2	1.0	0.9	0.7	0.6	0.5	
Be	4.0	2.9	1.9	1.7	1.6	1.4	1.2	1.0	0.9	0.7	0.6	0.5	
B⁺³	2.0	2.0	1.9	1.8	1.7	1.6	1.4	1.3	1.2	1.0	0.9	0.7	
B	5.0	3.5	2.4	1.9	1.7	1.5	1.4	1.2	1.2	1.0	0.9	0.7	
C	6.0	4.6	3.0	2.2	1.9	1.7	1.6	1.4	1.3	1.2	1.0	0.9	
N⁺⁵	2.0	2.0	2.0	1.9	1.9	1.8	1.7	1.6	1.5	1.4	1.3	1.2	
N⁺³	4.0	3.7	3.0	2.4	2.0	1.8	1.7	1.6	1.5	1.4	1.3	1.2	
N	7.0	5.8	4.2	3.0	2.3	1.9	1.7	1.5	1.5	1.4	1.3	1.2	
O	8.0	7.1	5.3	3.9	2.9	2.2	1.8	1.6	1.5	1.4	1.4	1.3	
O⁻²	10.0	8.0	5.5	3.8	2.7	2.1	1.8	1.5	1.5	1.4	1.4	1.3	
F	9.0	7.8	6.2	4.5	3.4	2.7	2.2	1.9	1.7	1.6	1.5	1.4	
F	10.0	8.7	6.7	4.8	3.5	2.8	2.2	1.9	1.7	1.6	1.5	1.4	
Ne	10.0	9.3	7.5	5.8	4.4	3.4	2.7	2.2	1.9	1.7	1.6	1.5	
Na⁺	10.0	9.5	8.2	6.7	5.3	4.1	3.2	2.7	2.3	2.0	1.8	1.6	
Na	11.0	9.7	8.2	6.7	5.3	4.1	3.2	2.7	2.3	2.0	1.8	1.6	
Mg⁺²	10.0	9.8	8.6	7.3	6.0	4.8	3.9	3.2	2.6	2.2	2.0	1.8	
Mg	12.0	10.5	8.6	7.3	6.0	4.8	3.9	3.2	2.6	2.2	2.0	1.8	
AI⁺³	10.0	9.7	8.9	7.8	6.7	5.5	4.5	3.7	3.1	2.7	2.3	2.0	
Al	13.0	11.0	9.0	7.8	6.6	5.5	4.5	3.7	3.1	2.7	2.3	2.0	
Si⁺⁴	10.0	9.8	9.2	8.3	7.2	6.1	5.1	4.2	3.4	3.0	2.6	2.3	
Si	14.0	11.4	9.4	8.2	7.2	6.1	5.1	4.2	3.4	3.0	2.6	2.3	
P⁺⁵	10.0	9.8	9.3	8.5	7.5	6.6	5.7	4.8	4.1	3.4	3.0	2.6	
P	15.0	12.4	10.0	8.5	7.5	6.5	5.7	4.8	4.1	3.4	3.0	2.6	
P⁻³	18.0	12.7	9.8	8.4	7.5	6.5	5.7	4.9	4.1	3.4	3.0	2.6	
S⁺⁶	10.0	9.9	9.4	8.7	7.9	6.9	6.1	5.3	4.5	3.9	3.4	2.9	
S	16.0	13.6	10.7	9.0	7.9	6.9	6.0	5.3	4.5	3.9	3.4	2.9	
S⁻²	18.0	14.3	10.7	8.9	7.9	6.9	6.0	5.3	4.5	3.9	3.4	2.9	
CI	17.0	14.6	11.3	9.3	8.1	7.3	6.5	5.8	5.1	4.4	3.9	3.4	
Cl⁻	18.0	15.2	11.5	9.3	8.1	7.3	6.5	5.8	5.1	4.4	3.9	3.4	
Ar	18.0	15.9	12.6	10.4	8.7	7.8	7.0	6.2	5.4	4.7	4.1	3.6	
K⁺	18.0	16.5	13.3	10.8	8.9	7.8	7.1	6.4	5.9	5.3	4.8	4.2	

(continued)

$\frac{\sin\theta}{\lambda}$ (nm^{-1})	0.0	1.0	2.0	3.0	4.0	5.0	6.0	7.0	8.0	9.0	10.0	11.0	12.0
K	19.0	16.5	13.3	10.8	9.2	7.9	6.7	5.9	5.2	4.6	4.2	3.7	3.3
Ca^{+2}	18.0	16.8	14.0	11.5	9.3	8.1	7.4	6.7	6.2	5.7	5.1	4.6	
Ca	20.0	17.5	14.1	11.4	9.7	8.4	7.3	6.3	5.6	4.9	4.5	4.0	3.6
Sc^{+3}	18.0	16.7	14.0	11.4	9.4	8.3	7.6	6.9	6.4	5.8	5.4	4.9	
Sc	21.0	18.4	14.9	12.1	10.3	8.9	7.7	6.7	5.9	5.3	4.7	4.3	3.9
Ti^{+4}	18.0	17.0	14.4	11.9	9.9	8.5	7.9	7.3	6.7	6.2	5.7	5.1	
Ti	22.0	19.3	15.7	12.8	10.9	9.5	8.2	7.2	6.3	5.6	5.0	4.6	4.2
V	23.0	20.2	16.6	13.5	11.5	10.1	8.7	7.6	6.7	5.9	5.3	4.9	4.4
Cr	24.0	21.1	17.4	14.2	12.1	10.6	9.2	8.0	7.1	6.3	5.7	5.1	4.6
Mn	25.0	22.1	18.2	14.9	12.7	11.1	9.7	8.4	7.5	6.6	6.0	5.4	4.9
Fe	26.0	23.1	18.9	15.6	13.3	11.6	10.2	8.9	7.9	7.0	6.3	5.7	5.2
Co	27.0	24.1	19.8	16.4	14.0	12.1	10.7	9.3	8.3	7.3	6.7	6.0	5.5
Ni	28.0	25.0	20.7	17.2	14.6	12.7	11.2	9.8	8.7	7.7	7.0	6.3	5.8
Cu	29.0	25.9	21.6	17.9	15.2	13.3	11.7	10.2	9.1	8.1	7.3	6.6	6.0
Zn	30.0	26.8	22.4	18.6	15.8	13.9	12.2	10.7	9.6	8.5	7.6	6.9	6.3
Ga	31.0	27.8	23.3	19.3	16.5	14.5	12.7	11.2	10.0	8.9	7.9	7.3	6.7
Ge	32.0	28.8	24.1	20.0	17.1	15.0	13.2	11.6	10.4	9.3	8.3	7.6	7.0
As	33.0	29.7	25.0	20.8	17.7	15.6	13.8	12.1	10.8	9.7	8.7	7.9	7.3
Se	34.0	30.6	25.8	21.5	18.3	16.1	14.3	12.6	11.2	10.0	9.0	8.2	7.5
Br	35.0	31.6	26.6	22.3	18.9	16.7	14.8	13.1	11.7	10.4	9.4	8.6	7.8
Kr	36.0	32.5	27.4	23.0	19.5	17.3	15.3	13.6	12.1	10.8	9.8	8.9	8.1
Rb$^+$	36.0	33.6	28.7	24.6	21.4	18.9	16.7	14.6	12.8	11.2	9.9	8.9	
Rb	37.0	33.5	28.2	23.8	20.2	17.9	15.9	14.1	12.5	11.2	10.2	9.2	8.4
Sr	38.0	34.4	29.0	24.5	20.8	18.4	16.4	14.6	12.9	11.6	10.5	9.5	8.7
Y	39.0	35.4	29.9	25.3	21.5	19.0	17.0	15.1	13.4	12.0	10.9	9.9	9.0
Zr	40.0	36.3	30.8	26.0	22.1	19.7	17.5	15.6	13.8	12.4	11.2	10.2	9.3
Nb	41.0	37.3	31.7	26.8	22.8	20.2	18.1	16.0	14.3	12.8	11.6	10.6	9.7
Mo	42.0	38.2	32.6	27.6	23.5	20.8	18.6	16.5	14.8	13.2	12.0	10.9	10.0
Tc	43.0	39.1	33.4	28.3	24.1	21.3	19.1	17.0	15.2	13.6	12.3	11.3	10.3
Ru	44.0	40.0	34.3	29.1	24.7	21.9	19.6	17.5	15.6	14.1	12.7	11.6	10.6
Rb	45.0	41.0	35.1	29.9	25.4	22.5	20.2	18.0	16.1	14.5	13.1	12.0	11.0
Pd	46.0	41.9	36.0	30.7	26.2	23.1	20.8	18.5	16.6	14.9	13.6	12.3	11.3
Ag	47.0	42.8	36.9	31.5	26.9	23.8	21.3	19.0	17.1	15.3	14.0	12.7	11.7
Cd	48.0	43.7	37.7	32.2	27.5	24.4	21.8	19.6	17.6	15.7	14.3	13.0	12.0
In	49.0	44.7	38.6	33.0	28.1	25.0	22.4	20.1	18.0	16.2	14.7	13.4	12.3
Sn	50.0	45.7	39.5	33.8	28.7	25.6	22.9	20.6	18.5	16.6	15.1	13.7	12.7
Sb	51.0	46.7	40.4	34.6	29.5	26.3	23.5	21.1	19.0	17.0	15.5	14.1	13.0
Te	52.0	47.7	41.3	35.4	30.3	26.9	24.0	21.7	19.5	17.5	16.0	14.5	13.3
I	53.0	48.6	42.1	36.1	31.0	27.5	24.6	22.2	20.0	17.9	16.4	14.8	13.6
Xe	54.0	49.6	43.0	36.8	31.6	28.0	25.2	22.7	20.4	18.4	16.7	15.2	13.9
Cs	55.0	50.7	43.8	37.6	32.4	28.7	25.8	23.2	20.8	18.8	17.0	15.6	14.5
Ba	56.0	51.7	44.7	38.4	33.1	29.3	26.4	23.7	21.3	19.2	17.4	16.0	14.7
La	57.0	52.6	45.6	39.3	33.8	29.8	26.9	24.3	21.9	19.7	17.9	16.4	15.0
Pr	59.0	54.5	47.4	40.9	35.2	31.1	28.0	25.4	22.9	20.6	18.8	17.1	15.7
Nd	60.0	55.4	48.3	41.6	35.9	31.8	28.6	25.9	23.4	21.1	19.2	17.5	16.1
Pm	61.0	56.4	49.1	42.4	36.6	32.4	29.2	26.4	23.9	21.5	19.6	17.9	16.4
Sm	62.0	57.3	50.0	43.2	37.3	32.9	29.8	26.9	24.4	22.0	20.0	18.3	16.8
Eu	63.0	58.3	50.9	44.0	38.1	33.5	30.4	27.5	24.9	22.4	20.4	18.7	17.1
Gd	64.0	59.3	51.7	44.8	38.8	34.1	31.0	28.1	25.4	22.9	20.8	19.1	17.5
Tb	65.0	60.2	52.6	45.7	39.6	34.7	31.6	28.6	25.9	23.4	21.2	19.5	17.9
Dy	66.0	61.1	53.6	46.5	40.4	35.4	32.2	29.2	26.3	23.9	21.6	19.9	18.3
Ho	67.0	62.1	54.5	47.3	41.1	36.1	32.7	29.7	26.8	24.3	22.0	20.3	18.6

(continued)

$\dfrac{\sin\theta}{\lambda}(\text{nm}^{-1})$	0.0	1.0	2.0	3.0	4.0	5.0	6.0	7.0	8.0	9.0	10.0	11.0	12.0
Er	68.0	63.0	55.3	48.1	41.7	36.7	33.3	30.2	27.3	24.7	22.4	20.7	18.9
Tm	69.0	64.0	56.2	48.9	42.4	37.4	33.9	30.8	27.9	25.2	22.9	21.0	19.3
Yb	70.0	64.9	57.0	49.7	43.2	38.0	34.4	31.3	28.4	25.7	23.3	21.4	19.7
Lu	71.0	65.9	57.8	50.4	43.9	38.7	35.0	31.8	28.9	26.2	23.8	21.8	20.0
Hf	72.0	66.8	58.6	51.2	44.5	39.3	35.6	32.3	29.3	26.7	24.2	22.3	20.4
Ta	73.0	67.8	59.5	52.0	45.3	39.9	36.2	32.9	29.8	27.1	24.7	22.6	20.9
W	74.0	68.8	60.4	52.8	46.1	40.5	36.8	33.5	30.4	27.6	25.2	23.0	21.3
Re	75.0	69.8	61.3	53.6	46.8	41.1	37.4	34.0	30.9	28.1	25.6	23.4	21.6
as	76.0	70.8	62.2	54.4	47.5	41.7	38.0	34.6	31.4	28.6	26.0	23.9	22.0
Ir	77.0	71.7	63.1	55.3	48.2	42.4	38.6	35.1	32.0	29.0	26.5	24.3	22.3
Pt	78.0	72.6	64.0	56.2	48.9	43.1	39.2	35.6	32.5	29.5	27.0	24.7	22.7
Au	79.0	73.6	65.0	57.0	49.7	43.8	39.8	36.2	33.1	30.0	27.4	25.1	23.1
Hg	80.0	74.6	65.9	57.9	50.5	44.4	40.5	36.8	33.6	30.6	27.8	25.6	23.6
Tl	81.0	75.5	66.7	58.7	51.2	45.0	41.1	37.4	34.1	31.1	28.3	26.0	24.1
Pb	82.0	76.5	67.5	59.5	51.9	45.7	41.6	37.9	34.6	31.5	28.8	26.4	24.5
Bi	83.0	77.5	68.4	60.4	52.7	46.4	42.2	38.5	35.1	32.0	29.2	26.8	24.8
Po	84.0	78.4	69.4	61.3	53.5	47.1	42.8	39.1	35.6	32.6	29.7	27.2	25.2
At	85.0	79.4	70.3	62.1	54.2	47.7	43.4	39.6	36.2	33.1	30.1	27.6	25.6
Rn	86.0	80.3	71.3	63.0	55.1	48.4	44.0	40.2	36.8	33.5	30.5	28.0	26.0
Fr	87.0	81.3	72.2	63.8	55.8	49.1	44.5	40.7	37.3	34.0	31.0	28.4	26.4
Ra	88.0	82.2	73.2	64.6	56.5	49.8	45.1	41.3	37.8	34.6	31.5	28.8	26.7
Ac	89.0	83.2	74.1	65.5	57.3	50.4	45.8	41.8	38.3	35.1	32.0	29.2	27.1
Th	90.0	84.1	75.1	66.3	58.1	51.1	46.5	42.4	38.8	35.5	32.4	29.6	27.5
Pa	91.0	85.1	76.0	67.1	58.8	51.7	47.1	43.0	39.3	36.0	32.8	30.1	27.9
U	92.0	86.0	76.9	67.9	59.6	52.4	47.7	43.5	39.8	36.5	33.3	30.6	28.3
Np	93.0	87.0	78.0	69.0	60.0	53.0	48.0	44.0	40.0	37.0	34.0	31.0	29.0
Po	94.0	88.0	79.0	69.0	61.0	54.0	49.0	44.0	41.0	38.0	34.0	31.0	29.0
Am	95.0	89.0	79.0	70.0	62.0	55.0	50.0	45.0	42.0	38.0	35.0	32.0	30.0
Cm	96.0	90.0	80.0	71.0	62.0	55.0	50.0	46.0	42.0	39.0	35.0	32.0	30.0
Bk	97.0	91.0	81.0	72.0	63.0	56.0	51.0	46.0	43.0	39.0	36.0	33.0	30.0
Cf	98.0	92.0	82.0	73.0	64.0	57.0	52.0	47.0	43.0	40.0	36.0	33.0	31.0

Appendix 6
Electron Binding Energies

	1s$_{1/2}$ K	2s$_{1/2}$ L$_1$	2p$_{1/2}$ L$_2$	2p$_{3/2}$ L$_3$	3s$_{1/2}$ M$_1$	3p$_{1/2}$ M$_2$	3p$_{3/2}$ M$_3$	3d$_{3/2}$ M$_4$	3d$_{3/2}$ M$_5$	4s$_{1/2}$ N$_1$	4p$_{1/2}$ N$_2$	4p$_{3/2}$ N$_3$	4d$_{3/2}$ N$_4$	4d$_{3/2}$ N$_5$	4f$_{3/2}$ N$_6$	4f$_{1/2}$ N$_7$
1 H	14															
2 He	25															
3 Li	55															
4 Be	111															
5 B	188	5														
6 C	284	7														
7 N	399	9														
8 O	532	24	7													
9 F	686	31	9													
10 Ne	867	45	18													
11 Na	1072	63	31		1											
12 Mg	1305	89	52		2											
13 Al	1560	118	74	73	1											
14 Si	1839	149	100	99	8	3										
15 P	2149	189	136	135	16	10										
16 S	2472	229	165	164	16	8										
17 Cl	2823	270	202	200	18	7										
18 Ar	3203	320	247	245	25	12										
19 K	3608	377	297	294	34	18										
20 Ca	4038	438	350	347	44	26		5								
21 Sc	4493	500	407	402	54	32		7								
22 Ti	4965	564	461	455	59	34		3								

(continued)

	$1s_{1/2}$ K	$2s_{1/2}$ L_1	$2p_{1/2}$ L_2	$2p_{3/2}$ L_3	$3s_{1/2}$ M_1	$3p_{1/2}$ M_2	$3p_{3/2}$ M_3	$3d_{3/2}$ M_4	$3d_{3/2}$ M_5	$4s_{1/2}$ N_1	$4p_{1/2}$ N_2	$4p_{3/2}$ N_3	$4d_{3/2}$ N_4	$4d_{3/2}$ N_5	$4f_{3/2}$ N_6	$4f_{1/2}$ N_7
23 V	5465	628	520	513	66	38		2								
24 Cr	5989	695	584	575	74	43		2								
25 Mr	6539	769	652	641	84	49		4								
26 Fe	7114	846	723	710	95	56		6								
27 Co	7709	926	794	779	101	60		3								
28 Ni	8333	1008	872	855	112	68		4								
29 Cu	8979	1096	951	931	120	74		2								
30 Zn	9659	1194	1044	1021	137	87		9								
31 Ga	10367	1298	1143	1116	158	107	103	18				1				
32 Ge	11104	1413	1249	1217	181	129	122	29				3				
33 As	11867	1527	1359	1323	204	147	141	41				3				
34 Se	12658	1654	1476	1436	232	168	162	57				6				
35 Br	13474	1782	1596	1550	257	189	182	70	69	27		5				
36 Kr	14326	1921	1727	1675	289	223	214	89		24	15	11				
37 Rb	15200	2065	1864	1805	322	248	239	112	111	30		14				
38 Sr	16105	2216	2007	1940	358	280	269	135	133	38		20				
39 Y	17039	2373	2155	2080	395	313	301	160	158	46		26				
40 Zr	17998	2532	2307	2223	431	345	331	183	180	52		29				
41 Nb	18986	2698	2465	2371	469	379	363	208	205	58		34	4			
42 Mo	20000	2866	2625	2520	505	410	393	230	227	62		35	2			
43 Tc	21044	3043	2793	2677	544	445	425	257	253	68		39	2			
44 Ru	22117	3224	2967	2838	585	483	461	284	279	75		43	2	3		
45 Rh	23220	3412	3146	3004	627	521	496	312	307	81		48	3	3		
46 Pd	24350	3605	3331	3173	670	559	531	340	335	86		51	1			
47 Ag	25514	3806	3524	3351	717	602	571	373	367	95	62	56	3			
48 Cd	26711	4018	3727	3538	770	651	617	411	404	108	67		9			
49 In	27940	4238	3938	3730	826	702	664	451	443	122	77		16			
50 Sn	29200	4465	4156	3929	884	757	715	494	485	137	89		24			

(continued)

Z		$1s_{1/2}$ K	$2s_{1/2}$ L_1	$2p_{1/2}$ L_2	$2p_{3/2}$ L_3	$3s_{1/2}$ M_1	$3p_{1/2}$ M_2	$3p_{3/2}$ M_3	$3d_{3/2}$ M_4	$3d_{5/2}$ M_5	$4s_{1/2}$ N_1	$4p_{1/2}$ N_2	$4p_{3/2}$ N_3	$4d_{3/2}$ N_4	$4d_{5/2}$ N_5	$4f_{5/2}$ N_6	$4f_{7/2}$ N_7
51	Sb	30 491	4 699	4 381	4 132	944	812	766	537	528	152	99		32			
52	Te	31 814	4 939	4 612	4 341	1 006	870	819	582	572	168	110		40			
53	I	33 170	5 188	4 852	4 557	1 072	931	875	631	620	186	123		50			
54	Xe	34 561	5 453	5 104	4 782	1 145	999	937	685	672	208	147		63			
55	Cs	35 985	5 713	5 360	5 012	1 217	1 065	998	740	726	231	172	162	79	77		
56	Ba	37 441	5 987	5 624	5 247	1 293	1 137	1 063	796	781	253	192	180	93	90		
57	La	38 925	6 267	5 891	5 483	1 362	1 205	1 124	849	832	271	206	192		99		
58	Ce	40 444	6 549	6 165	5 724	1 435	1 273	1 186	902	884	290	224	208		111	1	
59	Pr	41 991	6 835	6 441	5 965	1 511	1 338	1 243	951	931	305	237	218		114	2	
60	Nd	43 569	7 126	6 722	6 208	1 576	1 403	1 298	1 000	978	316	244	225		118	2	
61	Pm	45 185	7 428	7 013	6 460	1 650	1 472	1 357	1 052	1 027	331	255	237		121	4	
62	Sm	46 835	7 737	7 312	6 717	1 724	1 542	1 421	1 107	1 081	347	267	249		130	7	
63	Eu	48 519	8 052	7 618	6 977	1 800	1 614	1 481	1 161	1 131	360	284	257		134	0	
64	Gd	50 239	8 376	7 931	7 243	1 881	1 689	1 544	1 218	1 186	376	289	271		141	0	
65	Tb	51 996	8 708	8 252	7 515	1 968	1 768	1 612	1 276	1 242	398	311	286		148	3	
66	Dy	53 788	9 047	8 581	7 790	2 047	1 842	1 676	1 332	1 295	416	332	293		154	4	
67	Ho	55 618	9 395	8 919	8 071	2 128	1 923	1 741	1 391	1 351	436	343	306		161	4	
68	Er	57 486	9 752	9 265	8 358	2 207	2 006	1 812	1 453	1 409	449	366	320	177	168	4	
69	Tm	59 390	10 116	9 618	8 648	2 307	2 090	1 885	1 515	1 468	472	386	337		180	5	
70	Yb	61 332	10 488	9 978	8 943	2 397	2 172	1 949	1 576	1 527	487	396	343	197	184	6	
71	Lu	63 314	10 870	10 349	9 244	2 491	2 264	2 024	1 640	1 589	506	410	359	205	195	7	
72	Hf	65 351	11 272	10 739	9 561	2 601	2 365	2 108	1 716	1 662	538	437	380	224	214	19	18
73	Ta	67 417	11 680	11 136	9 881	2 708	2 469	2 194	1 793	1 735	566	465	405	242	230	27	25
74	W	69 525	12 099	11 544	10 205	2 820	2 575	2 281	1 872	1 810	595	492	426	259	246	37	34
75	Re	71 677	12 527	11 959	10 535	2 932	2 682	2 367	1 949	1 883	625	518	445	274	260	47	45
76	Os	73 871	12 968	12 385	10 871	3 049	2 792	2 458	2 031	1 960	655	547	469	290	273	52	50
77	Ir	76 111	13 419	12 824	11 215	3 174	2 909	2 551	2 116	2 041	690	577	495	312	295	63	60
78	Pt	78 395	13 880	13 273	11 564	3 298	3 027	2 646	2 202	2 121	724	608	519	331	314	74	70
79	Au	80 725	14 353	13 734	11 918	3 425	3 150	2 743	2 291	2 206	759	644	546	352	334	87	83
80	Hg	83 103	14 839	14 209	12 284	3 562	3 279	2 847	2 385	2 295	800	677	571	379	360	103	99

(continued)

		$1s_{1/2}$ K	$2s_{1/2}$ L$_1$	$2p_{1/2}$ L$_2$	$2p_{3/2}$ L$_3$	$3s_{1/2}$ M$_1$	$3p_{1/2}$ M$_2$	$3p_{3/2}$ M$_3$	$3d_{3/2}$ M$_4$	$3d_{5/2}$ M$_5$	$4s_{1/2}$ N$_1$	$4p_{1/2}$ N$_2$	$4p_{3/2}$ N$_3$	$4d_{3/2}$ N$_4$	$4d_{5/2}$ N$_5$	$4f_{5/2}$ N$_6$	$4f_{7/2}$ N$_7$
81	Tl	85 531	15 347	14	12 657	3 704	3 416	2 957	2 485	2 390	846	722	609	407	386	122	118
82	Pb	88 005	15 861	15	13 035	3 851	3 554	3 067	2 586	2 484	894	764	645	435	413	143	138
83	Bi	90 526	16 388	15	13 418	3 999	3 697	3 177	2 688	2 580	939	806	679	464	440	163	158
84	Po	93 105	16 939	16	13 814	4 149	3 854	3 302	2 798	2 683	995	851	705	500	473	184	
85	At	95 730	17 493	16	14 214	4 317	4 008	3 426	2 909	2 787	1 042	886	740	533	507	210	
86	Rn	98 404	18 049	17	14 619	4 482	4 159	3 538	3 022	2 892	1 097	929	768	567	541	238	
87	Fr	101 137	18 639	17	15 031	4 652	4 327	3 663	3 136	3 000	1 153	980	810	603	577	268	
88	Ra	103 922	19 237	18	15 444	4 822	4 490	3 792	3 248	3 105	1 208	1 058	879	636	603	299	
89	Ac	106 755	19 840	19	15 871	5 002	4 656	3 909	3 370	3 219	1 269	1 080	890	675	639	319	
90	Th	109 651	20 472	19	16 300	5 182	4 831	4 046	3 491	3 332	1 330	1 168	968	714	677	344	335

Appendix 7
X-Ray Wavelengths (nm)

Element		$E(K_\alpha)$ (keV)	$K_{\alpha 1}$	$K_{\alpha 2}$	K_β	K Edge	$L_{\alpha 1}$	L_{III} Edge
1	H							
2	He							
3	Li	0.05	22.8000			22.6500		
4	Be	0.11	11.4000			11.1000		
5	B	0.18	6.7600			6.5957		
6	C	0.28	4.4700			4.3680		
7	N	0.39	3.1600			3.0990		
8	O	0.52	2.3620			2.3320		
9	F	0.68	1.8320			1.8076		
10	Ne	0.85	1.4610		1.4452	1.4302		
11	Na	1.04	1.1910		1.1575	1.1569		40.5000
12	Mg	1.25	9.8900		0.9521	0.9512		25.0700
13	Al	1.49	0.8339	0.8342	0.7960	0.7948		17.0400
14	Si	1.74	0.7125	0.7128	0.6753	0.6738		12.3000
15	P	2.01	0.6157	0.6160	0.5796	0.5784		9.4000
16	Si	2.31	0.5372	0.5375	0.5032	0.5019		7.5610
17	Cl	2.62	0.4728	0.4731	0.4403	0.4397		6.2000
18	Ar	2.96	0.4192	0.4195	0.3886	0.3871		5.0612
19	K	3.31	0.3741	0.3745	0.3454	0.3437		4.2100
20	Ca	3.69	0.3358	0.3362	0.3090	0.3070	3.6330	3.5490
21	Sc	4.09	0.3031	0.3034	0.2780	0.2762	3.1350	3.0846
22	Ti	4.51	0.2749	0.2752	0.2514	0.2497	2.7420	2.7290
23	V	4.95	0.2504	0.2507	0.2284	0.2269	2.4250	2.4172
24	Cr	5.41	0.2290	0.2294	0.2085	0.2070	2.1640	2.0700
25	Mn	5.90	0.2102	0.2106	0.1910	0.1896	1.9450	1.9345
26	Fe	6.40	0.1936	0.1940	0.1757	0.1743	1.7590	1.7525
27	Co	6.93	0.1790	0.1793	0.1621	0.1608	1.5972	1.5915
28	Ni	7.47	0.1658	0.1662	0.1500	0.1488	1.4561	1.4525
29	Cu	8.04	0.1541	0.1544	0.1392	0.1381	1.3336	1.3288
30	Zn	8.63	0.1435	0.1439	0.1295	0.1283	1.2254	1.2131

(*continued*)

Element		$E(K_\alpha)$ (keV)	$K_{\alpha1}$	$K_{\alpha2}$	K_β	K Edge	$L_{\alpha1}$	L_{III} Edge
31	Ga	9.24	0.1340	0.1344	0.1208	0.1196	1.1292	1.1100
32	Ge	9.88	0.1254	0.1258	0.1129	0.1117	1.0436	1.0187
33	As	10.53	0.1176	0.1180	0.1057	0.1045	0.9671	0.9367
34	Se	11.21	0.1105	0.1109	0.0992	0.0980	0.8990	0.8646
35	Br	11.91	0.1040	0.1044	0.0933	0.0920	0.8375	0.7984
36	Kr	12.63	0.0980	0.0984	0.0879	0.0866	0.7817	0.7392
37	Rb	13.38	0.0926	0.0930	0.0829	0.0816	0.7318	0.6862
38	Sr	14.14	0.0875	0.0879	0.0783	0.0770	0.6863	0.6387
39	Y	14.93	0.0829	0.0833	0.0741	0.0728	0.6449	0.5962
40	Zr	15.75	0.0786	0.0790	0.0702	0.0689	0.6071	0.5579
41	Mb	16.58	0.0746	0.0750	0.0666	0.0653	0.5724	0.5230
42	Mo	17.44	0.0709	0.0714	0.0632	0.0620	0.5407	0.4913
43	Tc	18.33	0.0675	0.0679	0.0601	0.0589	0.5115	0.4630
44	Ru	19.24	0.0643	0.0647	0.0572	0.0561	0.4846	0.4369
45	Rh	20.17	0.0613	0.0618	0.0546	0.0534	0.4597	0.4130
46	Pd	21.12	0.0585	0.0590	0.0521	0.0509	0.4368	0.3907
47	Ag	22.11	0.0559	0.0564	0.0497	0.0486	0.4154	0.3700
48	Cd	23.11	0.0535	0.0539	0.0475	0.0464	0.3956	0.3505
49	In	24.14	0.0512	0.0517	0.0455	0.0444	0.3772	0.3324
50	Sn	25.20	0.0491	0.0495	0.0435	0.0425	0.3600	0.3156
51	Sb	26.28	0.0470	0.0475	0.0417	0.0407	0.3439	0.3000
52	Te	27.38	0.0451	0.0456	0.0400	0.0390	0.3289	0.2856
53	I	28.51	0.0433	0.0438	0.0384	0.0374	0.3149	0.2720
54	Xe	29.67	0.0416	0.0421	0.0369	0.0358	0.3017	0.2593
55	Cs	30.86	0.0400	0.0405	0.0354	0.0345	0.2892	0.2474
56	Ba	32.07	0.0385	0.0390	0.0341	0.0331	0.2776	0.2363
57	La	33.31	0.0371	0.0375	0.0328	0.0318	0.2666	0.2261
58	Ce	34.57	0.0357	0.0362	0.0316	0.0306	0.2562	0.2166
59	Pr	35.87	0.0344	0.0349	0.0304	0.0295	0.2463	0.2079
60	Nd	37.19	0.0332	0.0336	0.0293	0.0285	0.2370	0.1997
61	Pm	38.54	0.0320	0.0325	0.0283	0.0274	0.2282	0.1919
62	Sm	39.92	0.0309	0.0314	0.0273	0.0265	0.2200	0.1846
63	Eu	41.33	0.0298	0.0303	0.0264	0.0256	0.2121	0.1776
64	Gd	42.77	0.0288	0.0293	0.0255	0.0247	0.2047	0.1712
65	Tb	44.24	0.0279	0.0283	0.0246	0.0238	0.1977	0.1650
66	Dy	45.73	0.0270	0.0274	0.0238	0.0230	0.1909	0.1592
67	Ho	47.26	0.0261	0.0265	0.0230	0.0223	0.1845	0.1537
68	Er	48.83	0.0252	0.0257	0.0223	0.0216	0.1784	0.1484
69	Tm	50.42	0.0244	0.0249	0.0216	0.0209	0.1727	0.1433
70	Yb	52.04	0.0237	0.0241	0.0209	0.0202	0.1672	0.1386
71	Lu	53.70	0.0229	0.0234	0.0202	0.0196	0.1620	0.1341
72	Hf	55.40	0.0222	0.0227	0.0196	0.0190	0.1570	0.1297
73	Ta	57.11	0.0215	0.0220	0.0190	0.0184	0.1522	0.1255
74	W	58.87	0.0209	0.0214	0.0184	0.0178	0.1476	0.1216
75	Re	60.67	0.0203	0.0208	0.0179	0.0173	0.1432	0.1177

(continued)

Element		$E(K_\alpha)$ (keV)	$K_{\alpha1}$	$K_{\alpha2}$	K_β	K Edge	$L_{\alpha1}$	L_{III} Edge
76	Os	62.50	0.0197	0.0202	0.0174	0.0168	0.1391	0.1141
77	Ir	64.36	0.0191	0.0196	0.0169	0.0163	0.1351	0.1106
78	Pt	66.26	0.0186	0.0190	0.0164	0.0158	0.1313	0.1072
79	Au	68.20	0.0180	0.0185	0.0159	0.0154	0.1276	0.1040
80	Hg	70.18	0.0175	0.0180	0.0154	0.0149	0.1241	0.1009
81	Tl	72.19	0.0170	0.0175	0.0150	0.0145	0.1207	0.0979
82	Pb	74.25	0.0165	0.0170	0.0146	0.0141	0.1175	0.0951
83	Bi	76.34	0.0161	0.0166	0.0142	0.0137	0.1144	0.0923
84	Po	78.48	0.0156	0.0161	0.0138	0.0133	0.1114	0.0898
85	At	80.66	0.0152	0.0157	0.0134	0.0130	0.1085	0.0872
86	Rn	82.88	0.0148	0.0153	0.0131	0.0126	0.1057	0.0848
87	Fr	85.14	0.0144	0.0149	0.0127	0.0123	0.1030	0.0825
88	Ra	87.46	0.0140	0.0145	0.0124	0.0119	0.1005	0.0803
89	Ac	89.81	0.0136	0.0141	0.0121	0.1160	0.0980	0.0781
90	Th	92.22	0.0133	0.0138	0.0117	0.0113	0.0956	0.0761

Appendix 8
Mass Absorption Coefficients (μ/ρ)

		Mo		Cu		Co		Cr	
		k_α	k_β	k_α	k_β	k_α	k_β	k_α	k_β
Absorber	Density (gm/cm³)	0.0711 nm	0.0632 nm	0.1542 nm	0.1392 nm	0.1790 nm	0.1162 nm	0.2291 nm	0.20895 nm
1	H	0.08×10^{-3}	0.3727	0.3699	0.3912	0.3882	0.3966	0.3928	0.4116	0.4046
2	He	0.17×10^{-3}	0.2019	0.1972	0.2835	0.2623	0.3288	0.2966	0.4648	0.4001
3	Li	0.53	0.1968	0.1866	0.4770	0.3939	0.6590	0.5283	1.243	0.9639
4	Be	1.85	0.2451	0.2216	1.007	0.7742	1.522	1.152	3.183	2.388
5	Be	2.47	0.3451	0.2928	2.142	1.590	3.357	2.485	7.232	5.385
6	C	2.27 (graphite)	0.5348	0.4285	4.219	3.093	6.683	4.916	14.46	10.76
7	N	1.17×10^{-3}	0.7898	0.6054	7.142	5.215	11.33	8.33	24.42	18.23
8	O	1.3×10^{-3}	1.147	0.8545	11.03	8.06	17.44	12.85	37.19	27.88
9	F	1.70×10^{-3}	1.584	1.154	15.95	11.66	25.12	18.57	53.14	39.99
10	Ne	0.84×10^{-3}	2.209	1.597	22.13	16.24	34.69	25.72	72.71	54.91
11	Na	0.97	2.939	2.098	30.30	22.23	47.34	35.18	98.48	74.66
12	Mg	1.74	3.979	2.825	40.88	30.08	63.54	47.38	130.8	99.62
13	Al	2.70	5.043	3.585	50.23	37.14	77.54	58.08	158.0	120.7
14	Si	2.33	6.533	4.624	65.32	48.37	100.4	75.44	202.7	155.6
15	P	1.82 (yellow)	7.870	5.569	77.28	57.44	118.0	89.05	235.5	181.6
16	S	2.09	9.625	6.835	92.53	68.90	141.2	106.6	281.9	217.2
17	Cl	3.2×10^{-3}	11.64	8.261	109.2	81.79	164.7	125.3	321.5	250.2
18	Ar	1.67×10^{-3}	12.62	8.949	119.5	89.34	180.9	137.3	355.5	275.8
19	K	0.86	16.20	11.51	148.4	111.7	222.0	169.9	426.8	334.2
20	Ca	1.53	19.00	13.56	171.4	129.0	257.4	196.4	499.6	389.3

(continued)

Absorber		Density (gm/cm³)	Mo		Cu		Co		Cr	
			k_α 0.0711 nm	k_β 0.0632 nm	k_α 0.1542 nm	k_β 0.1392 nm	k_α 0.1790 nm	k_β 0.1162 nm	k_α 0.2291 nm	k_β 0.20895 nm
21	Sc	2.99	21.04	15.00	186.0	140.8	275.5	212.2	520.9	410.7
22	Ti	4.51	23.25	16.65	202.4	153.2	300.5	231.0	571.4	449.0
23	V	6.09	25.24	18.07	222.6	168.0	332.7	254.7	75.1	501.0
24	Cr	7.19	29.25	20.99	252.3	191.1	375.0	288.1	85.7	65.8
25	Mn	7.47	31.86	22.89	272.5	206.7	405.1	311.2	96.1	73.8
26	Fe	7.87	37.74	27.21	304.4	233.6	56.25	345.5	113.1	86.8
27	Co	8.80	41.02	29.51	338.6	258.7	62.86	47.71	124.6	96.1
28	Ni	8.91	47.24	34.18	48.83	282.8	73.75	56.05	145.7	112.5
29	Cu	8.93	49.34	35.77	51.54	38.74	78.11	59.22	155.7	119.5
30	Zn	7.13	55.46	40.26	59.51	45.30	88.71	68.00	171.7	133.5
31	Ga	5.91	56.90	41.69	62.13	46.65	94.15	71.39	186.9	144.0
32	Ge	5.32	60.47	44.26	67.92	51.44	102.0	77.79	199.9	154.5
33	As	5.78	65.97	48.57	75.65	57.01	114.0	86.76	224.0	173.3
34	Se	4.81	68.82	51.20	82.89	62.32	125.1	95.11	246.1	190.4
35	Br	3.12 (liquid)	74.68	55.56	90.29	68.07	135.8	103.5	266.2	206.2
36	Kr	3.488 × 10⁻³	79.10	5864	97.02	73.22	145.7	111.2	284.6	220.7
37	Rb	1.53	83.00	62.07	106.3	80.16	159.6	121.8	311.7	241.8
38	Sr	2.58	88.04	65.59	115.3	8677	173.5	132.2	339.3	263.4
39	Y	4.48	97.56	72.57	127.1	96.19	190.2	145.4	368.9	286.9
40	Zr	6.51	16.10	75.2	136.8	1033	204.9	156.6	398.6	309.7
41	Nb	8.58	16.96	81.22	148.8	112.3	222.9	170.4	431.9	336.4
42	Mo	10.22	18.44	13.29	158.3	119.7	236.6	181.0	457.4	356.5
43	To	11.50	19.78	14.30	167.7	126.9	250.8	191.9	485.5	378.0
44	Ru	12.36	21.33	15.40	180.8	137.0	269.4	206.6	517.9	404.4
45	Rh	12.42	23.05	16.65	194.1	147.1	289.0	221.8	555.2	433.7
46	Pd	12.00	24.42	17.63	205.0	155.6	304.3	234.0	580.9	455.1
47	Ag	10.50	26.38	19.10	218.1	165.8	323.5	248.9	617.4	483.5

(continued)

Absorber		Density (gm/cm³)	Mo k_α 0.0711 nm	Mo k_β 0.0632 nm	Cu k_α 0.1542 nm	Cu k_β 0.1392 nm	Co k_α 0.1790 nm	Co k_β 0.1162 nm	Cr k_α 0.2291 nm	Cr k_β 0.20895 nm
48	Cd	8.65	27.73	20.13	229.3	174.0	341.8	262.1	658.8	513.5
49	In	7.29	29.13	21.18	242.1	183.3	362.7	277.1	705.8	548.0
50	Sn	7.29	31.18	22.62	253.3	193.1	374.1	2887	708.8	556.6
51	Sb	6.69	33.01	23.91	266.5	203.6	391.3	303.1	733.4	578.8
52	Te	6.25	33.92	24.67	273.4	208.4	404.4	311.7	768.9	602.7
53	I	4.95	36.33	26.53	291.7	221.9	434.0	333.2	835.2	650.8
54	Xe	5.495×10^{-3}	38.31	27.86	309.8	235.9	459.0	353.4	755.4	685.2
55	Cs	1.91 (−10°C)	40.44	29.51	325.4	247.5	483.8	371.5	802.7	725.1
56	Ba	3.59	42.37	31.00	336.1	256.0	499.0	383.6	587.3	644.6
57	La	6.17	45.34	33.10	353.5	270.8	519.0	401.9	222.9	661.5
58	Ce	6.77	4856	35.54	378.8	289.6	559.1	431.5	240.4	509.4
59	Pr	6.78	50.78	37.09	402.2	306.8	596.2	458.8	260.5	205.3
60	Nd	7.00	53.28	38.88	417.9	319.8	531.7	475.9	271.3	213.4
61	Pm		55.52	40.52	441.1	336.4	401.4	503.2	284.7	223.8
62	Sm	7.54	57.96	42.40	453.5	346.6	411.8	446.3	295.0	231.5
63	Eu	5.25	61.18	44.74	417.9	369.0	165.2	476.9	312.7	244.9
64	Gd	7.87	62.79	45.95	426.7	377.2	169.5	346.7	318.9	250.3
65	Tb	8.27	66.77	48.88	321.9	399.9	178.7	367.1	338.9	265.2
66	Dy	8.53	68.89	50.38	336.6	360.2	184.9	142.9	351.7	275.0
67	Ho	8.80	72.14	52.76	128.4	272.4	189.8	146.3	363.3	283.4
68	Er	9.04	75.61	55.07	134.3	291.7	198.4	153.0	379.7	296.1
69	Tm	9.33	78.98	57.94	140.2	288.5	207.4	159.8	397.0	309.6
70	Yb	6.97	80.23	59.22	144.7	110.9	214.0	164.9	409.6	319.4
71	Lu	9.84	84.18	62.04	152.0	116.5	224.6	173.2	429.5	335.1
72	Hf	13.28	86.33	64.15	157.7	121.0	232.9	179.6	445.0	347.2
73	Ta	16.67	89.51	66.07	161.5	123.9	238.3	183.9	454.7	3550
74	W	19.25	95.76	70.57	170.5	131.5	249.7	193.7	470.4	369.1
75	Re	21.02	98.74	72.47	178.3	137.3	261.8	202.7	495.5	388.0

(continued)

Absorber		Density (gm/cm³)	Mo		Cu		Co		Cr	
			k_α 0.0711 nm	k_β 0.0632 nm	k_α 0.1542 nm	k_β 0.1392 nm	k_α 0.1790 nm	k_β 0.1162 nm	k_α 0.2291 nm	k_β 0.20895 nm
76	Os	22.58	100.2	74.13	183.8	141.3	270.3	209.0	512.4	401.1
77	Ir	22.55	103.4	77.20	192.2	147.6	283.4	218.8	539.6	421.6
78	Pt	21.44	108.6	80.23	198.2	161.2	295.2	226.4	571.6	443.9
79	Au	19.28	111.3	82.33	207.8	160.6	303.3	235.7	568.0	446.7
80	Hg	13.55	114.7	85.30	216.2	166.6	317.0	245.7	597.9	468.9
81	Tl	11.87	119.4	88.25	222.2	171.1	326.3	252.7	616.9	483.3
82	Pb	11.34	122.8	90.55	232.1	178.6	340.8	263.8	644.5	504.9
83	Bi	9.80	125.9	93.50	242.9	187.5	355.3	275.8	667.2	524.2
86	Rn	4.4 (−62°C)	117.2	100.7	263.7	203.0	387.1	299.7	731.4	573.2
90	Th	11.72	99.46	73.34	306.8	236.6	449.0	348.4	844.1	663.0
92	U	19.05	96.67	72.63	305.7	236.2	446.3	346.9	774.0	657.1
94	Pu	19.81	48.84	78.99	352.9	271.2	519.6	401.6	803.2	771.6

Appendix 9
KLL Auger Energies (eV)

		$2s^0 2p^6$	$2s^1 2p^5$			$2s^2 2p^4$				
		$^1S_0$	$^1P_1$	$^3P_0$	$^3P_2$	$^1S_0$	$^1D_2$	$^3P_0$	$^3P_0$	$^3P_2$
		KL_1L_1	KL_1L_2	KL_1L_2	KL_1L_2	KL_1L_3	KL_2L_2	KL_2L_3	KL_3L_3	KL_3L_3
C	6	0.243	0.252	0.258	0.258	0.258	0.265	0.266	0.267	0.267
N	7	0.356	0.362	0.369	0.369	0.369	0.373	0.375	0.377	0.377
O	8	0.474	0.486	0.495	0.495	0.495	0.504	0.507	0.509	0.509
F	9	0.610	0.627	0.638	0.638	0.638	0.650	0.654	0.657	0.657
Ne	10	0.761	0.781	0.794	0.794	0.794	0.808	0.813	0.816	0.816
Na	11	0.928	0.952	0.967	0.967	0.967	0.984	0.989	0.993	0.993
Mg	12	0.105	1.135	1.151	1.151	1.151	1.172	1.179	1.183	1.183
Al	13	0.301	1.336	1.354	1.354	1.354	1.379	1.387	1.392	1.392
Si	14	0.516	1.554	1.574	1.574	1.575	1.602	1.611	1.616	1.617
P	15	0.742	1.784	1.805	1.806	1.806	1.835	1.845	1.851	1.852
S	16	0.982	2.034	2.057	2.058	2.059	2.096	2.107	2.114	2.115
Cl	17	2.249	2.305	2.329	2.330	2.331	2.370	2.382	2.389	2.391
Ar	18	2.527	2.586	2.612	2.613	2.614	2.656	2.669	2.677	2.679
K	19	2.815	2.881	2.909	2.910	2.912	2.959	2.973	2.981	2.984
Ca	20	3.122	3.195	3.224	3.225	3.227	3.279	3.294	3.303	3.306
Sc	21	3.456	3.533	3.563	3.564	3.567	3.622	3.638	3.647	3.651
Ti	22	3.799	3.886	3.916	3.919	3.922	3.985	4.002	4.011	4.016
V	23	4.168	4.259	4.290	4.293	4.298	4.362	4.381	4.391	4.397
Cr	24	4.557	4.651	4.683	4.687	4.692	4.757	4.778	4.788	4.795
Mn	25	4.956	5.056	5.089	5.094	5.100	5.169	5.191	5.202	5.211
Fe	26	5.374	5.480	5.514	5.519	5.527	5.598	5.622	5.634	5.644
Co	27	5.808	5.923	5.957	5.964	5.972	6.049	6.075	6.088	6.099
Ni	28	6.264	6.384	6.419	6.426	6.436	6.514	6.542	6.556	6.568
Cu	29	6.732	6.861	6.896	6.905	6.916	7.000	7.030	7.045	7.059
Zn	30	7.214	7.348	7.384	7.394	7.407	7.493	7.526	7.543	7.558
Ga	31	7.712	7.852	7.888	7.900	7.915	8.000	8.037	8.057	8.073
Ge	32	8.216	8.365	8.401	8.416	8.433	8.523	8.563	8.586	8.603
As	33	8.749	8.903	8.939	8.957	8.975	9.063	9.107	9.133	9.152
Se	34	9.283	9.447	9.483	9.504	9.524	9.616	9.665	9.695	9.715
Br	35	9.840	10.014	10.049	10.074	10.096	10.189	10.244	10.279	10.300
kr	36	10.412	10.594	10.630	10.658	10.682	10.777	10.837	10.877	10.899
Rb	37	10.995	11.186	11.221	11.255	11.280	11.376	11.442	11.487	11.511

(continued)

		$2s^0 2p^6$	$2s^1 2p^5$			$2s^2 2p^4$				
		$^1S_0$	$^1P_1$	$^3P_0$	$^3P_2$	$^1S_0$	$^1D_2$	$^3P_0$	$^3P_0$	$^3P_2$
		KL_1L_1	KL_1L_2	KL_1L_2	KL_1L_2	KL_1L_3	KL_2L_2	KL_2L_3	KL_3L_3	KL_3L_3
Sr	38	11.595	11.795	11.830	11.870	11.897	11.992	12.066	12.118	12.143
Y	39	12.213	12.422	12.457	12.503	12.532	12.626	12.708	12.767	12.793
Zr	40	12.851	13.069	13.104	13.157	13.188	13.279	13.370	13.437	13.464
Mb	41	13.505	13.731	13.766	13.827	13.860	13.948	14.049	14.125	14.153
Mo	42	14.179	14.414	14.449	14.519	14.554	14.639	14.750	14.836	14.865
Te	43	14.867	15.111	15.146	15.226	15.263	15.343	15.466	15.563	15.593
Ru	44	15.574	15.827	15.862	15.952	15.991	16.066	16.202	16.310	16.341
Rh	45	16.298	16.560	16.595	16.697	16.738	16.806	16.956	17.077	17.109
Pd	46	17.040	17.312	17.347	17.462	17.504	17.565	17.729	17.864	17.897
Ag	47	17.797	18.078	18.113	18.242	18.286	18.339	18.519	18.668	18.702
Cd	48	18.568	18.857	18.892	19.037	19.082	19.125	19.322	19.488	19.523
In	49	19.354	19.653	19.688	19.849	19.896	19.930	20.144	20.327	20.364
Sn	50	20.157	20.465	20.501	20.680	20.728	20.750	20.984	21.185	21.223
Sb	51	20.977	21.295	21.331	21.529	21.579	21.588	21.844	22.065	22.104
Te	52	21.814	22.142	22.179	22.398	22.449	22.444	22.722	22.965	23.005
I	53	22.668	23.006	23.043	23.284	23.338	23.316	23.618	23.884	23.925
Xe	54	23.527	23.879	23.916	24.182	24.237	24.201	24.530	24.822	24.863
Cs	55	24.426	24.783	24.820	25.111	25.167	25.109	25.463	25.781	25.823
Ba	56	25.330	25.697	25.735	26.053	26.111	26.033	26.416	26.762	26.805
La	57	26.251	26.631	26.669	27.018	27.077	26.978	27.393	27.769	27.813
Ce	58	27.201	27.590	27.628	28.009	28.069	27.945	28.393	28.802	28.847
Pr	59	28.171	28.572	28.610	29.024	29.086	28.936	29.420	29.863	29.909
Nd	60	29.163	29.574	29.612	30.063	30.126	29.947	30.468	30.948	30.995
Pm	61	30.170	30.592	30.631	31.120	31.184	30.976	31.537	32.056	32.104
Sm	62	31.199	31.631	31.671	32.200	32.266	32.024	32.627	33.186	33.235
Eu	63	32.247	32.690	32.730	33.303	33.370	33.092	33.740	34.345	34.395
Gd	64	33.315	33.769	33.809	34.429	34.497	34.182	34.877	35.528	35.579
Tb	65	34.402	34.868	34.909	35.576	35.646	35.291	36.036	36.736	36.788
Dy	66	35.512	35.988	36.029	36.749	36.820	36.421	37.220	37.972	38.025
Ho	67	36.640	37.127	37.169	37.944	38.016	37.570	38.425	39.234	39.287
Er	68	37.788	38.287	38.329	39.162	39.236	38.740	39.655	40.522	40.576
Tm	69	38.958	39.469	39.512	40.406	40.481	39.934	40.911	41.840	41.895
Yb	70	40.151	40.674	40.716	41.675	41.752	42.192	41.149	43.186	43.242
Lu	71	41.361	41.897	41.940	42.967	43.045	42.383	43.496	44.559	44.617
Hf	72	42.589	43.137	43.181	44.280	44.359	43.635	44.821	45.957	46.015
Ta	73	43.831	44.391	44.436	45.611	45.691	44.900	46.164	47.377	47.436
W	74	45.097	45.671	45.715	46.971	47.053	46.193	47.538	48.831	48.891
Re	75	46.385	46.972	47.018	48.357	48.440	47.507	48.938	50.315	50.376
Os	76	47.690	48.291	48.337	49.767	49.851	48.839	50.361	51.830	51.892
Fr	77	49.022	49.636	49.682	51.205	51.291	50.195	51.812	53.375	53.437
Pt	78	50.375	51.003	51.050	52.672	52.759	51.575	53.292	54.954	55.017
Au	79	51.752	52.393	52.440	54.167	54.255	52.978	54.801	56.568	56.633
Hg	80	53.149	53.802	53.849	55.685	55.774	54.397	56.330	58.206	58.272
Tl	81	54.554	55.227	55.275	57.225	57.316	55.840	57.890	59.882	59.948
Pb	82	55.992	56.677	56.726	58.799	58.891	57.302	59.476	61.591	61.658
Bi	83	57.451	58.155	58.205	60.402	60.495	58.799	61.098	63.338	63.406
Po	84	58.918	59.640	59.690	62.026	62.120	60.299	62.739	65.118	65.187
At	85	60.427	61.163	61.213	63.689	63.784	61.836	64.416	66.935	67.005
Rn	86	61.980	62.720	62.771	65.392	65.489	63.397	66.124	68.789	68.860
Fr	87	63.523	64.286	64.337	67.114	67.212	64.983	67.868	70.690	70.762

(*continued*)

		$2s^0 2p^6$	$2s^1 2p^5$			$2s^2 2p^4$				
		$^1S_0$	$^1P_1$	$^3P_0$	$^3P_2$	$^1S_0$	$^1D_2$	$^3P_0$	$^3P_0$	$^3P_2$
		KL_1L_1	KL_1L_2	KL_1L_2	KL_1L_2	KL_1L_3	KL_2L_2	KL_2L_3	KL_3L_3	KL_3L_3
Ra	88	65.103	65.887	65.939	68.879	68.978	66.604	69.654	72.640	72.712
Ac	89	66.720	67.509	67.562	70.673	70.774	68.232	71.453	74.611	74.684
Th	90	68.341	69.153	69.207	72.498	72.600	69.898	73.302	76.640	76.714
Pu	91	70.016	70.842	70.896	74.373	74.476	71.599	75.190	78.714	78.789
U	92	71.704	72.550	72.604	76.280	76.384	73.327	77.116	80.839	80.916
Np	93	73.437	74.297	74.351	78.236	78.342	75.085	79.086	83.019	83.096
Pa	94	75.204	76.080	76.135	80.237	80.344	76.884	81.103	85.254	85.332
Am	95	77.060	77.930	77.985	82.317	82.425	78.727	83.177	87.558	87.637
Cm	96	78.867	79.590	79.646	84.386	84.495	80.240	85.099	89.888	89.968
Bk	97	80.594	81.528	81.585	86.408	86.518	82.388	87.331	92.204	92.284
Cf	98	83.286	84.187	84.245	89.453	89.565	85.017	90.348	95.607	95.688
Es	99	85.219	86.146	86.204	91.701	91.814	86.997	92.617	98.165	98.248
Fm	100	87.205	88.144	88.203	93.998	94.113	89.006	94.926	100.774	100.857
Md	101	89.221	90.192	90.251	96.356	96.471	91.085	97.315	103.472	103.556
No	102	91.267	92.260	92.320	98.763	98.880	93.173	99.744	106.240	106.325
Lr	103	93.373	94.388	94.448	101.250	101.368	95.322	102.252	109.108	109.194
Ku	104	95.518	96.555	96.615	103.796	103.915	97.510	104.820	112.055	112.142

Appendix 10
Table of the Elements

ELEMENT	AT. # (Z)	ISOTOPIC MASS (amu)	RELATIVE ABUNDANCE	ATOMIC WEIGHT (amu)	ATOMIC DENSITY (atom/cm^3)	SPECIFIC GRAVITY
H	1	1.007825	0.9999	1.008		
He	2	4.002603	1.0000	4.003		
Li	3	6.015125	0.0756	6.94	4.70E + 22	0.542
		7.016004	0.9244			
Be	4	9.012186	1.0000	9.012	1.21E + 23	1.82
B	5	10.012939	0.1961	10.814	1.30E + 23	2.47
		11.009305	0.8039			
C	6	12.000000	0.9889	12.011	1.76E + 23	3.516
		13.003354	0.0111			
N	7	14.003074	0.9963	14.007		
		15.000108	0.0037			
C	8	15.994915	0.9976	15.999		
		16.999133	0.0004			
		17.999160	0.0020			
F	9	18.998405	1.0000	18.998		
Ne	10	19.992441	0.9092	20.171	4.36E + 22	1.51
		20.993849	0.0026			
		21.991385	0.0882			
Na	11	22.989771	1.0000	22.99	2.65E + 22	1.013
Mg	12	23.985042	0.7870	24.31	4.30E + 22	1.74
		24.985839	0.1013			
		25.982593	0.1117			
Al	13	26.981539	1.0000	26.982	6.02E + 22	2.7
Si	14	27.976929	0.9221	28.086	5.00E + 22	
		28.976496	0.0470			
		29.973763	0.0309			
P	15	30.973765	1.0000	30.974		
S	16	31.972074	0.9500	32.061		
		32.971462	0.0076			
		33.967865	0.0422			
		35.967090	0.0001			
Cl	17	34.968851	0.7577	35.453		
		36.965899	0.2423			

(*continued*)

ELEMENT	AT. # (Z)	ISOTOPIC MASS (amu)	RELATIVE ABUNDANCE	ATOMIC WEIGHT (amu)	ATOMIC DENSITY (atom/cm^3)	SPECIFIC GRAVITY
Ar	18	35.967545	0.0034	39.948	2.66E + 22	1.77
		37.962728	0.0006			
		39.962384	0.9960			
K	19	38.963710	0.9310	39.097	1.40E + 22	0.91
		39.964000	0.0001			
		40.961832	0.0688			
Ca	20	39.962589	0.9697	40.081	2.30E + 22	1.53
		41.958625	0.0064			
		42.958780	0.0015			
		43.955491	0.0206			
		47.952531	0.0019			
Sc	21	44.955919	1.0000	44.956	4.27E + 22	2.99
Ti	22	45.952632	0.0793	47.879	5.66E + 22	4.51
		46.951769	0.0728			
		47.947950	0.7394			
		48.947870	0.0551			
		49.944786	0.0534			
V	23	49.947164	0.0024	50.942	7.22E + 22	6.09
		50.943961	0.9976			
Cr	24	49.946055	0.0435	51.996	8.33E + 22	7.19
		51.940513	0.8376			
		52.940653	0.0951			
		53.983882	0.0238			
Mn	25	54.938050	1.0000	54.938	8.18E + 22	7.47
Fe	26	53.939617	0.0582	55.847	8.50E + 22	7.87
		55.934936	0.9166			
		56.935398	0.0219			
		47.933282	0.0033			
Co	27	58.933189	1.0000	58.933	8.97E + 22	8.9
Ni	28	57.935342	0.6788	58.728	9.14E + 22	8.91
		59.930787	0.2623			
		60.931056	0.0119			
		61.928342	0.0366			
		63.927958	0.0108			
Cu	29	62.929592	0.6917	63.546	8.45E + 22	8.93
		64.927786	0.3083			
Zn	30	63.929145	0.4889	65.387	6.55E + 22	7.13
		65.926052	0.2781			
		66.927145	0.0411			
		67.924857	0.1857			
		69.925334	0.0062			
Ga	31	68.925574	0.6040	69.717	5.10E + 22	5.91
		70.924706	0.3960			
Ge	32	69.924252	0.2052	72.638	4.42E + 22	5.32
		71.922082	0.2743			
		72.923463	0.0776			
		73.921181	0.3654			
		75.921405	0.0776			
As	33	74.921596	1.0000	74.922		

(*continued*)

ELEMENT	AT. # (Z)	ISOTOPIC MASS (amu)	RELATIVE ABUNDANCE	ATOMIC WEIGHT (amu)	ATOMIC DENSITY (atom/cm^3)	SPECIFIC GRAVITY
Se	34	73.922476	0.0087	78.99	3.67E + 22	4.81
		75.919207	0.0902			
		76.919911	0.0758			
		77.917314	0.2352			
		79.916527	0.4982			
		81.916707	0.0919			
Br	35	78.918329	0.5069	79.904	2.36E + 22	4.05
		80.916292	0.4931			
Kr	36	77.920403	0.0035	83.801	2.17E + 22	3.09
		79.916380	0.0227			
		81.913482	0.1156			
		82.914131	0.1155			
		83.911503	0.5690			
		85.910616	0.1737			
Rb	37	84.911800	0.7215	85.468	1.15E + 22	1.629
		86.909187	0.2785			
Sr	38	83.913430	0.0056	87.616	1.78E + 22	2.58
		85.909285	0.0986			
		86.908892	0.0702			
		87.905641	0.8256			
Y	39	88.905872	1.0000	88.906	3.02E + 22	4.48
Zr	40	89.904700	0.5146	91.224	4.29E + 22	6.51
		90.905642	0.1123			
		91.905031	0.1711			
		93.906313	0.1740			
		95.908286	0.0280			
Nb	41	92.906382	1.0000	9.2906	5.56E + 22	8.58
Mo	42	91.906810	0.1584	95.89	6.42E + 22	10.22
		93.905090	0.0904			
		94.905839	0.1572			
		95.904674	0.1653			
		96.906022	0.0946			
		97.905409	0.2378			
		99.907475	0.0963			
Tc	43	0.000000	0.0000		7.04E + 22	11.5
Ru	44	95.907598	0.0551	1.01E+02	7.36E + 24	12.36
		97.905289	0.0187			
		98.905636	0.1272			
		99.904218	0.1262			
		100.905577	0.1707			
		101.904348	0.3161			
		103.905430	0.1858			
Rh	45	102.905511	1.0000	102.906	7.26E + 22	12.42
Pd	46	101.905609	0.0096	106.441	6.80E + 22	12
		103.904011	0.1097			
		104.905064	0.2223			
		105.903479	0.2733			
		107.903891	0.2671			
		109.905164	0.1181			
Ag	47	106.905094	0.5183	107.868	5.85E + 22	10.5
		108.904756	0.4817			

(continued)

ELEMENT	AT. # (Z)	ISOTOPIC MASS (amu)	RELATIVE ABUNDANCE	ATOMIC WEIGHT (amu)	ATOMIC DENSITY (atom/cm^3)	SPECIFIC GRAVITY
Cd	48	105.906463	0.0122	112.434	4.64E + 22	8.65
		107.904187	0.0088			
		109.903012	0.1239			
		110.904188	0.1275			
		111.902763	0.2407			
		112.904409	0.1226			
		113.903360	0.2886			
		115.904762	0.0758			
In	49	112.904089	0.0428	114.818	3.83E + 22	7.29
		114.903871	0.9572			
Sn	50	111.904835	0.0086	118.734	3.62E + 22	5.76
		113.902773	0.0066			
		114.903346	0.0035			
		115.901745	0.1430			
		116.902958	0.0761			
		117.901606	0.2403			
		118.903313	0.0858			
		119.902198	0.3285			
		121.903441	0.0472			
		123.905272	0.0594			
Sb	51	120.903816	0.5725	121.759	3.31E + 22	6.69
		122.904213	0.4275			
Te	52	119.904023	0.0009	127.628	2.94E + 22	6.25
		121.903066	0.0246			
		122.904277	0.0087			
		123.902842	0.0461			
		124.904418	0.0699			
		125.903322	0.1871			
		127.904476	0.3179			
		129.906238	0.3448			
I	53	126.904470	1.0000	126.904	2.36E + 22	4.95
Xe	54	123.906120	0.0010	131.305	1.64E + 22	3.78
		125.904288	0.0009			
		127.903540	0.0192			
		128.904784	0.2644			
		129.903509	0.0408			
		130.905085	0.2118			
		131.904161	0.2689			
		133.905367	0.1044			
		135.907221	0.0887			
Cs	55	132.905355	1.0000	132.905	9.05E + 21	1.997
Ba	56	129.906245	0.0010	137.327	1.60E + 22	3.59
		131.905120	0.0010			
		133.904612	0.0242			
		134.905550	0.0659			
		135.904300	0.0781			
		136.905500	0.1132			
		137.905000	0.7166			
La	57	137.906910	0.0009	138.905	2.70E + 22	6.17
		138.906140	0.9991			

(*continued*)

ELEMENT	AT. # (Z)	ISOTOPIC MASS (amu)	RELATIVE ABUNDANCE	ATOMIC WEIGHT (amu)	ATOMIC DENSITY (atom/cm^3)	SPECIFIC GRAVITY
Ce	58	135.907100	0.0019	140.101	2.91E + 22	6.77
		137.905830	0.0025			
		139.905392	0.8848			
		141.909140	0.1107			
Pr	59	140.907596	1.0000	140.908	2.92E + 22	6.78
Nd	60	141.907663	0.2711	144.241	2.93E + 22	7
		142.909779	0.1217			
		143.910039	0.2385			
		144.912538	0.0830			
		145.913086	0.1722			
		147.916869	0.0573			
		149.920915	0.0562			
Pm	61	0.000000	0.0000			
Sm	62	143.911989	0.0309	150.363	3.03E + 22	7.54
		146.914867	0.1497			
		147.914791	0.1124			
		148.917180	0.1383			
		149.917276	0.0744			
		151.919756	0.2672			
		153.922282	0.2271			
Eu	63	150.919838	0.4782	151.964	2.04E + 22	5.25
		152.921242	0.5218			
Gd	64	151.919794	0.0020	157.256	3.02E + 22	7.89
		153.920929	0.0215			
		154.922664	0.1473			
		155.922175	0.2047			
		156.924025	0.1568			
		157.924178	0.2487			
		159.927115	0.2190			
Tb	65	158.925351	1.0000	158.925	3.22E + 22	8.27
Dy	66	155.923930	0.0005	162.484	3.17E + 22	8.53
		157.924449	0.0009			
		159.925202	0.0229			
		160.926945	0.1888			
		161.926803	0.2553			
		162.928755	0.2497			
		163.929200	0.2818			
Ho	67	164.930421	1.0000	164.93	3.22E + 22	8.8
Er	68	161.928740	0.0014	167.261	3.26E + 22	9.04
		163.929287	0.0156			
		165.930307	0.3341			
		166.932060	0.2294			
		167.932383	0.2707			
		169.935560	0.1488			
Tm	69	168.934245	1.0000	168.934	3.32E + 22	9.32
Yb	70	167.934160	0.0014	173.036	3.02E + 22	6.97
		169.935020	0.0303			
		170.936430	0.1431			
		171.936360	0.2182			
		172.938060	0.1613			
		173.938740	0.3184			
		175.942680	0.1273			

(continued)

ELEMENT	AT. # (Z)	ISOTOPIC MASS (amu)	RELATIVE ABUNDANCE	ATOMIC WEIGHT (amu)	ATOMIC DENSITY (atom/cm^3)	SPECIFIC GRAVITY
Lu	71	174.940640	0.9741	174.967	3.39E + 22	9.84
		175.942660	0.0255			
Hf	82	173.940360	0.0018	178.509	4.52E + 22	13.2
		175.941570	0.0520			
		176.943400	0.1850			
		177.943880	0.2714			
		178.946030	0.1375			
		179.946820	0.3524			
Ta	73	179.947544	0.0001	180.948	5.55E + 22	16.66
		180.948007	0.9999			
W	74	179.947000	0.0014	183.842	6.30E + 22	19.25
		181.948301	0.2641			
		182.950324	0.1440			
		183.951025	0.3064			
		185.954440	0.2841			
Re	75	184.953059	0.3707	186.213	6.80E + 22	21.03
		186.955833	0.6293			
Cs	76	183.952750	0.0002	190.333	7.14E + 22	22.58
		185.953870	0.0159			
		186.955832	0.0164			
		187.956081	0.1330			
		188.958300	0.1610			
		189.958630	0.2640			
		191.961450	0.4100			
Ir	77	190.960640	0.3730	192.216	7.06E + 22	22.55
		192.963012	0.6270			
Pt	78	189.959950	0.0001	195.081	6.62E + 22	21.47
		191.961150	0.0078			
		193.962725	0.3290			
		194.964813	0.3380			
		195.964967	0.2530			
		197.967895	0.0721			
Au	79	196.966541	1.0000	196.967	5.90E + 22	19.28
Hg	80	195.965820	0.0015	200.617	4.26E + 22	14.26
		197.966756	0.1002			
		198.968279	0.1684			
		199.968327	0.2313			
		200.970308	0.1322			
		201.970642	0.2980			
		203.973495	0.0685			
Tl	81	202.972353	0.2950	204.384	3.50E + 22	11.87
		204.974442	0.7050			
Pb	82	203.973044	0.0148	207.177	3.30E + 22	11.34
		205.974468	0.2360			
		206.975903	0.2260			
		207.976650	0.5230			
Bi	83	208.980394	1.0000	208.98	2.82E + 22	9.8

Appendix 11
Table of Fluoresence Yields
for *K*, *L*, and *M* Shells

Z Element	ω_K		$\overline{\omega}_L$	$\overline{\omega}_M$	Z Element	ω_K		$\overline{\omega}_L$	$\overline{\omega}_M$
4 Be	4.5	−4			45 Rh	8.07	−1		
5 B	10.1	−4			46 Pd	8.19	−1		
6 C	2.0	−3			47 Ag	8.30	−1	5.6	−2
7 N	3.5	−3			48 Cd	8.40	−1		
8 0	5.8	−3			49 In	8.50	−1		
9 F	9.0	−3			50 Sn	8.59	−1		
10 Ne	1.34	−2			51 Sb	8.67	−1	1.2	−1
11 Na	1.92	−2			52 Te	8.75	−1	1.2	−1
12 Mg	2.65	−2			53 1	8.82	−1		
13 Al	3.57	−2			54 Xe	8.89	−1	1.1	−1
14 Si	4.70	−2			55 Cs	8.95	−1	8.9	−2
15 P	6.04	−2			56 Ba	9.01	−1	9.3	−2
16 S	7.61	−2			57 La	9.06	−1	1.0	−1
17 Cl	9.42	−2			58 Ce	9.11	−1	1.6	−1
18 Ar	1.15	−1			59 Pr	9.15	−1	1.7	−1
19 K	1.38	−1			60 Nd	9.20	−1	1.7	−1
20 Ca	1.63	−1			61 Pm	9.24	−1		
21 Sc	1.90	−1			62 Sm	9.28	−1	1.9	−1
22 Ti	2.19	−1			63 Eu	9.31	−1	1.7	−1
23 V	2.50	−1	2.4	.03	64 Gd	9.34	−1	2.0	−1
24 Cr	2.82	−1	3.0	.03	65 Tb	9.37	−1	2.0	−1
25 Mn	3.14	−1			66 Dy	9.40	−1	1.4	−1
26 Fe	3.47	−1			67 Ho	9.43	−1		
27 Co	3.81	−1			68 Er	9.45	−1		
28 Ni	4.14	−1			69 Tm	9.48	−1		
29 Cu	4.45	−1	56	.03	70 Yb	9.50	−1		
30 Zn	4.79	−1			71 Lu	9.52	−1	2.9	−1
31 Ga	5.10	−1	6.4	.03	72 Hf	9.54	−1	2.6	−1
32 Ge	5.40	−1			73 Ta	9.56	−1	2.3	−1
33 As	5.67	−1			74 W	9.57	−1	3.0	−1
34 Se	5.96	−1			75 Re	9.59	−1		
35 Br	6.22	−1			76 Os	9.61	−1	3.5	−1
36 Kr	6.46	−1	1.0	.02	77 Ir	9.62	−1	3.0	−1
37 Rb	6.69	−1	1.0	.02	78 Pt	9.63	−1	3.3	−1
38 Sr	6.91	−1			79 Au	9.64	−1	3.9	−1

(continued)

Z Element	ω_K		$\overline{\omega}_L$		$\overline{\omega}_M$		Z Element	ω_K		$\overline{\omega}_L$		$\overline{\omega}_M$	
39 Y	7.11	−1	3.2	−2			80 Hg	9.66	−1	3.9	−1		
40 Zr	7.30	−1					81 Tl			4.6	−1		
41 Nb	7.48	−1					82 Pb	9.68	−1	3.8	−1	2.9	−2
42 Mo	7.64	−1	6.7	−2			83 Bi			4.1	−1	3.6	−2
43 Tc	7.79	−1					92 U	9.76	−1	5.2	−1	6	−2
44 Ru	7.93	−1											

Note that the multipliers in specific columns are powers of 10 (e.g., ω_k and ω_L for Ag are 8.3×10^{-1} and 3×10^{-3}, respectively). The fluorescence yield values ω_L and ω_M indicate average values for the L subshells and M subshells.

Appendix 12
Physical Constants, Conversions, and Useful Combinations

Physical constants

Avogadro constant	$N_A = 6.022 \times 10^{23}$ particles/mole
Boltzmann constant	$k_B = 8.617 \times 10^{-5}$ eV/K
Elementary charge	$e = 1.602 \times 10^{-19}$ Coulombs
Planck constant	$h = 4.136 \times 10^{-15}$ eV $-$ s $= 6.626 \times 10^{-34}$ J-s
	$\hbar = h/2\pi = 6.582 \times 10^{-16}$ eV-s
Speed of light	$c = 2.998 \times 10^8$ cm/s

Useful Combinations

Electron rest mass	$mc^2 = 0.511$ MeV
Bohr radius	$\alpha_o = \hbar^2/me^2 = 0.0529$ nm
Bohr velocity	$v_o = e^2/\hbar = 2.188 \times 10^8$ cm/s
Fine structure constant	$\alpha = e^2/\hbar c = 7.297 \times 10^{-3} \approx 1/137$
Hydrogen binding energy	$e^2/2\alpha_o = 13.606$ eV
Classical electron radius	$r_e = e^2/mc^2 = 2.818 \times 10^{-13}$ cm
Electron charge	$e^2 = 1.4395$ eV-nm
Electron Compton wavelength	$\bar{\lambda} = \hbar/mc = 3.861 \times 10^{-11}$ cm
Photon energy-wavelength	$hc = 1239.85$ eV-nm

Conversions

$1\,\text{Å} = 10^{-8}$ cm $= 10^{-1}$ nm

1 eV $= 1.602 \times 10^{-19}$ J

1 eV/particle $= 23.06$ kcal/mol

m_p (proton mass) $= 1.6726 \times 10^{-27}$ kg m

(electron mass) $= 9.1095 \times 10^{-31}$ kg m_p/m

$\quad m_p/m = 1836.1$

Appendix 13
Acronyms

AEM	Analytical Electron Microscopy
AES	Auger Electron Spectroscopy
AFM	Atomic Force Microscopy
ED	Electron Diffraction
EDS	Energy-Dispersive Spectroscopy
EELS	Electron Energy-loss Spectroscopy
EMA	Electron Microprobe Analysis
ESCA	Electron Spectroscopy for Chemical Analysis
EXAFS	Extended X-ray Absorption Fine Structure
FRS	Forward Recoil Spectroscopy
GAXRD	Glancing Angle X-Ray Diffraction
HEED	High-Energy Electron Diffraction
HEIS	High-Energy Ion Scattering
IIXS	Ion-Induced X-ray Spectroscopy
IMMA	Ion Microprobe Mass Analysis
LEED	Low-Energy Electron Diffraction
LEIS	Low-Energy Ion Scattering Neutron Activation Analysis
NRA	Nuclear Reaction Analysis
PES	Photoelectron Spectroscopy
PIXE	Particle-Induced X-ray Emission
PRA	Prompt Reaction Analysis
RBS	Rutherford Backscattering Spectrometry
RHEED	Reflection High Energy
SEM	Scanning Electron Microscopy
SEXAF	Surface Extended X-ray Absorption Fine Structure
SIMS	Secondary Ion Mass Spectroscopy
SNMS	Secondary Neutral Mass Spectroscopy
SNOM	Scanning Near-Field Optical Microscopy
SPM	Scanning Probe Microscopy
STM	Scanning Tunneling Microscopy
TEM	Transmission Electron Microscopy
TRXRF	Total Reflection X-Ray Fluorescence

UPS	Ultraviolet Photoelectron Spectroscopy
WDS	Wavelength Dispersive Spectroscopy
XPS	X-ray Photoelectron Spectroscopy
XRD	X-Ray Diffraction
XRF	X-Ray Fluorescence

Index